COMPUTER AUTOMATED MANUFACTURING

JOHN H. POWERS JR.

McGraw-Hill Book Company

*New York Atlanta Dallas St. Louis
San Francisco Auckland Bogotá Guatemala
Hamburg Lisbon London
Madrid Mexico Milan Montreal New Delhi
Panama Paris San Juan São Paulo
Singapore Sydney Tokyo Toronto*

Sponsoring Editor: D. Eugene Gilmore
Editing Supervisor: Suzette Andre
Design and Art Supervisor: Caryl Valerie Spinka
Production Supervisor: Priscilla Taguer

Text Designer: Suzanne Bennett
Cover Designer: Edward A. Butler

Library of Congress Cataloging-in-Publication Data

Powers, John H.
  Computer-automated manufacturing.

  Bibliography: p. 303
  Includes index.
  1. CAD/CAM systems. I. Title.
TS155.6.P69 1987   670'.285   86-21305
ISBN 0-07-050601-9

The manuscript for this book was prepared electronically.

Computer - Automated Manufacturing

1 2 3 4 5 6 7 8 9 0   SEMSEM   8 9 4 3 2 1 0 9 8 7

ISBN 0-07-050601-9

CONTENTS

Preface   v
About the Author   vii

**Part I Introduction to Computer-Automated
Manufacturing**   1

Chapter 1 What is Computer-Automated Manufacturing? 2
    Introduction 2   Basic Elements of CAM 2   The
    Computer's Role in Manufacturing 4   The Evolution
    of CAM 5   The Nature of the Manufacturing Environ-
    ment 5   Summary 8   Review Questions 9

Chapter 2 Why Do We Use Computers in
Manufacturing?   10
    Introduction 10   Types of Data Used in
    Manufacturing 10   Sources of Data Used in
    Manufacturing 12   Complexity of Products in
    Manufacturing 12   Nature of the Manufacturing
    Process 14   Interdependencies 15   General Needs
    and Trends in the Use of Computers in Manufact-
    uring 16   Summary 17   Review Questions 17

Chapter 3 Computer Control   18
    Introduction 18   Numerical Control 18   Extensions
    of Numerical Control 22   Manufacturing Control
    Systems 24   Summary 27   Review Questions 27

Chapter 4 Computer Applications in Manufacturing   28
    Introduction 28   Tool Control Applications 28
    Design 29   Shop Floor Control 30   Materials
    Handling 31   Automation 33   Testing and Mea-
    surement 34   Process Control 35   Other Applica-
    tions 36   Summary 37   Review Questions 37

Chapter 5 Trends in the Use of Computers in
Manufacturing   38
    Introduction 38   Product Technology 38   Computer
    Capabilities 40   Systems Architecture 40   Manufac-
    turing Technologies 41   Computer Applications in
    Manufacturing 43   Other Considerations 44
    Summary 45   Review Questions 46

Part II Computer Technologies   47

Chapter 6 Computer Hardware   48
    Introduction 48   Computer Basics 48   Types of
    Data Processing Equipment 51   Data Communica-
    tions 55   Computer Hardware Trends 57
    Summary 62   Review Questions 62

Chapter 7 Computer Software   63
    Introduction 63   Software Basics 63   System Soft-
    ware 64   Application Software 67   Programming
    Languages 68   Programming 69   Trends in Computer
    Software 72   Summary 75   Review Questions 76

Chapter 8 Micro- and Minicomputers   77
    Introduction 77   Microprocessors 77   Microcom-
    puters 79   I/O Communications 80   Minicompu-
    ters 81   Programmable Controllers 82   Programming
    Micros and Minis 83   Micro- and Minicomputer
    Trends 87   Summary 90   Review Questions 91

Chapter 9 Artificial Intelligence   92
    Introduction 92   Basics of Artificial Intelligence 92
    Expert Systems 94   Software for AI Systems 100
    Computer Hardware for AI Systems 101   AI Applica-
    tions in Manufacturing 102   Trends in Artificial Intel-
    ligence 107   Summary 109   Review Questions 109

Part III Computer-Automated Engineering   111

Chapter 10 Computer Graphics Technology   112
    Introduction 112   Basics of Computer Graphics 112
    Graphics Hardware 116   Graphics Software 120
    Geometric Modeling 123   Applications of Computer
    Graphics 128   Trends in Computer Graphics 129
    Summary 131   Review Questions 131

Chapter 11 Computer-Automated Design   132
    Introduction 132   Automating the Design Pro-
    cess 132   Basic CAD Systems 134   Mech-
    anical Design Systems 137   Electronic Design
    Systems 140   Trends in Computer-Automated
    Design 144   Summary 146   Review
    Questions 147

Chapter 12 Computer Tools for Engineering Analysis   148
  Introduction 148   Design Analysis 148   Manufac-
  turing Analysis 152   Manufacturing Modeling and
  Simulation 154   Integrated CAE Systems 157
  Trends in CAE 160   Summary 161   Review
  Questions 161

Part IV Robotics   **163**

Chapter 13 Basic Robotics Technology   164
  Introduction 164   What Is a Robot? 164   Types
  of Robots 165   Performance Capabilities 169
  Programming Robots 171   Robot Operation 174
  Summary 175   Review Questions 175

Chapter 14 Intelligent Robotics Systems   176
  Introduction 176   What Is an Intelligent Robot? 176
  End-of-Arm Tooling 177   Sensors 180   Robot
  Vision 184   Control Systems and Software for Intelligent
  Robots 188   Artificial Intelligence in Robotics 190
  Trends in Robotics 192   Summary 193   Review
  Questions 193

Chapter 15 Robot Applications   195
  Introduction 195   Materials Handling Opera-
  tions 195   Tool Handling Operations 196   Other
  Robot Tasks 198   Integrating the Robot Task and the
  Workplace 200   Metalworking Applications 201
  Plastics Manufacturing Applications 202   Assembly
  Applications 204   Process Operations 205
  Summary 206   Review Questions 206

Chapter 16 Implementing Robotics in Manufacturing   207
  Introduction 207   Deciding Where and When to
  Use a Robot 207   Implementing a Robot Application 210   Robot Safety 211   Simulating Robot Applications 213   Summary 215   Review Questions 216

Part V Manufacturing Systems   **217**

Chapter 17 System Architecture   218
  Introduction 218   What Is a Manufacturing
  System? 219   Hierarchical Systems 220   Local Area
  Networks 222   Control Systems 224   Trends in
  System Architecture 230   Summary 232   Review
  Questions 233

Chapter 18 Management Systems   234
  Introduction 234   Technical Data Systems 234
  Logistical Data Systems 240   Administrative Data
  Systems 245   Designing for Manufacturing 248
  Optimizing Manufacturing Operations 249
  Summary 252   Review Questions 253

Chapter 19 Integrated Manufacturing Systems   254
  Introduction 254   Integrated Database
  Systems 255   Data Collection Systems 258   Mater-
  ials Handling Systems 259   Flexible Manufacturing
  Systems 264   Integrating Manufacturing Sys-
  tems 268   Summary 272   Review Questions 272

Part VI Computer-Automated Manufacturing   **273**

Chapter 20 Automated Manufacturing   274
  Introduction 274   Machining 275   Assembly 278
  Process 286   Summary 289   Review
  Questions 289

Chapter 21 Implementing CAM   290
  Introduction 290   Why Implement CAM? 290
  Barriers to CAM 290   Changes Caused by CAM 291
  Keys to Successful Implementation of CAM 292
  Justifying CAM 292   Planning for CAM 293
  The Implementation Process 295   Summary 296

Glossary   297

References   303

Index   307

PREFACE

Computer-automated manufacturing (CAM) is the key to competitive production operations. CAM encompasses the advanced technologies used to automate the physical tasks in manufacturing as well as the handling of the data that drives the process. The tools of CAM include computer technologies, computer-automated engineering (CAE), and robotics. CAM uses these technologies to integrate the design process with automated production machines, materials handling equipment, and control systems. Modern manufacturing has progressed beyond the use of individual computer-aided design and computer-aided manufacturing techniques to fully integrate and automate factories. CAM ties together all the pieces of advanced, computer-based tools; the result is a total manufacturing system.

This book is intended to provide the reader with a basic introduction to the concept of CAM. Although the book is broad in scope, it covers all the key elements involved. To understand and use CAM requires many specialized topics to be covered. *Computer-Automated Manufacturing* offers a comprehensive, yet fundamental treatment of the subject. The object is to help the reader develop a working knowledge of what exists, how it works, and how it can be used. There is an emphasis on the state of the art and on future trends. The book is intended to be used to establish a basic understanding of CAM for practical application in industry, as well as to provide a basis for advanced study. It is designed to be useful as an introduction to CAM for both students and professionals.

The book begins each topic by providing a general overview. Then an understanding of the basics involved is developed from the use of fundamental concepts and actual practical examples. Each topic is then summarized and reviewed. The coverage of the subject progresses by continually building upon key elements. References are provided for further study or research, and there is a comprehensive glossary of terms.

Some of the book's unique features include:

☐ Easy-to-read-and-understand explanations that avoid some of the technical details and mathematics that are characteristic of more specialized textbooks

☐ Up-to-date information on all elements of the subject including the state of the art and future trends

☐ Entire chapters and sections on the latest technologies, including artificial intelligence, expert systems, intelligent robotics, microprocessors, local area networks, interactive graphics, solid modeling, machine vision, group technology, flexible manufacturing systems, and factories of the future

☐ Numerous examples and illustrations of actual industrial applications—including fabrication, assembly, and process-type manufacturing operations

☐ Extensive lists of up-to-date references for each subject

☐ Chapters explaining how to implement CAM with practical recommendations of "dos and don'ts"

☐ An emphasis on the integration and automation of both the physical and data handling processes in manufacturing

The use of the computer is the key to the future of manufacturing. Computers have become the most important tool for industry and are used in an ever-increasing variety of applications. Computers do not replace people; they help people to be more productive and to do things that may not otherwise be possible. This book, for example, was written, edited, and composed on computers. Some of the "high-technology" products of today would not exist if it were not for computer-automated design and manufacturing techniques.

ACKNOWLEDGMENTS

I would like to express my gratitude to those who supported the writing of this book and to the companies and publishers that contributed up-to-date information and high quality illustrations, particularly my employer, the International Business Machines Corporation. I am also indebted to many of my friends and associates in the technical community for their ideas, suggestions, and sources of information. My publisher and editors, too, were extremely helpful in the planning, reviewing, and editing of the book. Of course the patience, encouragement, and support of my family was essential to the completion of this book.

John H. Powers Jr.

ABOUT THE AUTHOR

John H. Powers Jr. draws from his more than 20 years' experience in the development and manufacturing of electronic components and systems for this book. As Manager of Equipment and System Engineering at IBM's semiconductor manufacturing facility in East Fishkill, New York, Mr. Powers oversees the design, fabrication, installation, and maintenance of production machines, automation, and manufacturing control systems for advanced integrated circuit manufacturing. His prior assignments as Manager of the Semiconductor Production Management Center at East Fishkill and Manager of Technical Services on IBM's Corporation Manufacturing staff directly supported key IBM executives who had worldwide manufacturing responsibilities.

Mr. Powers authored numerous technical papers, publications, and presentations on electronics and manufacturing technologies. His leadership in the technical community is evidenced by numerous professional affiliations: he has served as general chairman and has served on the executive committees of several international conferences; he is a senior member of the Institute of Electrical and Electronics Engineers (IEEE) and past president of its Components, Hybrids, and Manufacturing Technology Society (CHMT) as well as chairman of its technical committee on Manufacturing Technology. Mr. Powers is a recipient of the CHMT Outstanding Contribution Award and the IEEE Centennial Medal. He holds a bachelor of engineering degree from Stevens Institute of Technology, Hoboken, New Jersey, and a masters of science degree from Union College, Schenectady, New York.

PART ONE

INTRODUCTION TO COMPUTER-AUTOMATED MANUFACTURING

CHAPTER 1
WHAT IS COMPUTER-AUTOMATED MANUFACTURING?

CHAPTER 2
WHY DO WE USE COMPUTERS IN MANUFACTURING?

CHAPTER 3
COMPUTER CONTROL

CHAPTER 4
COMPUTER APPLICATIONS IN MANUFACTURING

CHAPTER 5
TRENDS IN THE USE OF COMPUTERS IN MANUFACTURING

The basic objective of this book is to provide an understanding of the fundamental concepts of computer-automated manufacturing (CAM). This includes the key elements and technologies involved, the typical applications that exist in industry today, and the major trends that will affect the future of manufacturing. This book is not intended or expected to create experts in CAM. However, it is intended to develop an understanding and knowledge of the subject area so that the reader will be able to operate in a modern, high-technology manufacturing environment.

To maintain a practical approach to the subject, examples are used frequently to illustrate fundamental concepts, as well as specific, real examples from industry. Most of the concepts involved are not unique to any particular company or industry. Therefore, examples are drawn from a number of different sources to show the variety and potential for the application of CAM techniques.

The first part of the book provides a broad introduction to the total scope of the subject. It addresses why we use computers in manufacturing and how CAM evolved historically. Some examples of typical types of applications and an overview of the major trends for the future are also presented. Parts 2 through 4 deal with the key technologies involved in making CAM possible: computers (i.e., hardware, software, and their capabilities), computer-automated engineering (CAE—e.g., graphics and simulation), and robotics. Part 5 covers systems concepts (e.g., architecture and types of systems) and how manufacturing operations can be integrated with the use of the computer. Part 6 discusses a variety of typical types of automated manufacturing operations.

The general approach taken in this book is to first establish a basic context for understanding the subject. Next, the subject is broken down into its essential parts to a reasonable level of technical depth. Thus a fundamental working knowledge is established for practical application or further development. Finally, all the elements are tied together to demonstrate they all relate, how CAM works, and is applied in industry. The emphasis is on the state of the art today rather than historical evolution or common practice. Special focus is put on future trends for the reader to learn about what is possible with CAM and be better prepared for the future.

Each chapter includes review questions to aid in comprehension and retention. A reference section listing sources used and available for further information is provided at the back of the book, together with a glossary of terms.

WHAT IS COMPUTER-AUTOMATED MANUFACTURING?

1-1 INTRODUCTION

This first chapter will introduce and explain the basic concepts behind computer-automated manufacturing (CAM). Automation may have a variety of different meanings to different people. In this book, we will focus on how computers are used to automate manufacturing activities. This includes the physical tasks in the manufacturing process as well as the handling of data. To understand and use the capabilities of CAM, one must be familiar with the technologies involved and the applications that can use them. It is first necessary to understand the basics of computers and manufacturing.

We will start by defining some of the basic elements of CAM. This should clarify some of the more commonly used terms and concepts involved. Then we will discuss the role of the computer in manufacturing and the evolution of CAM. To provide a basic understanding of how manufacturing operates, we will then describe the basic nature of the manufacturing environment. This will set the stage in the remaining chapters of Part 1 for discussing how computers are used in manufacturing.

1-2 BASIC ELEMENTS OF CAM

For those not familiar with computers and automation, some of the basic elements of CAM should first be clarified. The most basic concepts and common terms used in this book will be explained to provide an understanding of some of the key elements of the subject. These are general explanations to introduce these concepts and are not intended to be formal or thorough descriptions or definitions. Several similar terms that are often used interchangeably will also be put into context relative to each other. A complete glossary of terms is included at the back of the book for your reference and convenience.

COMPUTER

The computer is the building block for the entire concept of CAM. Computers are so commonplace today that it may seem unnecessary to explain what they are. However, the term "computer" is used so frequently to refer to

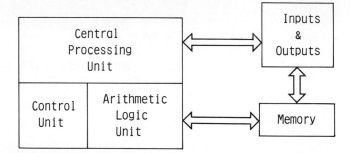

Fig. 1-1 Basic elements of a digital computer.

so many things that it is probably worthwhile to clarify the basic functions of a computer and the forms it can take. A computer is a machine that consists of an arithmetic logic unit (ALU), a memory, a control unit, and input/output (I/O) devices that, through the use of a stored program, can process data (Fig. 1-1). Many types of machines can fit this general definition. As long as they have all the basic features and functions, they can be considered to be computers. The range in size and capability is substantial, however. A variety of terms describe a number of different types of computers.

At the high end of the spectrum are large-scale data processing systems made up of one or more separate machines to perform each of the basic functions (i.e., central processing unit or CPU, main memory, I/O devices such as disk and tape drives and printers, etc.). Such large systems can usually store millions of bytes (characters) of information in main memory and can process millions of instructions per second. Intermediate systems are similar in design and have all the same elements as the large systems. However, they do not have as much memory or operate as fast.

At the low end of the spectrum, however, a wide variety of small computers have emerged. They provide computer power to individuals or control other machines. The minicomputer is usually a machine that contains all the processing and storage functions in a small package with limited capabilities. The microcomputer, however, evolved from the development of the microprocessor, which is a single semiconductor integrated circuit (IC) device that contains all the arithmetic logic function of a CPU. This small, low-cost computer building block made possible the portable, personal, and compact computers and controllers that are used throughout business and industry today (Fig. 1-2). For our purposes here, it is important to remember that in manufacturing we use the computer's ability to collect, store, process, and transmit data, and we choose a particular type of machine depending on the magnitude and complexity of the data handling job involved.

PROGRAM

Computers are usually thought of in terms of the physical "hardware" of electronics equipment that can be seen. However, computers could not operate at all without the programs that are not visible. "Program" is another commonly used and generally understood term. Because of its fundamental importance to the subject and the number of related terms involved, however, it should be explained. A computer program is a set of coded instructions in an ordered sequence which tells the computer to carry out certain arithmetic or logical operations. "Software" is a term commonly used to refer to programs that have been developed to accomplish specific tasks. This term is used to distinguish the programs from the hardware of the computer equipment.

Programs are written in "languages" which communicate with the computer logic through the use of a symbolic code. The most elementary language, that is, "machine language," uses a binary code which directly relates to the on/off switching operation of the computer logic circuits. "High-level languages" use simpler codes similar to normal speech to communicate complex instructions. One highly advanced stage of programming is called "artificial intelligence" (AI). AI is a set of software that permits the computer to deal with very high-level languages, adapt to sensory inputs, interpret data, and "learn" from experience. Programming is a tool as power-

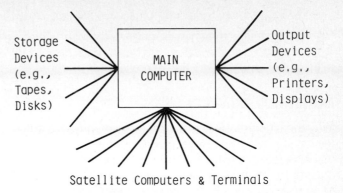

Fig. 1-3 Elements of a computer system.

ful and important as the computer hardware itself in providing capabilities which industry can use to increase its productivity.

SYSTEM

A data processing or computer system consists of all the hardware, software, and data that is required to perform a particular set of tasks. In its simplest form, a system is made up of a CPU, main memory, input and output peripheral equipment (e.g., displays, printers, and disk drives), and operating system software and applications programs (Fig. 1-3). In manufacturing, such a system might be used for data collection and reporting on the production floor (e.g., job status, labor claiming, work in process (WIP), and throughput). Some systems in large manufacturing operations can be extremely large and complex. They can involve many computers and terminals tied together with support equipment to help run the factory. Systems are designed to perform specific tasks. The appropriate hardware and software is arranged in a way that will get the job done efficiently and economically.

There are many ways to arrange, or "configure," systems. A variety of other terms are used in relation to systems design:

☐ The "architecture" of a computer system is the basic structure of the data flow. For example, a "hierarchical architecture" is made up of several levels of processors and controllers between the points at which data is collected or generated and the central "host" computer which stores the database.

☐ "Networks" are groups of computer systems which are tied together with data communication links in a particular configuration that permits them to exchange or share data.

☐ "Distributed data processing" is a popular approach to configuring systems which provides processing and storage capability at a low level in the system hierarchy. It uses small computers to reduce the dependency and demand on the large central host computer.

☐ A "manufacturing control system" is usually a computer system that handles technical data for process or tool control applications.

Fig. 1-2 Typical small business computer built from a microprocessor. *(International Business Machines Corp.)*

☐ The "integrated manufacturing system" is the ultimate form of a system in a manufacturing environment. Such a system controls all the different types of data needed to manage the entire production operation.

NUMERICAL CONTROL

Numerical control (NC) is a form of programmable automation that was first developed in the early 1950s for machine tool applications. An NC system is composed of a control program of coded instructions (which in some early systems was in the form of paper tape), a control unit, and a machine or tool. "Direct numerical control" (DNC) is a later form in which the control program is loaded by a direct link to a computer. "Computer numerical control" (CNC) is the most recent and powerful form of NC. It uses a dedicated computer as the tool controller.

Many applications do not require the power of a computer. They use "programmable controllers" (PCs) which are sequential logic devices with limited functions (e.g., counting, timing, signal generation). Advances in electronics and computer technology have blurred the lines that distinguish PCs and microcomputers. They now often have similar capabilities. Both may be tied into a manufacturing control system to run the tools or process.

AUTOMATION

In a manufacturing environment, automation is usually associated with the mechanization and control of the physical movement or fabrication operations in production rather than the data handling tasks. Automated manufacturing can be either fixed or programmable. "Fixed automation" is the control of a fixed or repetitive sequence of operations. "Programmable automation," through the use of computer control, can be reprogrammed to change the sequence or control of operations. Programmable automation is obviously more flexible and includes more sophisticated forms of automation, such as robotics.

The term "robot" means different things to different people. It has been used in science fiction for years. To the manufacturing industry, however, robots are a new form of automated tool (Fig. 1-4). They can be programmed to perform a wide variety of tasks using human-like capabilities. The Robot Institute of America (a large industry group) has developed the following definition: "A robot is a reprogrammable, multifunctional manipulator designed to move material, parts, tools or specialized devices through variable programmed motions for the performance of a variety of tasks."

COMPUTER-AUTOMATED MANUFACTURING

CAM is a concept that will be used to describe the general category of advanced approaches to manufacturing which use the power of the computer to automate the handling of data as well as the physical operations in the

Fig. 1-4 Industrial robot. *(Gilman Engineering and Manufacturing Co.)*

process. CAM employs many modern manufacturing technologies, each of which will be addressed at length in this text. Some of the topics are computer-aided design (CAD), CNC, robotics, distributed processing, computer modeling and simulation, and data communications.

A number of other terms are also used to describe this subject. "Computer-aided manufacturing" is an earlier and more commonly used term. It has been formally defined by an industry organization, CAM-I (Computer-Aided Manufacturing-International, Inc.), as "the effective utilization of computer technology in the management, control, and operations of the manufacturing facility through either direct or indirect computer interface with the physical and human resources of the company."

Other, more recent terms used to express a larger, more active role of the computer are "computer-integrated manufacturing" (CIM) and "flexible manufacturing systems" (FMS). The common concept behind these terms is the use of the computer to tie together, or "integrate," all the movement of data and product in the factory under the control of one complete manufacturing system. In addition, such systems make possible the automation of small-batch and customized production. Without computer control, they would use more labor and be more costly.

These are not all the elements of CAM. However, this should clarify some of the more basic concepts and commonly used terms found in this book. Each subject area will, of course, be dealt with in more depth in later chapters.

1-3 THE COMPUTER'S ROLE IN MANUFACTURING

Computers are a major product of the manufacturing industry. They are also one of its most valuable tools. Modern manufacturing is a data-driven operation; therefore, computers are used to process all that data efficiently. As product technologies and manufacturing processes have become increasingly more complex, the amount of data involved has also increased substantially to the point that computers must be used to handle it.

The use of computers in manufacturing has become increasingly widespread and necessary as computers serve in a wider variety of roles in the production environment. Indeed, some products and technologies today could not be manufactured at all without them. Many of the "high-technology" wonders of modern living, which we often take for granted, would not exist because it would not be possible to design or manufacture them without the use of computers and data-driven tools. Automation and computer control, in these applications at least, have become a necessity.

Computers are not new to manufacturing. Early versions of computers were being applied to manufacturing tasks ever since they were first commercially available. Computers are basically machines that can process data very efficiently. Therefore, they have found their way into a variety of uses or applications that require large amounts of data to be handled quickly. There are many such applications in manufacturing, such as controlling automated tools and processes, tracking parts and materials, testing and inspection, and generating management reports. Today, computers are a key to the total manufacturing process—from product design to process control to product assembly, administrative management, and distribution. Computers have been used to help in every part of the manufacturing process.

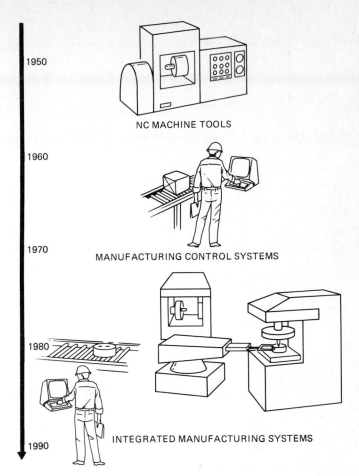

Fig. 1-5 Evolution of computer-automated manufacturing (CAM).

1-4 THE EVOLUTION OF CAM

CAM evolved from relatively simple stand-alone applications, such as tool control, to completely integrated manufacturing systems (Fig. 1-5). Each of the stages in that evolution will be discussed in later chapters. However, this book will focus mainly on how computers are being used in manufacturing today and on the major trends for the future. Over time, the computer's role in manufacturing has changed from just controlling individual operations to being the building block of a system that manages the factory.

There are many forms of CAM and degrees of automation even today. The traditional fixed or hard automation that is often used to make large volumes of identical parts may use computer control. This differs significantly from programmable control and FMS. They can be used in a batch manufacturing operation which involves small quantities and a variety of types of parts or products.

The major trend for the future in the use of computers in manufacturing is to integrate all the key functions of production operations into a computer-controlled system. Many people see such "factories of the future" as the key to the next industrial revolution. It is sometimes thought that computers displace people in manufacturing jobs. The trend for the future seems to be automated, "people-less" factories where computers and robots will do all the work. Computers, we will see, are not merely machines

that replace direct labor. There are many jobs in those future factories that humans will not be able to do. However, such factories will also require many people to design, build, operate, program, maintain, repair, and manage those sophisticated tools and processes. Historically, in industry, major technological advances in manufacturing productivity result in greater production and more jobs. These jobs, however, often require different skills. The computer is a key productivity tool for industry. It will therefore continue to play a major role in making new products and technologies possible and economical.

1-5 THE NATURE OF THE MANUFACTURING ENVIRONMENT

To form a basic understanding of CAM, we should start with a general discussion of the typical manufacturing environment where computers are used today. Manufacturing operations are managed by the flow of information and data about when to do what and how as well as the actual process of doing it. We can best trace that flow by examining the major elements involved and understanding their interaction. These elements can be thought of in terms of three categories that together describe the basic manufacturing system. They are organization, operation, and data.

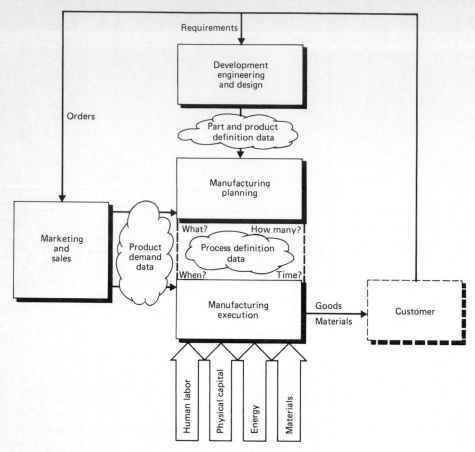

Fig. 1-6 Organizational elements of manufacturing. *(M. M. Kutcher, "Automating It All," IEEE Spectrum, May 1983; © 1983 IEEE)*

ORGANIZATIONAL STRUCTURE

Figure 1-6 depicts the major organizational elements of an industrial enterprise. Simplistically, we can think of the roles of these elements as:

1. Development, which is responsible for the design and specification of new products and processes compatible with customer requirements

2. Marketing, which must identify the customer requirements as well as sell the product

3. Manufacturing, which produces the product

It takes all three organizational elements in an industrial enterprise together, not the manufacturing organization alone, to describe the total process which involves the source and use of data that affects manufacturing operations. As can be seen, the data which is transferred between these organizations ultimately drives the manufacturing process. Data from development describes the product and processes for manufacturing. Marketing data identifies the demand for the product in terms of sales forecasts as well as actual orders. Within manufacturing, the planning organization uses the development and marketing data to determine when how many of what products have to be manufactured. The "true" manufacturing organization then applies resources such as labor, capital, and materials with the direction of the manufac-

turing planning data to "execute" the manufacturing process and produce the product.

MANUFACTURING OPERATIONS

The manufacturing operation is itself made up of a number of elements which generate and use data. A number of control and data collection functions are involved as the product moves through the manufacturing process (Fig. 1-7). The exact nature and form of these functions vary with the particular type of product and process involved. However, the general categories are common to most manufacturing operations.

The control functions usually draw data from other sources, such as development engineering, marketing, or planning, to direct what product is to be manufactured, when, and how. The data collection functions, however, draw data from the manufacturing process itself to monitor its performance and provide management reports and records of what actually occurred. Together, these functions make up what is sometimes referred to as the "execution" portion of manufacturing. It usually involves many of the principal organizations that one normally finds in a plant.

Production control, for example, takes customer order and forecast data and develops detailed production schedules for manufacturing. It also monitors the actual

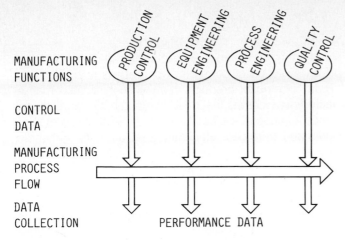

MANUFACTURING FUNCTIONS

PRODUCTION CONTROL   EQUIPMENT ENGINEERING   PROCESS ENGINEERING   QUALITY CONTROL

CONTROL DATA

MANUFACTURING PROCESS FLOW

DATA COLLECTION   PERFORMANCE DATA

Fig. 1-7 Manufacturing operations data flow.

flow of product in the process and measures manufacturing's performance against those schedules. Similarly, manufacturing engineering takes technical data which describes the product and process and develops detailed instructions to direct operators and tools. It then collects data from the production operations to monitor the performance of the process. Generally, the same type of relationship also applies to the other organizations, such as purchasing, maintenance, or quality control. Depending on the complexity of the process and the product involved, these functions can generate and collect large amounts of data.

TYPES OF MANUFACTURING DATA

The last category describing the manufacturing system is the data itself. The data can take many forms and can vary significantly according to the type of manufacturing operation involved. However, one can usually think of it in terms of several typical elements. The "how many" and "when" type of data that drives the flow of materials and products through the process is often referred to as "logistical data." The data which describes the product and processes, drives the tools, and directs the operators can be considered the "technical data." Finally, the data that provides records and reports of actual performance, such as labor and costs, is basically "administrative data." It takes all this data together to run a manufacturing operation.

TYPES OF MANUFACTURING

The manufacturing environment will, of course, be different for different types of manufacturing operations. This influences how and why computers are used. Generally, there are three principal types of manufacturing:

1. Process-type manufacturing. This involves a continuous flow of materials through a series of process steps that eventually forms a finished product, such as chemicals or semiconductors. Figure 1-8 shows an example of computer applications in that type of environment.

2. Fabrication. This involves the manufacture of individual parts by a series of operations, such as machining or

Fig. 1-8 A computer in a process-type manufacturing application: operator using a terminal in a semiconductor clean room. *(International Business Machines Corp.)*

welding. Computers also have many applications in that type of environment as illustrated in Fig. 1-9.

3. Assembly. This involves putting parts together into a complete product, such as a machine. This environment has historically been the least automated or computerized. However, with the introduction of robotics, applications for computers are increasing significantly (Fig. 1-10).

Let us put all these pieces of the generalized manufacturing environment together and look at what the future holds for CAM. We will see that the factory of the future will have an integrated manufacturing system as illustrated in Fig. 1-11. This concept, sometimes referred to as CIM, uses the computer to tie together all the operations and organizations in a plant which control the process and the flow of product. In complex manufacturing operations, even today, such an approach may be the only way to assure that a product can be manufactured at the

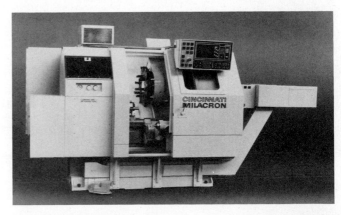

Fig. 1-9 A computer in a fabrication-type manufacturing application: computer numerical control (CNC) machining center. *(Cincinnati Milacron)*

Fig. 1-10 A computer in an assembly-type manufacturing application: robot welding in an automated automobile assembly line. *(Gilman Engineering and Manufacturing Co.)*

cost, quality, and schedule required. Note that CAM need not be represented by the physical motion of a manufacturing operation. It can also be the automation of the flow of data for control, monitoring, and reporting.

1-6 SUMMARY

Computer-automated manufacturing (CAM) is the result of an evolution of computer applications in the factory. Today's complex products require sophisticated manufacturing processes that need computers to control them. Since computers are basically machines that can manipulate and store data efficiently, they are used to handle the large amounts of data that drive modern manufacturing operations. The computer system in manufacturing manages the flow of information between the organizations and operations involved in making a product. There are several types of data which together describe what product is to be manufactured, when, and how. The amount and nature of this data will vary for different types of manufacturing operations.

There are a number of different technologies involved in CAM. Each is developing rapidly, and together they are making complex processes and computer-controlled factories possible.

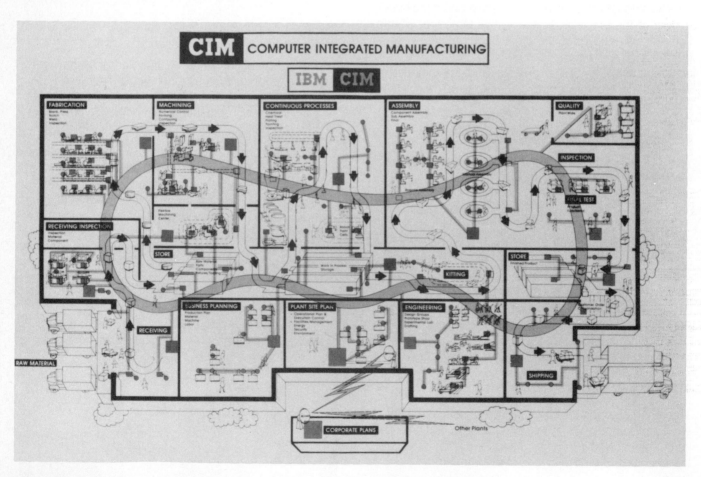

Fig. 1-11 A view of a computer-integrated manufacturing (CIM) operation. (*International Business Machines Corp.*)

REVIEW QUESTIONS

The answer to each question can be found in the section indicated at the end of the question.

1. Define computer-automated manufacturing. [1-2]

2. Describe the basic elements of a computer. [1-2]

3. Identify the major elements of a manufacturing operation in terms of organizations, operations, and data. [1-5]

4. Identify the principal types of manufacturing operations. [1-5]

5. Explain the basic functions of a computer in a manufacturing environment. [1-3]

6. What are programming languages? [1-2]

7. Explain what a computer "system" is and how can it be used in manufacturing. [1-2]

8. Define numerical control. [1-2]

9. Explain the difference between "fixed" and "programmable" automation. [1-2]

10. What is an industrial robot? [1-2]

WHY DO WE USE COMPUTERS IN MANUFACTURING?

2-1 INTRODUCTION

Before we study the technologies of computer-automated manufacturing (CAM), we should understand the reasons why computers are used in manufacturing in the first place. In order to understand this, we must start with the basic objectives of any manufacturing operation in terms of cost, quality, and schedule. Manufacturing operations make commitments to their customers regarding the quantities of specific products they will produce and deliver by certain dates at a stated cost and level of quality. Anything done in manufacturing should ultimately relate to one or more of these basic performance measurements. Therefore, computers, to be productive, must also help the manufacturing operation achieve its cost, quality, and schedule objectives. All three objectives are interactive. All the key elements of the manufacturing operation must be controlled in a balanced manner to achieve these objectives. This can be a very complicated job in a large production line. It usually requires a lot of data processing help.

For a company to be competitive in today's modern industrial environment, its manufacturing operation must not only meet its current commitments; it must also continue to improve its performance. For example, some of the essential areas for improving manufacturing efficiency are shortening production cycle times, increasing the use of equipment, improving product quality, and reducing inventory levels. In addition, for companies and industries to grow—in high technology areas such as electronics, for example—they must continue to offer better products at lower costs. In order to be constantly improving a manufacturing operation, it must be totally under control and continuously monitored and analyzed for opportunities to make improvements. Again, this can require a lot of data handling for control and analysis. This is why the computer is such a critical tool.

Computers can be used in many ways to help improve the performance of a manufacturing operation. The basic strength of a computer is efficient data handling. There are many opportunities in manufacturing to handle large amounts of data. The computer is basically a productivity tool that can help people be more efficient and do things

that would otherwise not be possible or practical. For example, computers can help reduce costs by doing things that will lessen the requirement for high-cost labor or materials, such as through automation. They can help improve product quality by providing process controls or inspections. They can also help to meet production schedules by controlling the flow of product on the manufacturing floor. Computers have therefore become essential to the manufacturing industry. They help to meet manufacturing's basic performance objectives and to keep it competitive. In this chapter, we will study those factors which influence why computers are used in manufacturing.

2-2 TYPES OF DATA USED IN MANUFACTURING

The computer's basic job is to handle data. Most manufacturing operations use three types of data:

1. Logistical data, such as quantities, schedules, and routings (i.e., manufacturing instructions)

2. Technical data, which describes the design, process or testing of the product

3. Administrative data, such as costs and payroll

In a large or complex manufacturing operation, this can amount to a lot of data handling. In most modern factories, the ability to handle all this data efficiently will have a significant influence on the ability to meet those basic manufacturing objectives we just discussed. These three types of data, although not unrelated, usually operate in three distinct data systems (Fig. 2-1). Together, they drive and manage the manufacturing process. Note that the logistical and technical data systems tend to be the more visible direct links to the operation of the manufacturing process. The administrative data systems usually are data collection functions that management uses for performance reporting and accounting purposes.

LOGISTICAL DATA

Logistical data controls the flow of materials and products through the manufacturing process from the original customer demand to the ultimate supply. It involves mainly

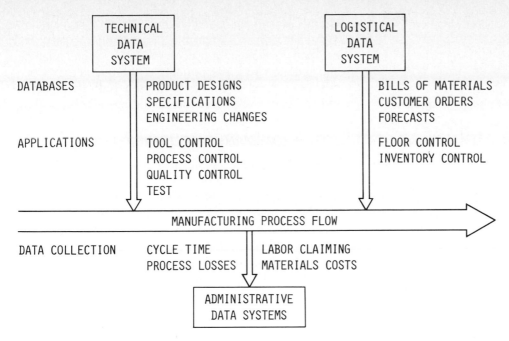

Fig. 2-1 The three main types of data used in a manufacturing operation: technical, logistical, and administrative.

the traditional production control function in manufacturing, which is responsible for scheduling and tracking the production floor. The data is derived primarily from product demands (either actual orders or forecasts), product and process descriptions (i.e., bills of materials and process routings), and actual manufacturing performance (e.g., throughput). The amount of logistical data in a manufacturing operation will depend a lot on the variety and complexity of the products and processes involved. It can also be influenced significantly by changes that may occur to the product design or demand. Such changes can be substantial in a new or growing product line. For large, high-technology manufacturing operations it is not unusual to find hundreds of different products. These products may have thousands of different design variations or features and millions of individual part types or part numbers; the data systems may have to handle hundreds of thousands of transactions or orders per week. Sophisticated computer systems are obviously essential in such a logistical environment.

TECHNICAL DATA

Technical data starts with the design of the product and process. Ultimately, this data will drive the tools and physical manufacturing activities on the factory floor. It usually includes the basic information on how a product or part is to be fabricated and tested. It often takes the form of computer instructions controlling the operation of a production tool [e.g., numerical control (NC) data]. The amount of technical data involved in a manufacturing operation will again depend heavily on the variety and complexity of the products and processes. In the world of

electronics manufacturing, for example, it is not unusual to find processes that involve hundreds of individual steps. Some of these steps use sophisticated tools that require millions of instructions per second. This is only possible through the use of computer control.

ADMINISTRATIVE DATA

Administrative data includes all the types of information used to manage the manufacturing operation as a business on a daily basis. This involves such data as the direct labor and materials costs of production, the efficiency of the production operation (in terms of such measurements as cycle time, throughput, and equipment utilization), and the indirect costs supporting production (labor, capital, and expenses, for example). The amount of administrative data depends on the complexity of the organization in terms of the number of people, the accounting structure, management reports, and so on, more than on the product itself. In a large, modern manufacturing operation many comprehensive files and reports are updated on a daily, weekly, and monthly basis for management and accounting purposes. Millions of pieces of information are involved that together will tell managers how well the business is being run. Computers are needed to process this data. Otherwise, a lot of manual paperwork would be necessary, which could not be as complete or timely.

Modern manufacturing is a data-driven process. It is this data handling job that makes computers essential. The logistical, technical, and administrative data are all necessary to run a production operation. And each depends on computer power.

2-3 SOURCES OF DATA USED IN MANUFACTURING

Most of the data that manufacturing has to handle comes from three principal sources:

1. Development, which generates design information

2. The customer, through the marketing or sales organization, who ultimately determines what is to be manufactured, in what amount, and when

3. Manufacturing, which is the source for a variety of operational data, such as logistics, test, and cost data

All these sources of data, which eventually drive the manufacturing process, must be tied together by a manufacturing data system (Fig. 2-2).

DATA FROM DEVELOPMENT

Development is the source for the data that describes the product and process, and the changes that occur to them, to manufacturing. Such engineering design data must be analyzed and reformatted by manufacturing to generate specific instructions for operators (e.g., process routings and process specifications) and tools (e.g., tool control programs and NC data). In some advanced CAM systems, the process of generating or changing designs through the generation of operator and tool control data is all done automatically by the computer. This saves a substantial amount of time and effort.

DATA FROM THE CUSTOMER

Customer requirements come in several forms and can change frequently. For planning purposes, the marketing organization will forecast demand so that manufacturing can provide for sufficient capacity. However, customer orders dictate specific requirements in terms of what products, with what features, are required in what quantities, and when. These requirements are often very dynamic, and manufacturing must be prepared to respond to frequent changes. It should also be noted that in a large manufacturing company the customer for a particu-

lar manufacturing operation may be another plant or manufacturing line within the same company. Its product, part, or material is used in the fabrication of another end product. The requirements of the ultimate external customer must then be translated and transmitted through both levels of manufacturing operations.

DATA FROM MANUFACTURING

The manufacturing organization itself is not merely a receiver and translator of data from these sources. It also generates additional data to run its operation. This includes a variety of controls for the process, the product and material logistics, and the product quality. There are also a number of operational support functions within manufacturing that generate data to perform their jobs. These include maintenance, distribution, purchasing, and finance. Therefore, manufacturing is both a major user and a source of data. Ultimately, it has the greatest data handling task in the company because of the magnitude, variety, and changes of data that affect its operations from all these sources.

2-4 COMPLEXITY OF PRODUCTS IN MANUFACTURING

One significant factor that influences the amount of data handled by manufacturing is the complexity of the products and processes involved. Such factors as the number of different parts and the process steps and changes that can occur to them can generate a great deal of data. Modern manufacturing operations, particularly those involved with high-technology products, have experienced a great increase in the complexity of what they have to manufacture. Data processing equipment, for example, is one high-technology product whose manufacture has become very complex and data-intensive.

FACTORS AFFECTING PRODUCT COMPLEXITY

Following is a list of general factors that have made data processing equipment very complex.

1. A wide variety of products
2. Many models and features
3. Large numbers of part types
4. Frequent engineering changes
5. Many process steps
6. Multilevel packaging hierarchy
7. Complex product technologies

Although this list was derived for data processing equipment, these factors are not unique to the computer industry. They probably can apply to most modern products, particularly those driven by technology (e.g., aerospace, communications, chemicals, medicine, and consumer electronics). Manufacturing data processing equipment can involve large numbers. A large company

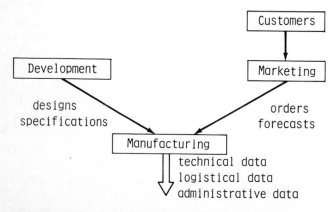

Fig. 2-2 The three main sources of data: development, marketing, and manufacturing.

in this field may manufacture hundreds of machine types. Each of these may have selective and optional features and configurations that can run into the hundreds of thousands. At any time, such a manufacturer may have to handle millions of individual active part numbers and hundreds of thousands of orders per week between its plants and customers. It is also important to recognize that this data is dynamic, and things like schedules and "mix" or variety are continually changing. The dynamics of the data in some operations may have a greater bearing on the total magnitude of the manufacturing data handling job than the absolute number of types or quantities of parts or products involved.

COMPLEXITY OF ELECTRONIC EQUIPMENT

To better understand the manufacturing process involved in a complex product, such as modern electronic equipment and computers, one can think in terms of a packaging structure or hierarchy, which is illustrated in Fig. 2-3. The packaging of this particular type of machine starts with semiconductor integrated circuit (IC) chips. They can each contain hundreds to thousands of individual electronic circuits. These chips are assembled onto carriers or "modules" which interconnect them. In turn, the modules are assembled onto printed circuit boards and finally into equipment frames or "gates" that may be made up of hundreds of thousands to millions of circuits. Each of the major elements of this packaging hierarchy involves a separate and quite complex production line. All of these lines depend on each other, at least in terms of technical and logistical data.

Electronic products and data processing equipment, in particular, can vary greatly in size and complexity. Products today can range from the small personal computer (Fig. 2-4) that must be produced in high volume at low cost, to the large processors and peripheral equipment in a "central electronics complex" or large system (Fig. 2-5). Such product variety, all with complex technologies in a packaging hierarchy, makes the total manufacturing job very complicated and involves a tremendous amount of data.

The state of the art in the technology used in the products to be manufactured also has a great influence on the magnitude and complexity of the data handling job in manufacturing. For example, in modern electronics equipment, such as computers, the product technologies used are extremely complex and the manufacturing techniques are very difficult. In fact, many of these products would be impossible to manufacture without the use of computers.

Perhaps the best examples of this are the IC and packaging technologies that are used today in computers and other electronic equipment. Figure 2-6 shows a bipolar logic device which contains over 11,000 high-speed circuits. Some of its features are as small as one-fortieth the width of a human hair. Hundreds of process steps are involved in manufacturing such devices. In the case of this particular device, four levels of metal wiring were used to interconnect all the circuits. The packaging for some of these devices can be just as complex, as illustrated by the multilayer ceramic module in Fig. 2-7. It incorporates over 30 internal layers of printed wiring and 1800 interconnecting pins to provide a package for over 100 IC

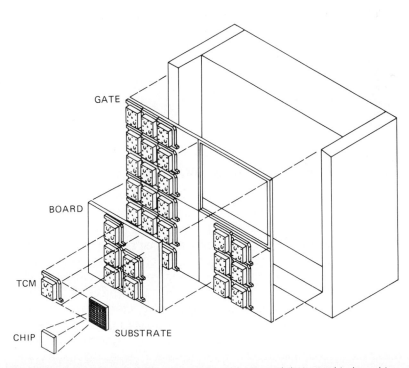

Fig. 2-3 Computer packaging hierarchy: chip to module to card to board to gate to frame. (*International Business Machines Corp.*)

Fig. 2-4 Personal computer. *(International Business Machines Corp.)*

Fig. 2-5 Large computer system. *(International Business Machines Corp.)*

Fig. 2-6 High-density bipolar integrated circuit device containing over 11,000 high-speed logic circuits. *(International Business Machines Corp.)*

devices. Together they contain over 45,000 high-speed logic circuits and 65,000 bits of high-performance bipolar memory.

Product complexity, therefore, involves the number, types, and nature of the products to be manufactured. It can be a very significant factor in why computers are used.

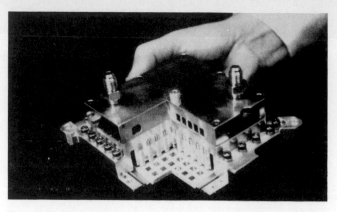

Fig. 2-7 High-performance large-scale integration circuit-packaging technology: thermal conduction module. *(International Business Machines Corp.)*

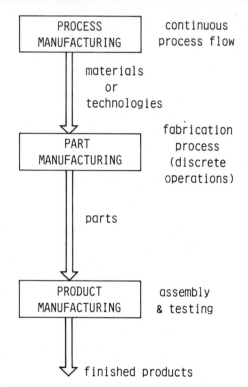

Fig. 2-8 Types of manufacturing operations.

2-5 NATURE OF THE MANUFACTURING PROCESS

"Manufacturing" is obviously a very broad term to categorize an industry. It includes many distinctly different types of production operations and products. Different types of manufacturing have different needs and problems, although their basic performance objectives are still usually the same. To be useful, computers must help satisfy the unique needs and problems of each type of manufacturing operation. As we know, a computer's principal function is data handling, and its ability to be programmed and organized into a wide variety of systems configurations makes it extremely flexible. This has made the computer successful in most types of manufacturing environments.

Figure 2-8 illustrates three major types of manufacturing. Each has its own distinctive characteristics. They may be part of the total manufacturing flow from raw materials to a complex end product, or they may stand alone to produce a finished product.

PROCESS MANUFACTURING

Process manufacturing involves a continuous flow of raw materials through a series of sequential operations. These operations transform the materials into a product, such as "manufacturing materials" or "technologies." Common examples of this type of manufacturing are chemicals, plastics, coatings, semiconductors, and metals. In this environment, there must be control of the process, equipment, and material at every step. Process manufacturing tends to involve a great deal of capital investment because of the sophisticated production equipment required. Therefore, the key operational objectives are to maximize the use of this equipment and minimize the in-process product losses. Computer systems are usually the only practical method of providing the controls and management tools needed in this type of environment.

PART MANUFACTURING

Part manufacturing, on the other hand, is a fabrication process that usually involves discrete operations on batches of objects to create parts or subassemblies. This type of manufacturing has been around much longer than the continuous process type. The numerous examples include machining, casting, molding, and metal fabrication. Also included in this category are "job shop" operations, which generally deal with small quantities of parts that may often be customized for specific applications. However, there are also many examples of automated high-volume part manufacturing. The common ingredients, from a control point of view, are the need to minimize the time used to set up tools for new jobs and the need to minimize the amount of in-process inventory. In both low-volume and high-volume operations, there are opportunities for computers to help optimize the operation.

PRODUCT MANUFACTURING

Product manufacturing can be thought of as the final process which assembles parts and subassemblies into a functional end product, such as a machine. This type of manufacturing usually involves a number of assembly and test operations, both in series and in parallel, which traditionally require a great deal of labor. Common products in this category are automobiles, airplanes, farm equipment, furniture, and even computers. In this environment, where many people are involved and the cost invested in the product is relatively high, it is desirable to minimize the labor costs and production cycle time. It is in this type of manufacturing where the power of the computer has, until recently, been least exploited. For

such industries to remain competitive by providing better products at lower costs, they had to introduce automation and control of the data and the physical assembly process.

2-6 INTERDEPENDENCIES

The magnitude of the data that manufacturing must handle is somewhat related to the complexity of the company itself and the interdependencies which exist between organizations and operations. The major business functions of a firm are highly dependent upon each other and have a great deal of influence on each other's activities (see Fig. 2-9). The marketing, development, manufacturing, and service organizations are closely involved with a product from its conception to the end of its market life. Throughout this life cycle, these organizations exchange data, place requirements, and make commitments to each other in order to keep the company's business running. In a large company with a lot of products, this can create an extremely large and complicated data handling job. At any time, for example, marketing should be able to find out the status of an order from manufacturing. Manufacturing, in turn, should be able to get the latest details of the product design from development. Service, at the same time, needs to be able to locate a product with any particular serial number, to order any necessary spare parts, and to have access to design and diagnostic information for maintenance or repair. These routine activities require a continuous interchange of information between these organizations. In most cases, this is made possible by communicating computer systems.

INTERDEPENDENCIES BETWEEN LOCATIONS

Most large companies also have interdependencies between the locations where they develop and manufacture products. A number of different plant and laboratory locations may be involved in developing and supplying materials, parts, assemblies, and data to each other (Fig. 2-10). Depending on how complicated the end product is and how vertically integrated the company is, a lot of locations and data exchange may be involved. Most companies with such interdependencies have established corporate database and data communication systems to tie all these operations together. In so doing, the company operates effectively as a single manufacturing entity.

INTERDEPENDENCIES BETWEEN ORGANIZATIONS

Even within a typical plant, there are many data generating functions or organizations. Each of these depends on the others for executing their part of the total manufacturing process. Production control, for example, translates the customer demand into detailed production schedules for manufacturing. The manufacturing organization, in turn, must then collect and report data on the status of

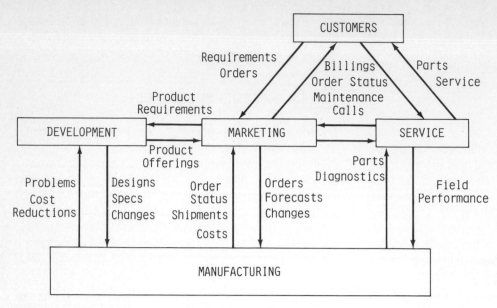

Fig. 2-9 Interdependencies between major business functions.

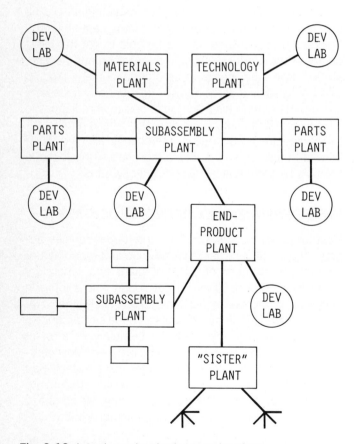

Fig. 2-10 Interdependencies between locations.

jobs in process. To do its job, manufacturing engineering needs performance data from manufacturing such as process and test losses and process and equipment problems. The quality control organization takes data from the process (e.g., inspections) and from the customer or user location relating to product quality and reliabil-

ity. This is fed back to manufacturing and manufacturing engineering for any necessary corrective action. Therefore, there must be a number of data systems within each plant that permit these organizations to communicate on a daily basis.

2-7 GENERAL NEEDS AND TRENDS IN THE USE OF COMPUTERS IN MANUFACTURING

Let's conclude this discussion of why computers are used in manufacturing by reflecting briefly on some of the general needs for computer use in the future. Consider which manufacturing operations, by their nature, lend themselves best to computer applications. It should now be clear that some of the major manufacturing activities with an increasing dependency on computer control are process data collection and monitoring, product scheduling, and the engineering functions of design and process control. To meet these needs, manufacturing must efficiently handle data collection and reporting and must be able to easily access large amounts of data from multiple sources.

Manufacturing operations in the future will need to be more flexible in providing for product, process, and schedule changes; in customizing products; and in responding to shorter product life cycles. Obviously, humans using manual techniques alone cannot satisfy all these needs. We must continue to expand our use of computers by exploiting their abilities to store, monitor, retrieve, process, and communicate data with increasing efficiency. Only computers can provide us with the capability to integrate all the elements necessary to control the total manufacturing operation efficiently.

2-8 SUMMARY

Computers are efficient data handling machines. Since modern manufacturing operations generate and use a great deal of data, there has been an increasing use of computers to help manage the factory. The need for computers is influenced by the amounts and types of data that must be handled by manufacturing. Data for production operations comes in several different forms from a number of sources (e.g., technical data from engineering). The complexity of the products to be manufactured and the processes involved also have a great deal of influence on the data handling job required for manufacturing. There are also different types of manufacturing operations which have different needs for data processing, communication, and control. Most large manufacturing companies have complex operations that require data communications between many organizations and locations. In summary, the use of computers in manufacturing gives us a tool to overcome the basic complexities and inefficiencies of production operations. Computers help make the manufacture of modern, complex products possible and integrate the management of the total factory.

REVIEW QUESTIONS

The answer to each question can be found in the section indicated at the end of the question.

1. Explain why computers are used in manufacturing. [2-1]
2. What are the basic performance objectives of a manufacturing operation? [2-1]
3. Identify the principal types of data found in manufacturing and give examples of each. [2-2]
4. Identify the principal sources of data in a manufacturing operation. [2-3]
5. Describe how product complexity affects the need for and use of computers in manufacturing. [2-4]
6. Describe the different types of manufacturing operations and how they differ in their need for computers. [2-5]
7. Identify the types of interdependencies that exist within a manufacturing company that can influence its need for computers. [2-6]
8. What are some of the future trends in manufacturing that will increase the need for and dependency on computers? [2-7]

COMPUTER
CONTROL

3-1 INTRODUCTION

The earliest applications for computers in manufacturing were for the control of production tools. This control function evolved from the development of numerical control (NC) technology, for the machine tool industry. Before practical electronic computers were developed, machine tools used less sophisticated control mechanisms such as switches and relays. The basic functions of a computer in a tool control application are to (Fig. 3-1):

1. Interpret a program of operating instructions
2. Generate signals which direct the operation of the tool
3. Monitor the performance of the tool
4. Compare performance against control limits
5. Adjust the direction of tool operation as needed

Depending on how complicated the tool and the operations are, these control functions can involve a lot of data handling and manipulation. As the use of computers in manufacturing evolved over the years, the applications they were used in became more complex. They covered not just the control of individual machine tools but eventually the control of entire manufacturing operations which included many types of production equipment and processes.

In this chapter we will trace the evolution of the computer's use in manufacturing by first establishing a basic

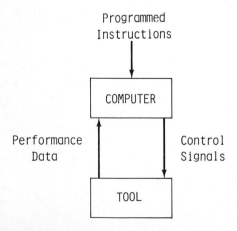

Fig. 3-1 Computer control.

understanding of NC and how it developed as a key manufacturing technology. Then we will follow that evolution to the use of computers in manufacturing control systems. This ultimately leads to the complex hierarchical and distributed systems which are emerging as the basis for computer-automated manufacturing (CAM) in the future.

3-2 NUMERICAL CONTROL

When NC was developed in the early 1950s, it was defined simply as "a technique to control machine tools through a program, coded in numerical form, and recorded on a suitable medium." (Mark Morgan, Numerical Control—The First Fifteen Years." IBM report, 1965.) But as NC technology evolved and computer technology was also introduced, it became much more than a technique to control machine tools. NC is basically a technique for data communication and control that can be applied throughout manufacturing. And, as we will see, it led to the development of many of the essential ingredients of CAM today.

Before there was NC, machine tools were controlled exclusively by the skill of operators who used their eyes for measurement and their judgment for feedback. With the development of NC, measurements taken by automatic instruments or gauges are fed back to the control system which directs the operation of the tool. This permits the optimization of the machining process while it is in operation and virtually eliminates errors and inaccuracies that could have been introduced by even the most skilled operators. NC did not eliminate the need for operators—it just made them more productive. In fact, the job of the machinist was expanded. It required additional skills to write and interpret control programs as well as operate the tool controller.

NC is a form of automation, but it differed from earlier forms of automation. It introduced the use of programmable control—the ability to perform a variety of tasks which can be changed or "reprogrammed." Traditional forms of automation, which emerged from the industrial revolution around the turn of the twentieth century, were fixed or hard automation systems which could only perform repetitive tasks. Traditional automation involves the high-volume production of individual parts or assemblies and is still common in manufacturing today. Such automation systems usually involve a number of tools in a fixed arrangement connected by materials handling and transfer mechanisms. They are usually very expensive to tool and are therefore most often used for high-volume, repetitive production operations (such as in the manufacture of containers, fasteners, and electrical parts). The typical characteristics of an NC application, however, include:

1. Similar, but not identical, parts and materials
2. Parts to be produced in a variety of sizes and shapes
3. Relatively small batches of work
4. A similar, but again not necessarily identical, sequence of operations
5. Machine movement in multiple axes simultaneously

NC therefore made possible, for the first time, the manufacture of small quantities of precision and even customized parts at relatively low costs. This technology, by its nature, became the mainstay of the machine tool industry and was then extended into a wider variety of manufacturing applications.

HISTORY OF NUMERICAL CONTROL

The earliest efforts to develop a numerically controlled machine tool began around 1950. Industrial feasibility was first demonstrated in 1955 on a milling machine, which is shown in Fig. 3-2. This "digital cam miller" had single-axis control from a program stored on paper tape which was programmed by an early digital computer. Further developments and refinements to this initial effort led to continuous-contouring millers, point-to-point positioning control, and automatic tool changers. An automatic jig borer developed in 1956 featured the indexing of a matrix of 30 different tools for selection, sensing of hole depths, selection of spindle feeds and speeds, and table positioning. All of these features were programmable. That meant that without operator intervention, holes could be bored in a workpiece at different diameters and depths at predetermined locations with great accuracy

and speed. This was a major breakthrough for production operations that needed small batches and short cycle times. It led to the evolution of a class of tool-changing machines that are the workhorses of the machining industry today, as illustrated by the modern example in Fig. 3-3.

NUMERICAL CONTROL SYSTEM

An NC system is comprised of three basic components (Fig. 3-4):

1. A program of instructions
2. A controller unit
3. The machine tool, equipment, or process which is being controlled

The program in early NC systems was coded on a punched tape, punch cards, or magnetic tape. It included two types of information: operational commands and dimensions. The operational commands generated signals to the switching systems in the machine, which actuated its control mechanisms. The dimensional data identified the desired position of the tool relative to the workpiece. This data was entered on the tape using a

Fig. 3-3 Modern tool changer: three-spindle machining center with automatic magazine for changing up to 36 different tools. *(Boston Digital Corp.)*

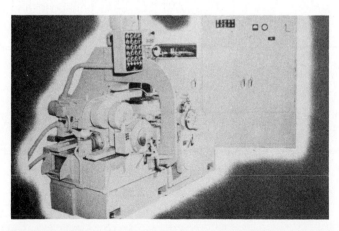

Fig. 3-2 Digital cam miller: the first computer-controlled machine tool, 1955. *(International Business Machines Corp.)*

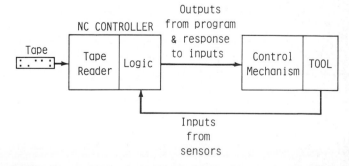

Fig. 3-4 Numerical control (NC) system.

standard perforation or punch coding system such as the one illustrated in Fig. 3-5. The controller read and interpreted the program of instructions and converted it into mechanical actions of the machine tool.

The principal elements of a traditional controller include:

1. A tape reader

2. A data buffer

3. Signal output channels to the machine tool

4. Feedback channels from the machine tool

5. Sequence controls to coordinate the overall operation of the machine

The tape reader reads the punched tape instructions into the data buffer and stores them in logical blocks representing the desired sequence of process steps. The signal output channels are connected to the control mechanisms on the tool, such as servomotors and solenoids. Feedback data comes from analog or digital devices (such as precision potentiometers or lead screws, variable transformers), which are coupled to the drives of the tool to assure that the instructions have been followed.

NUMERICAL CONTROL PROCESS

The NC process comprises three fundamental steps: planning, programming, and execution. "Process planning" is interpreting information about the design of a part to be fabricated, such as from an engineering drawing, into a description of manufacturing process steps. The process planning step is usually done by a "part programmer." Part programmers have special skills and experience with the machine tools involved; the programmer develops a detailed program which will direct a tool's operation. The programming job, therefore, involves planning and specifying every step and movement of the NC machine into a complete process sequence. This can be an extremely laborious task. It requires many calculations of tool position coordinates, feed rates, and speeds, which can get particularly complicated for complex designs. To machine curved surfaces, for example, requires variable feed rates and a continuous flow of position coordinate data to control the precise geometric path for the tool to follow. This task soon becomes constrained by the time and skills required.

Therefore, a more efficient approach was necessary, which led to the development of computer-assisted part programming. The use of the computer eliminated the manual computational work involved in defining geometries and directing tool movements. The computer's job is to translate the instructions of the programmer, perform the necessary calculations, and code the instructions for the machine tool. Early NC programming languages were first developed in the mid-1950s using English-like terms that were similar to those traditionally used in the machining environment and therefore familiar to machine operators. The basis for most of the symbolic languages used in NC today is the automatically programmed tool (APT) system. Over the years, APT has been enhanced and modified by a number of different machine tool and computer manufacturers to improve its efficiency and capabilities.

The part programmer prepares a program of instructions using the NC language, which is then entered into the computer (Fig. 3-6). After the computer translates the program and performs the necessary calculations, it codes and formats machine instructions that are punched onto an NC tape to be read by the machine tool. The NC program usually consists of two basic parts:

1. Defining the geometry of the workpiece

2. Specifying the sequence and path of the tool

In the APT language, these instructions take the form of relatively simple statements that use symbols to repre-

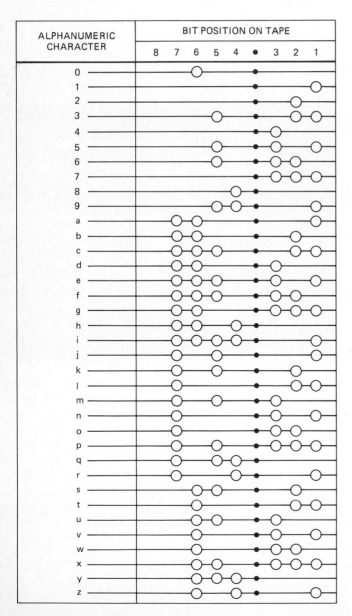

Fig. 3-5 Standard NC punched-tape code.

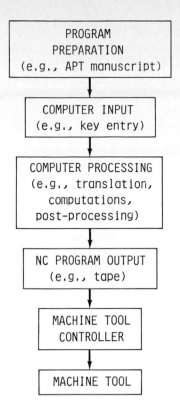

PROGRAM
PREPARATION
(e.g., APT manuscript)

↓

COMPUTER INPUT
(e.g., key entry)

↓

COMPUTER PROCESSING
(e.g., translation,
computations,
post-processing)

↓

NC PROGRAM OUTPUT
(e.g., tape)

↓

MACHINE TOOL
CONTROLLER

↓

MACHINE TOOL

Fig. 3-6 NC part programming.

sent the variables and words which provide instructions to the computer. A typical program uses four types of statements:

1. Geometry, to define a shape

2. Motion, to direct the movement of the tool

3. Post-processing, to control the operation of the features of the machine

4. Auxiliary statements, which define and identify parts and sizes

Such statements are shown in Fig. 3-7 as a general illustration.

TYPES OF NUMERICAL CONTROL SYSTEMS

NC systems can be classified by the different amounts of control they have over the relative motion between the workpiece and the cutting tool. There are three principal types of NC systems:

1. "Point-to-point" NC simply moves the tool to a predefined position; it has no control over the tool's speed or path. Examples of this type of NC system include drill presses and spot welding.

2. "Straight-cut" NC moves the cutting tool parallel to an axis at a controlled rate of speed, such as in an NC milling operation.

3. "Contouring" or "continuous path" NC requires the most control since it must be capable of simultaneously controlling more than one axis of motion and continuously controlling the path of the tool to generate the desired geometry of the workpiece. Examples of this include continuous milling and turning, which, in some cases, can involve more than two axes as well as both point-to-point and straight-cut moves.

Today, the machine tool industry has a class of tool called "machining centers," in which one machine tool is designed to perform a variety of operations, such as drilling, milling, boring, reaming, and tapping. These tools incorporate automatic tool changing, workpiece positioning, and loading, which provide efficiency as well as

Geometry statements

FORM: SYMBOL = GEOMETRY TYPE/DESCRIPTIVE DATA
e.g., P2 = POINT/1.0,2.0,3.0
OR L1 = LINE/P1,P2

Motion statements

FORM: MOTION COMMAND/DESCRIPTIVE DATA
e.g., GOTO/P2
OR FROM/1.0,2.0,3.0

Post—processor statements

FORM: FUNCTION/DESCRIPTIVE DATA
e.g., FEDRAT/30 (i.e., feed rate in inches per minute)
OR SPINDL/1000 (i.e., spindle speed in rpm)

Auxiliary statements

FORM: IDENTIFICATION/DESCRIPTIVE DATA
e.g., CUTTER/.100 (i.e., cutter diameter in inches)
OR OUTTOL/.001 (i.e., outside tolerance in inches)

Fig. 3-7 NC program statements—APT language.

flexibility for small-batch, precision machining applications (Fig. 3-8).

NC technology has therefore been a key contributor to the success and productivity of the machine tool industry. Many of the precision-machined parts and the products that they are used in (such as aerospace and automotive parts) would not be feasible or economical without NC. The advantages of NC include:

1. Reductions in both process and set-up time

2. Increased flexibility in terms of changes to design or schedules

3. Improved accuracy and repeatability

4. Less need for fixturing to hold and position the parts

NC has therefore proved to be most useful in applications that involve small lot sizes, complex and expensive designs with close tolerances, many machining operations, and changes in designs and schedules. This, however, is only the most obvious, and limited beginning of NC's role in manufacturing.

3-3 EXTENSIONS OF NUMERICAL CONTROL

PROGRAMMABLE CONTROLLERS

As the development of NC and computer technology progressed, opportunities presented themselves for increasing the capabilities and broadening the applications of NC. One step was the use of modern electronics technology to replace the "hard-wired" relay logic that had been used for tool controllers. This new generation of programmable controllers (PCs) are basically small, dedicated computers with logical but not computational capabilities. These units are usually mounted directly on the machine tool (Fig. 3-9). They have an operator control panel and can be reprogrammed either by physically replacing one of the internal electronic assemblies or tape memories or by transmitting a new control program from a central computer. A typical PC is comprised of four main sections:

1. The input section, which receives signals from sensors located on the tool

2. The output section, which generates signals that actuate the control mechanisms on the tool

3. The program section, which stores the control program

4. The operating section, which directs the logical operations of the controller itself

The basic functions of a PC include:

1. Control, often using the traditional relay-type logic to generate control signals

2. Timing, to control the duration or separation of signals

3. Counting, to generate an output based upon the completion of a predetermined number of steps

As electronics and computer technologies continue to progress, more and more function is being added to the PCs. This is making it difficult, if not irrelevant, to find the difference between them and minicomputers.

DIRECT NUMERICAL CONTROL

To overcome some of the difficulties and shortcomings of the earlier NC systems, such as programming errors, punched tape input, limited controller functions, and a lack of operational data, a computer was linked directly to the tool controller. This approach, called "direct numerical control" (DNC), loads the control programs for many tools from a central computer, eliminating the need for

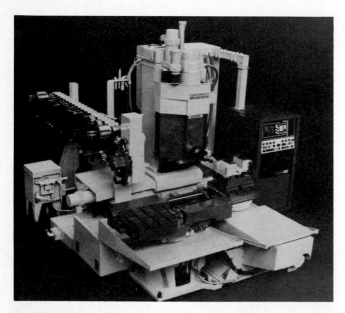

Fig. 3-8 Machining center provides multioperation production capability with a single, computer-controlled machining system. *(Cincinnati Milacron)*

Fig. 3-9 Programmable controller. *(Automation Intelligence Inc.)*

punched tape and readers. A DNC system is composed of a central computer with a main memory storing the NC programs, connected by data communication lines to a number of machine tools (Fig. 3-10).

There are generally two different types of DNC systems, on the basis of the means used to link the computer with the tool controller. In a "behind the tape reader" system, the tape reader is replaced by the data communication lines to the computer and the program instructions are stored in the controller's buffer memory. In the other type of system, a "special machine control unit" replaces the traditional "hard-wired" tool controller with a PC that is directly linked to the computer to improve communications efficiency and provide programming flexibility. Other configurations of more complex DNC systems can include communications controllers, satellite computers, and buffer memories. They minimize the burden on the main computer and reduce the system response time and exposure to downtime. DNC therefore has a number of advantages over the basic NC system, including:

1. Eliminating the tape and reader

2. Providing flexibility and increased capabilities in machine programming

3. Permitting the remote control of a large number of machines

In addition, DNC systems can provide reports to management, such as equipment performance and utilization data, by having access to a large central computer.

COMPUTER NUMERICAL CONTROL

With the invention of the minicomputer came the development of another form of NC called "computer numerical control" (CNC). In CNC, a dedicated stored-program computer is used to perform the NC functions for each tool. The major difference from conventional NC is the replacement of the tool controller by a computer. Relative to DNC, the main difference is the use of a single dedicated computer for each tool. It is typically located near the tool and programmed to optimize that particular tool's performance. A CNC system can provide many of the same advantages in flexibility and performance as does DNC, without the dependency and constraints of a general-purpose central computer system. CNC has therefore become an increasingly popular approach for small or specialized machining operations. The current trend in the machine tool industry is to capitalize on the advantages of both DNC and CNC by developing hybrid systems. Such systems use minicomputers for tool controllers as well as satellite and central computers to control an entire manufacturing process. We will discuss this in greater detail later in the chapter.

ADAPTIVE CONTROL

The power and the presence of the computer in the machining operation made possible a further advance in optimizing the manufacturing process, called "adaptive control." In adaptive control machining, real-time measurements are made of part parameters and fed back to the NC system that controls the operation of the tool. By responding to variations and changes through the control system, the machining process can be optimized while it is in operation, even in the continuous path mode. This permits the system to compensate for some of the common sources of variability that can occur in a machining operation, such as:

1. Wear of the tool

2. Hardness and rigidity of the workpiece

3. Position of the tool relative to the workpiece

4. The geometry of the cut

Adaptive control programs monitor process variables such as these, compare them to predetermined control limits, and manipulate the speeds and feeds of the machine to compensate and optimize its performance. Using adaptive control can increase production rates by:

1. Reducing machining time and increasing tool life

2. Avoiding unnecessary part losses

3. Reducing the need for operator intervention

4. Providing for easier part programming

OTHER NUMERICAL CONTROL APPLICATIONS

NC technology was fundamentally a technique to communicate with and control machines that could improve the efficiency of precision small-batch manufacturing operations. Therefore, it eventually found its way into many applications other than machining. One early application in the electronics industry was in the assembly of printed circuit boards where electronic components had to be selected and placed in specific positions for subsequent

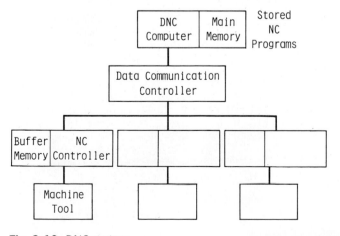

Fig. 3-10 DNC system.

soldering. The automation of this application evolved over the years from the earliest attempt to program the operation by the use of a template by the operator (Fig. 3-11). Then came the use of an NC tape to control the position of the table. Today, computer-controlled robotic assembly systems are used. Another early application in electronics that still exists in a variety of forms today is the wiring machine (Fig. 3-12). This machine combined the NC positioning and contouring concepts for routing and fastening wires between pins on the back panel of printed circuit boards.

NC technology was also applied to the development of plotters and automatic drafting machines, where the motion of the board relative to the pen is similar to that of the table of a machine tool relative to a cutter. Drafting systems continued to evolve with the use of higher-level languages for simpler programming, faster printers, video display terminals, and eventually mathematical functions to define geometries which made computer-aided design (CAD) systems feasible.

Finally, one of the most recent and fastest-growing applications of NC is in industrial robots. The complex motions of each of the robot axes must be controlled by an NC program. There are both point-to-point and continuous path robots which can be programmed by a variety of methods and controlled in an NC, DNC, or CNC mode. We will deal extensively with these and other applications of NC technology in subsequent chapters.

3-4 MANUFACTURING CONTROL SYSTEMS

Over the years, the use of the computer in manufacturing has evolved. It started with relatively simple stand-alone tool control applications, which used small dedicated computers. This eventually led to more complex manufacturing control systems. They have highly structured architectures involving host computers and satellites which exchange real-time data. This was a natural outgrowth of the DNC systems discussed in Sec. 3-3, which used a central computer to control the operation of multiple machine tools. By extending this concept to other manufacturing operations, early manufacturing control systems were able to integrate the handling of technical data which drove a variety of different types of tools (Fig. 3-13). Such systems often had an interface to an engineering database with design information, which provided the source of technical data for the operation of the tools (e.g., testers, design and drafting systems, NC tools).

FUNCTIONS OF A MANUFACTURING CONTROL SYSTEM

The basic functions of a computer control system are to:

1. Monitor (e.g., a process parameter)
2. Compare (e.g., performance versus control limits)

MANUAL PROGRAMMING BY TEMPLATE

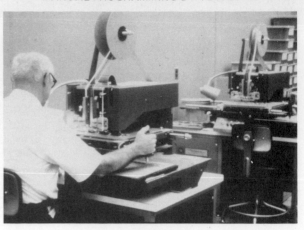

NC CONTROL BY PAPER TAPE

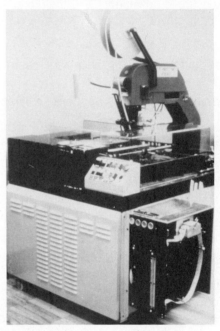

COMPUTER CONTROL WITH ROBOTICS

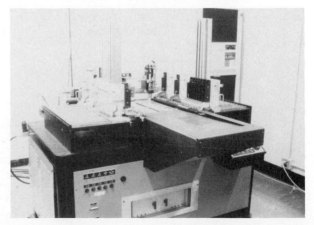

Fig. 3-11 Evolution of automating component insertion. *(International Business Machines Corp.)*

Fig. 3-12 Early automatic wiring machine. *(International Business Machines Corp.)*

3. Provide output (e.g., control signals, alarms, reports)

These functions can be applied to the control of tools, processes, and the logistics of product movement and scheduling on the factory floor. The nature of the control system can take several forms, for example:

☐ "Process monitoring" involves the observation of processes and equipment to collect and record data. This tracks the performance of the operation—but does not directly control it. The data collected may include process data (e.g., process variables), equipment data (e.g., utilization), and product data (e.g., quantities).

☐ "Direct digital control" (DDC) involves a more active role for the computer. It uses the process data that is monitored to control the operation of the tools. The computer may compare the performance of the tool or process against control limits. Or it may evaluate the effects of the performance using analytical calculations and feedback control signals to the tool or process to compensate or adjust its operation.

Sophisticated control systems may attempt to optimize the performance of an entire complex production process. The theory and operation of such control systems will be covered in greater depth in Chap. 17.

INTEGRATED MANUFACTURING SYSTEMS

Today, large manufacturing control systems are employed to tie many tools and production operations together in a hierarchical network of controllers, satellite computers, and large central host systems. Such systems may handle technical data for tool control, logistical data for factory floor control, and administrative data for production management. The data that drives these systems may come from multiple databases located in a number of different host computers, which could even be at different locations. The architecture and complexity of such systems will depend upon the nature of the manufacturing process and the data involved. As illustrated in Fig. 3-14, the stages of an assembly process controlled by a manufacturing control system may be driven by three types of data:

1. Customer order data

2. Engineering design data

3. Manufacturing data

In Fig. 3-14, the manufacturing control system drives the assembly processes in the manufacture of large computers through its packaging hierarchy, from integrated circuit (IC) modules to printed circuit card and board assemblies to the gates and mainframe of the computer. This particular system also permits the computers that are manufactured to be customized to the unique requirements of each customer. It is also continually updated to the latest design level by integrating the customer, engineering, and manufacturing data systems.

HIERARCHICAL SYSTEMS

Most early manufacturing control systems attempted to build total manufacturing process control around a large central computer to fully use the expensive computer system hardware involved. One advantage to this approach was the power and speed of a large computer to process large amounts of data efficiently or perform com-

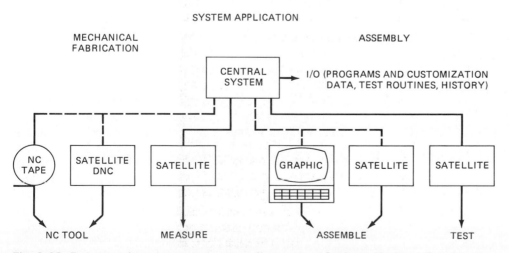

Fig. 3-13 Early manufacturing control system. *(International Business Machines Corp.)*

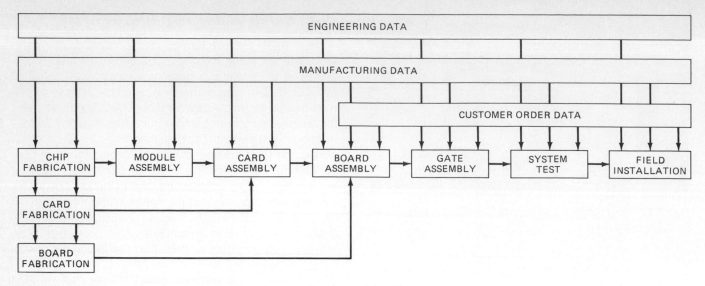

Fig. 3-14 Modern integrated manufacturing control system. *(International Business Machines Corp.)*

plex mathematical operations. However, it had the disadvantage of high programming and support costs as well as potentially slow responses to tool control due to the large amount of data transfer involved. Such systems were also limited in scope to handling only certain types of data—typically technical data for tool control. They were not linked to other sources of production information, such as logistical and administrative data. As manufacturing processes and production operations have become increasingly more complex, so have the systems which are used to control them. The continually increasing capabilities of computers which are available at lower costs make more sophisticated systems possible. In order to tie more tools and production operations together, it becomes necessary to employ additional levels of hierarchy in the systems through the use of "subhosts" and satellite computers.

Hierarchical systems (Fig. 3-15) are similar to the organization of a plant where each level performs a specific function. At the first level, a controller communicates directly with the process or tool. It performs most of the control tasks and provides constant monitoring. At the second level, an "area controller" provides supervisory control over other controllers. It coordinates the activities of many minicomputers in a particular area of production and provides for management reporting and data transfer. At the third level, the host computer manages or controls the central databases and provides high-level functions such as modeling and high-level management reports. The benefits of such hierarchical systems include:

1. Lower costs of hardware and software
2. Faster control
3. Modularity and ease of expandability
4. Limited exposure to system failure, with ease of backup

DISTRIBUTED SYSTEMS

In order to handle large amounts of data efficiently without creating an unmanageable dependency on a large central host computer, communications networks were established. Together with the subhost architecture, they permit the computer control to be "distributed" throughout the manufacturing operation. "Distributed system" architectures employ concepts which will be dealt with in later chapters, such as peer-to-peer communications, loosely coupled computers, common data buses, central databases, local controllers, and local area networks (LANs). Such systems have a multilevel, hierarchical structure, but the computer power is distributed throughout the system (Fig. 3-16). Therefore, there is not as great a dependency on the central host computer to perform the control functions at a lower level.

Control functions which do not require very complex

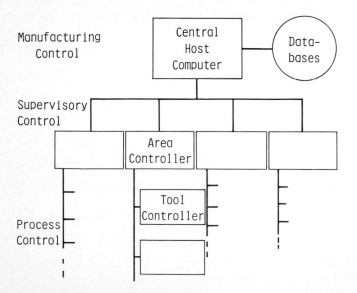

Fig. 3-15 Hierarchical control system.

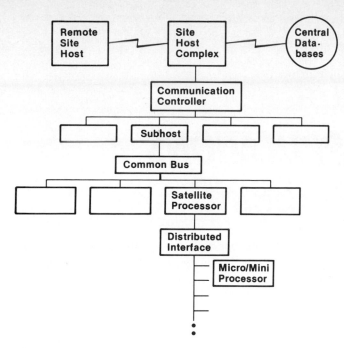

Fig. 3-16 Distributed hierarchical system.

calculations or large amounts of data can be done more efficiently on small systems. The minicomputer permits faster feedback and communication to the process or tool than could be provided by the large central host. The distributed system approach permits us to integrate the control of complex manufacturing operations that involve multiple functions and draw on remote data sources, which is essential in the management of the factory of the future.

One form of this, which is already occurring as an outgrowth of NC machining systems, is called "flexible manufacturing systems" (FMS). In such systems, multiple NC machining operations and automated materials handling are integrated into a total factory control system. These flexible systems make it economical to fabricate small batches of precision parts by automating the controls of the tools and the movement of the product between multiple operations. By optimizing the total process and using advanced manufacturing technologies (such as robotics and CNC), set-up times, delays, inventories, and errors can be reduced. These and other examples of CAM will be explored in depth in later chapters.

3-5 SUMMARY

Computer control is the basic technology that makes computer-automated manufacturing (CAM) possible. This control function evolved from the numerical control (NC) of machine tools where controllers were used to direct and adjust the operation of the tools. As the NC technology developed, it gained additional capabilities and applications. Direct numerical control (DNC) and computer numerical control (CNC) provided improvements in flexibility and performance as well as the ability to control multiple tools. These technologies eventually led to the development of manufacturing control systems which could tie together the control of a variety of manufacturing equipment and operations. Today, such systems are often very large and complex since they may control an entire production line or process.

The NC technology was also extended to other applications which themselves have become important technologies for CAM. These include automated assembly tools (such as component insertion and wiring machines), computer-automated design (CAD) and graphics, and robotics. The basic principles of NC provided the technical foundation for these and other manufacturing applications.

REVIEW QUESTIONS

The answer to each question can be found in the section indicated at the end of the question.

1. Identify the basic functions of a computer in a control system. [3-1]

2. What are the main advantages of NC? [3-2]

3. Explain how DNC is different from NC; what are the advantages of DNC? [3-3]

4. How is CNC different from DNC and NC; what are the advantages of CNC? [3-3]

5. Define NC part programming. [3-2]

6. Describe the differences between the types of NC systems. [3-2]

7. Explain what a programmable controller is and its basic functions. [3-3]

8. What is adaptive control and what are its advantages? [3-3]

9. Identify some manufacturing applications (other than machining) which use NC technology. [3-3]

10. Describe some of the different forms and functions of manufacturing control systems. [3-4]

11. What are hierarchical and distributed control systems and how do they differ from NC systems? [3-4]

12. Identify the advantages of distributed processing over centralized computer control systems. [3-4]

COMPUTER APPLICATIONS IN MANUFACTURING

4-1 INTRODUCTION

In Chap. 2 we discussed why we use computers in manufacturing. As you remember, the computer's use was driven by such factors as the types and sources of data, the complexity of the products, the nature of the manufacturing process, and the interdependencies that exist in the manufacturing operation. In Chap. 3 we addressed the basics of computer control technology, which established the foundation for computer applications in manufacturing. What started as numerical control (NC) of machine tools evolved into manufacturing control systems and a wide variety of applications that used NC technology. This chapter will focus on where computers are used in manufacturing. That is, it will identify some of the typical applications in industry today.

There are many different types of manufacturing industries, all of which use computers in their operations. Some of the major categories include both large and small companies engaged in important parts of the economy: automobiles, aerospace, computers, chemicals, machine tools, metals, packaging, appliances, consumer electronics, electronic components, and communications. Within such industries, there are many manufacturing applications in which computers can be used. Several major groups are design, tool control, floor control or logistics, automation, testing, materials handling, process control, facilities, and service. During the course of the book we will address some of these applications in greater depth. However, here we will attempt to provide a broad overview of the scope and variety of computer applications in modern manufacturing.

4-2 TOOL CONTROL APPLICATIONS

MACHINING CENTERS

As we saw in Chap. 3, tool control was one of the earliest applications for the computer in manufacturing, particularly in machining operations. NC technology evolved from the control of individual machine tools to entire machining systems. Machine tools themselves became

very complex and versatile, such as the modern "machining centers" which can perform a variety of precision machining operations on a workpiece at a single station. The principal features of a machining center are:

1. Multiple functions
2. Work and tool changes
3. NC control
4. Unattended operation

Machining centers find their greatest use in small job shops where they can:

1. Reduce machining time per part
2. Reduce nonproductive time (e.g., loading, tool changing)
3. Increase flexibility
4. Improve safety and reduce noise
5. Reduce operator involvement
6. Provide compatibility with control systems

NC machine tool technology took a further step with the help of the computer in the development of automated machining systems. These are essentially production lines for the fabrication of complex machined parts, based on the concept of a machining "cell." Machining cells consist of a set of complementary, computer-controlled machine tools that together perform a series of machining operations on a workpiece. These tools are connected by automated handling systems or robots for loading, unloading, and transporting the product between operations (Fig. 4-1).

FLEXIBLE MANUFACTURING SYSTEMS

With the machine cell as a base, advanced factories have now been developed which employ flexible manufacturing systems (FMS). They tie together multiple cells into an integrated production operation that controls the entire process in the fabrication of a finished product. The control of the individual tools and the materials handling and logistics is handled by a computer system. The Machine Tool Task Force (an industry group) defined FMS

Fig. 4-1 Machining cell. *(Cincinnati Milacron)*

as "a series of automatic machine tools or items of fabrication equipment linked together with an automatic material-handling system, a common hierarchical digital preprogrammed computer control, and providing for randomly fabricating parts or assemblies that fall within predetermined families."

FMS systems evolved from machining centers and cells by adding materials handling systems that transported workpieces between sets of tools and integrated control systems. Large and advanced FMS systems include multiple machine cells tied together with automated materials handling systems which employ carousels, conveyors, rotary tables, and robots (Fig. 4-2). FMS systems are ideal for batch manufacturing operations with midrange volumes and variety. They can reduce the cost penalty of product diversity, reduce inventory, reduce lead times, and provide the ability to quickly change the product mix and respond to market changes. The machining industry is again at the forefront of the application of computers in manufacturing, being one of the first to actually reduce to practice integrated manufacturing systems. Now there are also other applications for FMS's, including welding, assembly, molding, and sheet-metal working.

Fig. 4-2 Flexible manufacturing system (FMS). *(Hughes Aircraft Co.)*

TOOL CONTROL IN ELECTRONICS APPLICATIONS

There are many tool control applications for computers in manufacturing other than machining. For example, many of the sophisticated tools employed in manufacturing electronic components and assemblies require computer control. One common application that has been automated over the years, due to the increasing complexity, precision, and labor content involved, is wire bonding. In many electronics products, a hierarchical packaging system involves several stages which employ wire bonding to make electrical interconnections. Typical applications include wire bonding of chips to packages (modules), wires to substrates, and wires to boards. As we saw in Chap. 3, automatic wire bonders were a natural extension of NC technology and were first used for wiring between pins on printed circuit boards or back panels used in computers and other electronic equipment.

Figure 4-3 illustrates a modern automatic wire bonder which attaches small wires to the surface of multilayer ceramic substrates to provide additional interconnections between chips and permit engineering design changes. The surface of this particular substrate has 12,500 wiring pads with provision for 2000 wiring channels and 10 wiring levels. This wiring job, which employs wires that are about the same diameter as a human hair, is obviously very complex and could not be accomplished manually. The tool automatically strips, forms, and ultrasonically bonds each wire. To ensure that the wires are bonded to the exact positions required, the entire surface interconnection pattern of each substrate is first mapped and stored in the control computer. The control system also performs calculations based on the map and wiring needs in order to optimize the wire routing.

4-3 DESIGN

The manufacturing process starts with the design of the product. Computerized design systems are sometimes referred to as computer-aided design (CAD) or computer-aided engineering (CAE). The use of computers for design applications started with numerically controlled plotters and automated drafting systems. This technology developed into systems that could design two-dimensional (2D) mechanical drawings very efficiently. To this was added surface-geometry graphics capability, in the form of three-dimensional (3D) "wire-frame" shapes. Such systems are interactive with the designer (Fig. 4-4). They can include additional features such as high resolution, zoom, and visualization/transformation—that is, moving views.

Large design systems also incorporate libraries of standard symbols, dimensioning systems, formats, and databases for the storage and retrieval of designs. Such systems can be used to support NC part programming and even generate the NC data that drives production tools. Some design systems also incorporate analytical capabilities for modeling and simulation. This may take the form

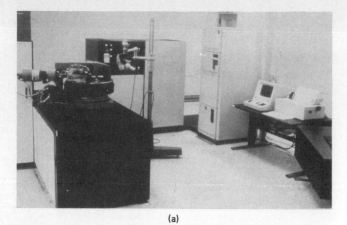

(a)

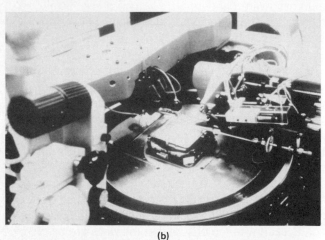

(b)

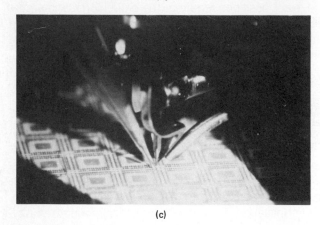

(c)

Fig. 4-3 Automated wire-bonding system. *(International Business Machines Corp.)*

of physical animation to test for interferences, or finite element analysis for stress or heat transfer. More recently, systems have been developed to provide solid geometric models and true 3D shapes (Fig. 4-5). More and more design systems are also using color displays, particularly to highlight surface contours, analytical data (e.g., stress and temperature), or multiple layers.

Automated design systems have also been developed for electrical and electronics applications, such as in the design of printed circuit boards, photo masks, and ceramic substrates, or even to drive electron-beam lithography tools. We will deal with these and other types of design systems in Chap. 11. For the time being, note that

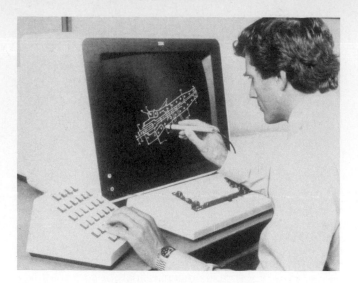

Fig. 4-4 Computer-aided design system. *(International Business Machines Corp.)*

some of the principal advantages of computer-aided design systems include:

1. Reduced design time
2. Fewer errors
3. Improved functionality
4. More standardization
5. Improved control of engineering changes
6. Higher designer productivity
7. Lower costs

4-4 SHOP FLOOR CONTROL

The control of the actual flow of product and materials on a factory floor can be automated by using logistics control systems. A shop floor control system involves quantities, types of parts, schedule dates and priorities, and the status of orders or jobs. Such a control system establishes

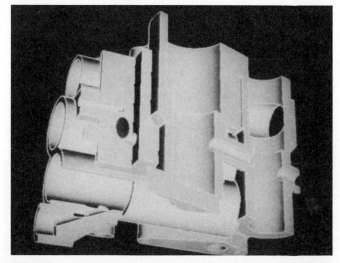

Fig. 4-5 3D solid geometrical modeling. *(McDonnell-Douglas Manufacturing Industry Systems Co.)*

standards, measures actual performance, reports results, and takes corrective actions. This is the classical job of production control and is really a part of the total manufacturing planning and control system that also includes forecasting, master scheduling, capacity planning, and materials requirements planning (Fig. 4-6).

A shop floor control system converts planned orders to actual job orders on the floor by:

1. Committing the parts and materials required
2. Selecting a manufacturing routing
3. Creating a shop order or job
4. Releasing it to the floor for execution

The control system then collects data from the floor and compares plan versus actual performance by work center or sector of the process in order to control the flow of product on the shop floor. The control system will attempt to optimize the loading of the line and the lead time through the process by monitoring job status, flow, queues, and work in process (WIP). Some of the types of data collected on the floor include:

Job order (number, status, location)

Priorities

Labor claiming

Equipment utilization

Many different methods of data collection are employed, such as terminals, key entry (batch), and optical or magnetic wands (Fig. 4-7). Such systems must be prepared to respond to changes in bills of material, product plans, or orders and schedules. Key objectives of a shop floor control system are to:

1. Ensure that delivery dates are met
2. Control the level of WIP
3. Control production lead times
4. Control the length of manufacturing queues
5. Prevent production bottlenecks

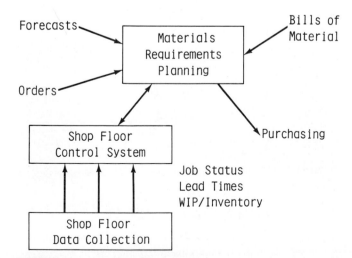

Fig. 4-6 Manufacturing planning and control system.

Fig. 4-7 Shop floor data collection: using magnetic-wand input. *(International Business Machines Corp.)*

6. Minimize idle time of people or tools
7. Manage priorities in production

The job of shop floor control is simply to manage the flow of product and materials through the manufacturing process. This job could be done manually, which has often been the case in many production operations. However, it requires a lot of time-consuming paperwork that is subject to errors and delays. Most modern production operations, especially those that are very complex, automate this activity by using computerized production control systems. The more complete job of managing the factory floor, however, involves more than scheduling and tracking production activity. It is also necessary to optimize the performance of the line in terms of throughput, labor costs, cycle times, and inventory levels. This can be achieved by monitoring and adjusting the production operations (e.g., machine scheduling and loading, operator scheduling and assignment, and defect and error detection). To do this efficiently requires an automated data collection and analysis system. We will learn more about materials logistics and floor control systems in subsequent chapters.

4-5 MATERIALS HANDLING

AUTOMATED MATERIALS HANDLING SYSTEMS

One of the keys to improving the efficiency of a manufacturing operation is to automate the handling of the product and materials involved. In traditional manufacturing processes, the product spends most of its time waiting rather than being worked on. This is due to inefficiencies in the movement and handling of materials and the scheduling of people and machines. By using advanced materials handling techniques, one can make substantial reductions in this waiting time, which results in lower WIP inventory and shorter manufacturing cycle times. These techniques include approaches to automating each step in the process that requires the movement or handling of the product or materials (e.g., machine loading and unloading, machine-to-machine transfer, in-line movement of WIP, and storage).

A variety of materials handling technologies are used in such applications, and they all lend themselves to the use of computers for control. The loading and unloading of machines is often the simplest application and can use traditional hard automation approaches or programmable robots. Materials can be moved between machines and process steps by a variety of techniques, including conveyors, transporter systems, transfer lines, carousels, and automatic guided vehicles (AGVs).

INTEGRATED MATERIALS HANDLING SYSTEMS

To assure an efficient flow of materials, computers are used as controllers to tie together an integrated materials handling system for the entire factory. An automated production line with an integrated materials handling system (Fig. 4-8):

1. Controls the movement of materials to optimize the utilization of equipment

2. Reduces the queues or waiting time

3. Increases overall throughput

To accomplish this, the system must be able to keep track of the material. This can be done by using sensors or readers (such as magnetic or optical devices) which provide inputs to the controller. Automated materials handling systems must also control the orientation and accurate placement of workpieces to assure the proper operation of production machines without operator intervention.

An integrated materials handling system can provide a number of significant benefits for a large or complex manufacturing operation, including:

1. A reduction in manual handling and labor

2. The control and reduction of inventory by tracking the WIP

3. Flexibility by programming changes in flow, sequences, quantities, or schedules

AUTOMATIC GUIDED VEHICLES

AGVs are battery-powered vehicles that can move and transfer materials by following prescribed paths around the manufacturing floor; they are neither physically tied to the production line nor driven by an operator like a forklift truck. They can take several different forms, but they are typically carts whose movements are controlled either by following guide paths buried beneath or painted on the floor, or by radio transmission (Fig. 4-9). Such vehicles have on-board controllers that can be programmed for complicated and varying routes as well as load and unload operations. The central computer for the materials handling system provides overall control functions, such as dispatching, routing, traffic control, and collision avoidance. AGVs usually complement an automated production line comprised of conveyor or transfer systems by providing the flexibility of complex and reprogrammable movement around the manufacturing process.

AUTOMATED STORAGE SYSTEMS

Of course, a key part of any materials handling system is storage. Major advances have been made in recent years to automate the storage and retrieval of product and materials by employing sophisticated materials handling machines, high-density storage techniques, and computer control. Such systems come in a variety of forms and sizes depending on the materials handling and storage job that has to be done. They often take the form of automated warehouses or materials distribution centers (MDCs), which use automatic storage and retrieval systems (ASRS's), conveyors, and computers to control the materials handling machines and to track and control the inventory (Fig. 4-10). The characteristics of such warehouses include:

1. High-density storage (in some cases, large, high-rise rack structures)

2. Automated handling systems (such as elevators, storage and retrieval machines, and conveyors)

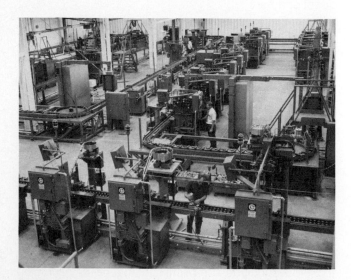

Fig. 4-8 Automated production line with integrated materials handling system. *(Gilman Engineering and Manufacturing Co.)*

Fig. 4-9 Automatic guided vehicle (AGV) loading products from a conveyor system on a manufacturing line. *(International Business Machines Corp.)*

Fig. 4-10 Automated warehouse. *(International Business Machines Corp.)*

3. Materials tracking systems (using optical or magnetic sensors)

In such a storage system, the computer can keep track of a large number of different parts, products, and materials and can assign bin locations to optimize the use of storage space. When such a system is tied into the production control system, parts and materials can be replenished as they are consumed on the factory floor, keeping the WIP to a minimum. An integrated materials storage and handling system can also improve the control of the inventory. It can provide control and reduce losses, delays, and WIP from the initial delivery of parts and material through the manufacturing process to the storage and shipment of the final product.

4-6 AUTOMATION

In the broadest sense, "automation" can mean almost everything we are talking about in this book. However, in this context of computer applications, we are concerned with a class of physical tasks in manufacturing which can be automated with the help of computers. The key is to use the control and programmability of the computer to replace manual or mechanized operations so that we can:

1. Increase throughput and productivity

2. Reduce costs

3. Improve quality and process control

4. Reduce manufacturing cycle times and delays

Such automation usually involves automating not only the physical manufacturing operation but also the materials handling and data handling jobs. One can automate individual tools or operations, groups of tools, or an entire manufacturing process tied together with a manufacturing control system. In the machining world automation evolved from the NC of individual tools composed of an entire machining process. In the world of assembly operations, however, there have traditionally been only two

extreme approaches to manufacturing: hard or fixed automation and manual assembly. This has now changed; the development of robotics and sophisticated materials handling systems offers a third alternative to fill the gap.

ROBOTIC ASSEMBLY

Robots can provide the complex motion, precision, and flexibility necessary in batch assembly operations. Robots come in a wide variety of forms and capabilities that must be matched to the task and workplace involved, which we will discuss extensively in subsequent chapters. Figure 4-11 illustrates how robots can be used in a flexible assembly operation—in this case, an automated line for the manufacture of electric motors. Such a line, by using sensors and vision systems in addition to programmable robots, can assemble a variety of different parts into several different types of motors. The robot's ability to detect and recover from errors or fault conditions permits it to respond to abnormal conditions (such as a missing part) without shutting the entire line down. To operate an automated assembly line like this requires a hierarchical computer control sytem which ties together the robot controllers, the materials handling system, and the sensor and inspection systems.

There are many different types of automated assembly applications in industry. Some use robots, such as in the welding operation in automotive assembly (Fig. 4-12). Others use automated tools that are tied together with materials handling systems and computer control, such as the electronic component assembly line illustrated in Fig. 4-13. This type of line evolved from the NC automation of individual machines that could eventually be linked efficiently into a total assembly process. In this

Fig. 4-11 Robotic assembly line: automated assembly of small electric motors. *(Westinghouse Electric Corp.)*

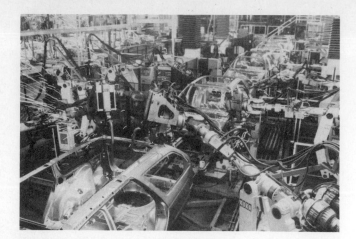

Fig. 4-12 Robotic welding line. *(Ford Motor Co.)*

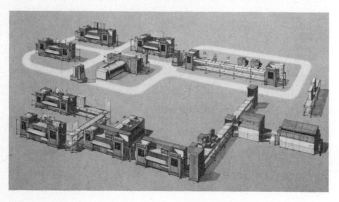

Fig. 4-13 Automated component assembly line. *(Universal Instruments Corp.)*

case it requires storage buffers for queuing at stations, automated parts-feeding mechanisms, and automatic verification and detection systems. The individual tool controllers must be tied into the control of the product flow by a computer system, which controls the entire manufacturing line.

4-7 TESTING AND MEASUREMENT

One of the major parts of almost any manufacturing operation is testing and measurement to assure that the product meets its specifications and functions properly. In precision and complex products the testing job itself can get extremely complicated. As product and manufacturing technologies have evolved, the test and measurement technologies have had to keep pace. There are now many applications which use computers. Three general areas which can be used to illustrate the role of the computer are:

1. Mechanical measurement and inspection
2. Electronic test systems and instruments
3. Quality control systems

MECHANICAL MEASUREMENT AND INSPECTION

In the world of fabricating precision mechanical parts, it has become increasingly more difficult and costly to perform measurements and inspections by manual techniques. One natural extension of the NC machine tool technology which is used to measure complex, precision-machined parts is the computer-controlled coordinate measurement machine (CMM). Such machines are actually NC tools that probe and record positions on a workpiece rather than perform machining operations (Fig. 4-14). However, they use the same basic motion control, sensing, and data logging functions. For smaller parts, particularly those being manufactured in large quantities, visual inspection is still necessary to detect defects. But as the complexities and quantities of the products increase, the effectiveness of the human eye decreases substantially. For this reason, automated optical inspection systems have been developed which use numerically controlled precision XY tables, electronic vision devices, and computer control systems. Such systems can read codes or numbers, detect the absence or presence of parts, and detect specific defects or voids with higher speed and lower incidence of error than human inspection techniques.

ELECTRONICS TEST SYSTEMS AND INSTRUMENTS

Electronics testing has become a large industry in and of itself and a major user of computers at the same time. Electronics testing covers a broad range of applications in electronics manufacturing, from testing individual integrated circuit (IC) chips to testing components, assemblies, and entire machines or systems. At each stage there is a need to remove defective parts and assure functionality at the next level of assembly. The types of tests that are usually performed include checking for electrical "opens" or "shorts," electrical parameters (e.g., resistance and capacitance), or a circuit function, such as a logic or memory function. As electronics technology

Fig. 4-14 Coordinate measurement machine (CMM) checking dimensions on an automobile body. *(General Motors Corp.)*

has become more complex, with increasingly higher-density circuits, it is sometimes impossible to perform these tests manually.

High-volume production of electronic parts and products also dictates that the testing operation have high throughput and relatively low cost. The answer to this has been the development of automated test equipment (ATE), such as the system illustrated in Fig. 4-15, which uses computer control to:

1. Draw test instructions from the product design system
2. Execute high-speed electronic tests on the product
3. Record test results
4. Conduct diagnostic analysis of failures
5. Generate reports

Automated test systems can reduce the number of defects and testing errors while also reducing the number of operators and testers required. To perform more complex electronic tests, very sophisticated instruments must often be used, such as logic analyzers, which today use controllers and commumicating interfaces to automate data collection, analysis, and reporting (Fig. 4-16).

QUALITY CONTROL SYSTEMS

To assure that a manufacturing process is under control and not making defective product, quality control procedures are used. These procedures are based on collecting and analyzing test and inspection data. The basic philosophy is to conduct tests and inspections in the process to detect defects early rather than to wait to find failures at the final test operation at the end of the line. Such procedures may involve tests or inspections that are conducted on every part or just on a sampling basis at several points in the process. In most cases this still involves a lot of data. Therefore, automated quality control systems have been developed to increase the efficiency of data collection, analysis, and reporting. Such in-line data may be collected by a variety of means, such as key entry by operators or inspectors, optical or magnetic coding, or direct data transfer from the test system.

Fig. 4-15 Automated electronic test system for testing integrated circuits. *(International Business Machines Corp.)*

Fig. 4-16 Automated electronic instrumentation system. *(John Fluke Manufacturing Co., Inc.)*

4-8 PROCESS CONTROL

PROCESS MANUFACTURING ENVIRONMENT

As discussed in Chap. 1, a process manufacturing environment differs from that of discrete part or product manufacturing operations. It involves a continuous flow of product, usually employs chemicals and materials which can have complex interactions with the process, and is subject to in-process "yield" losses. To operate successfully in this type of manufacturing environment, one must use process control techniques, which in many modern production operations are highly dependent on the use of computers. The basic functions of a process control system are to monitor, compare, and control. Process monitoring involves the acquisition, recording, and collection of process parameters, such as temperatures and times. The control function then compares this data to specified limits or calculates results predicted on the basis of process models, and then it directs control signals to the tools as required to adjust the process. For complex processes and tools, these functions can be automated by using computers for data collection, computation, and control.

PROCESS CONTROL COMPUTERS

The computers used in process control applications often differ from the standard data processing systems in ways that increase their efficiency. Process control computers are typically "sensor-based" and "event-driven," meaning that they are designed to respond directly to signals generated by control devices in the process or on the tools. They may also use specialized engineering languages for high-speed processing of technical data and efficient programming of complex functions. Process control applications can be found in many different manufacturing industries, such as electronics, chemicals, and materials, but they all have similar functions.

PROCESS CONTROL: PLATING

One manufacturing process that can be found in a number of industries is plating. Plating is a complex electrochemical process that must be carefully controlled to assure high-quality results. In precision plating operations, such as in the plating of small areas or complex patterns on electronic devices, each step in the process must be precisely controlled. In such cases one cannot rely on inspections to assure quality; defects must be prevented by controlling the process.

An example of automating a sophisticated plating operation by computer control is shown in Fig. 4-17, which involves the plating of nickel and gold pads on multilayer ceramic substrates which are used in complex electronic circuit packages. This particular process has both an automated handling system and a "clean-room" environment to reduce contamination from handling or operators. The computer control system records process variables, such as time, temperature, and the acidity of the plating solutions, to assure that they stay within prescribed control limits. This data can also be analyzed to find optimum points for improving quality and process yields. Computer control then permits one to modify the

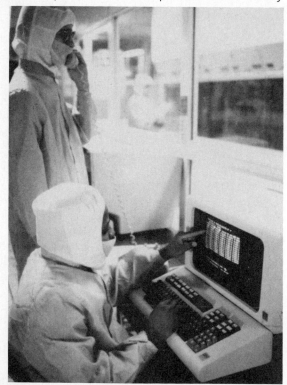

Fig. 4-17 Automated plating line: computer control system for plating multilayer ceramic substrates. *(International Business Machines Corp.)*

process by reprogramming the control limits of the appropriate process steps. Similar process control systems can be found in a wide variety of other applications where precise control of multiple process steps in an automated sequence is required.

4-9 OTHER APPLICATIONS

Although we have discussed the major and perhaps the most obvious applications for computers in manufacturing, there are many others that are also important. All of them exploit the basic capabilities of the computer, but some also have unique characteristics of their own. In many cases, these other applications involve peripheral or support activities that are not directly involved with the production operation itself. This does not mean, however, that they are not essential or critical to the success of the total manufacturing business.

FACILITIES APPLICATIONS

One large area of computer application is in facilities. This involves both the design and control of the physical plant, that is, the buildings and services. Within the realm of facilities-related activities there are several major areas of application for computers. Facilities design uses many of the same computer functions and tools as does product design. Buildings, along with their utilities and services (e.g., wiring and piping), can be designed by computer-aided design systems. This can include the internal physical layout of the manufacturing buildings—the location of tools, materials handling systems, and offices. Perhaps the greatest advantage of this approach, particularly with large, complex, and dynamic manufacturing operations, is that designs can be tried out and optimized without ever making any actual hard-copy drawings. Changes and modifications, such as frequent rearrangements, can also be made quickly without a major design effort.

Large manufacturing operations, particularly those involving high-technology products, are supported by large and complex utility systems that deliver heat, power, water, chemicals, and air-conditioning. To save money and prevent failures which can affect manufacturing operations, many plants use sophisticated computer control systems to manage their facilities services (Fig. 4-18). Such systems may also include special functions, such as energy management to minimize the consumption of electricity or environmental control to monitor air and water emissions from the processes for the protection and safety of employees. Some systems even include security controls, such as magnetic badge locks for critical areas.

MAINTENANCE APPLICATIONS

Another application area that has become increasingly dependent on computers is in the repair and maintenance of equipment. This can involve the products after

Fig. 4-18 Facilities control system. *(International Business Machines Corp.)*

they are produced or the manufacturing equipment used in the process. Both can be very complex and can require sophisticated testing and diagnostic analysis. The maintenance or repair of such equipment often involves the use of a portable computer or terminal which can communicate with a remote computer to use complex test and diagnostic routines to locate the source of the problem (Fig. 4-19).

4-10 SUMMARY

We have by no means discussed all the possible types of applications for computers in manufacturing. Nor do the examples deal with all the different types of manufacturing operations that exist. However, the categories of applications we covered are generic. That is, they can be found in many industries and are hopefully typical of the major computer application areas in manufacturing. In each case, the application is merely exploiting one or more of the basic strengths of a computer to make the manufacturing process more efficient, or in some instances even possible. We will cover some of these areas in much greater depth in later chapters.

Computers are used wherever there is a need for efficient data collection, control, and reporting. As manu-

Fig. 4-19 Portable computer service system for equipment diagnostics and maintenance. *(International Business Machines Corp.)*

facturing and product technologies become more complex, so do the processes and tools involved, which creates new applications for computers. In addition, as computer capabilities continue to improve and costs continue to go down, the scope of their application is extended as more functions become feasible, practical, and economical. Chapter 5 will deal with some of these trends that will influence how computers are used in manufacturing.

REVIEW QUESTIONS

The answer to each question can be found in the section indicated at the end of the question.

1. Identify the benefits of using computers in tool control applications. [4-2]
2. Define flexible manufacturing systems. [4-2]
3. Describe some of the functions and capabilities of computer-aided design systems. [4-3]
4. What is a shop floor control system? [4-4]
5. Identify some of the benefits of computerized shop floor control. [4-4]
6. Describe some of the different types of automated materials handling systems. [4-5]

7. Identify some of the benefits of computerized materials handling. [4-5]
8. Explain what flexible automation is and the role played by robotics. [4-6]
9. Identify some general areas of application for computers in testing and measurement. [4-7]
10. Explain why computers are necessary in automated electronics test systems. [4-7]
11. Identify the principal functions of the computer in process control applications. [4-8]
12. What are some of the uses for computers in facilities applications? [4-9]

5

TRENDS IN THE USE OF COMPUTERS IN MANUFACTURING

5-1 INTRODUCTION

The previous chapters addressed the basics of where we have been and where we are today in the use of computers in manufacturing. Before we begin to study the technologies of computer-automated manufacturing (CAM) in detail, let us focus first on some of the general trends that indicate where we are going in the future. Several major factors will have a significant influence on how the computer will be used in manufacturing:

1. The technology of the products to be manufactured
2. The capabilities of the computers themselves
3. The architecture of the computer systems
4. The technologies employed in manufacturing
5. The applications in manufacturing
6. Other environmental and economic considerations

In this chapter we will give a brief overview of each of these factors in order to gain an appreciation of the general direction that CAM is taking.

5-2 PRODUCT TECHNOLOGY

TRENDS IN ELECTRONICS TECHNOLOGY

Perhaps the most obvious trend in this high-technology world of ours is toward increasingly more complex product technologies. One area in particular that is very visible and will continue to expand its role in the products in our lives is electronics. Nowhere has this expansion been more dramatic than in "solid-state" (i.e., transistor) technology. The number of circuits that can be integrated on a silicon chip has increased by more than two orders of magnitude over the past 10 years (Fig. 5-1). Large-scale integration (LSI) has become a fact of life, and there seems to be no end in sight to this trend. Electronics technology will continue to strive to provide high-level function at low cost and in small packages.

Electronics technology has come a long way during the past few decades. If we look back to the 1950s, we find that even the most sophisticated electronics products used vacuum-tube technology and relatively crude packaging schemes (Fig. 5-2). However, this began to change rapidly with the development of the transistor and solid-state electronics. During much of the 1960s, electronics technology was typically comprised of discrete solid-state components mounted on relatively simple printed circuit boards, which are still used in some applications today. This evolved into much higher-density electronics with the development and application of integrated circuits

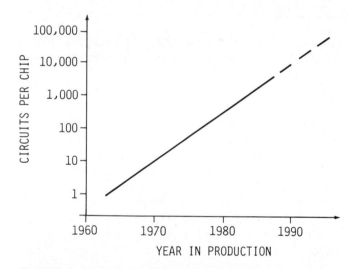

Fig. 5-1 Trends in integrated circuit chip density.

Fig. 5-2 Electronics technology of the 1950s: vacuum-tube assembly. *(International Business Machines Corp.)*

(ICs) during the 1970s and 1980s. Some of the more sophisticated electronics manufacturing today uses high-density, high-speed devices in complex packages to optimize performance (Fig. 5-3).

To see what the future holds for electronics technology, we need only look at what is coming out of the laboratories today. Memory devices are emerging at ever increasing densities (typically quadrupling every three to four years) and ever lower costs. For example, 1-million-bit (1-Mb) devices like the one shown in Fig. 5-4 have already been introduced into production. Advanced manufacturing techniques (e.g., photolithography) will permit a continued reduction in the dimensions of circuit patterns, which will make increased chip densities possible. The dimensions of devices have been decreasing steadily for years; the trend is shown in Fig. 5-5. Devices with features of less than one micron (i.e., one millionth of a

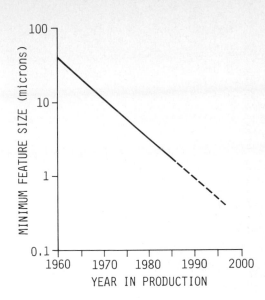

Fig. 5-5 Trends in feature size in integrated circuit devices.

meter or less than one-hundredth the diameter of a human hair) have already been fabricated (Fig. 5-6).

Such complex technologies will require sophisticated process controls and complex manufacturing tools that will be driven by large amounts of technical data. Computer-automated design and manufacturing provide the only way such advanced technologies can be produced. Figure 5-7 illustrates just how complex some of the manufacturing processes have become already. Since the introduction of hybrid ICs in the mid-1960s, the number of steps in a semiconductor manufacturing process have increased by approximately a factor of 10! To achieve economical yields with such complex processes, sophisticated controls and advanced manufacturing technologies are essential. This trend is not unique to electronics; it also can be found in most high-technology products, for example, chemicals and pharmaceuticals. Advanced product technologies will demand computer control to make their manufacture feasible and economical.

Fig. 5-3 Modern electronics technology: high-density multichip integrated circuit package. *(International Business Machines Corp.)*

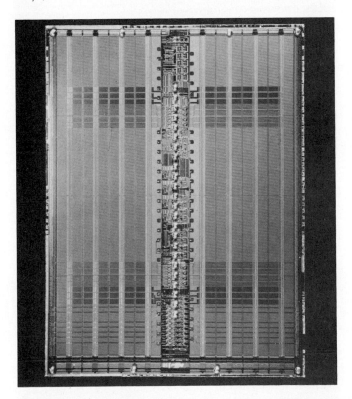

Fig. 5-4 One-million-bit FET memory device. *(International Business Machines Corp.)*

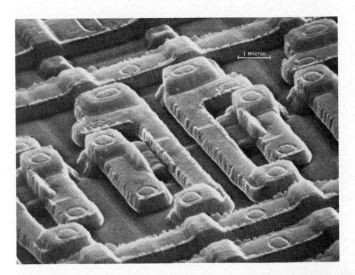

Fig. 5-6 Submicron lines on an integrated circuit chip. *(International Business Machines Corp.)*

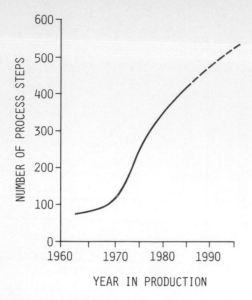

Fig. 5-7 Trends in semiconductor process complexity.

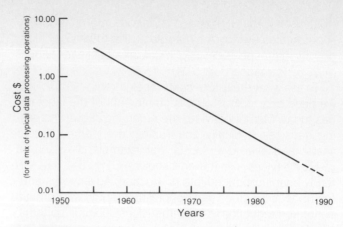

Fig. 5-8 Trends in data processing costs. *(International Business Machines Corp.)*

5-3 COMPUTER CAPABILITIES

TRENDS IN COMPUTER COSTS AND PERFORMANCE

Perhaps the most significant factor affecting the use of computers is the continued improvement in their cost and performance. The cost of data processing has come down dramatically over the past 20 years, making more and more applications economical (Fig. 5-8). A major contributor to this trend has been the dramatic reduction in logic and memory costs as a result of the continued advances in IC technology which we just discussed. These advances in technology have been so dramatic that it is sometimes difficult to appreciate how significant they have been. Figure 5-9, for example, shows a microprocessor chip being held next to a popular computer of the early 1960s having the same processing power!

TRENDS IN COMPUTER TYPES AND FUNCTIONS

Computers in the past were large, complex systems run by specialists and isolated from the average worker and from manufacturing operations. Today, computers come in a wide variety of sizes and functions to suit the needs of many applications. Personal and portable computers and "intelligent workstations" are now commonplace, bringing computer power to more people to improve their capability and productivity (Fig. 5-10). In the future, computer capabilities will continue to expand, offering not just lower cost and smaller size but increased functions, many varieties of features, and higher-level languages which will permit use by nonprofessionals. There will also be an increased ease of access and manipulation of data through the use of terminals, displays, and other I/O equipment which will be found throughout the workplace. Therefore, the advances in computer technology itself,

Fig. 5-9 Microprocessor chip compared to computer of the 1960s. *(International Business Machines Corp.)*

Fig. 5-10 Industrial computer on the factory floor. *(International Business Machines Corp.)*

which we will address in depth in the next few chapters, will be one of the most significant factors influencing the computer's use in manufacturing in the future.

5-4 SYSTEMS ARCHITECTURE

Over the years, since the first uses of computers in manufacturing, there has been an evolution toward more complex systems to control manufacturing operations. As discussed in Chap. 3, many of the manufacturing control systems found in industry today have a highly structured architecture of hosts and satellites which exchange real-time production data. Such systems have already become very complex. They may include a multiple-level hierarchy, multiple hosts, a number of central databases serving several functions, and even direct ties to systems at other locations (Fig. 5-11). Systems such as these often comprise hundreds of satellite computers and controllers as well as hundreds of terminals.

In the future, the increasing availability of computer power, and particularly the lower cost of memory and microprocessors, will promote the development of more distributed systems, using additional levels of hierarchy or subhosts that can be loosely coupled to a central database management system. Such systems, as discussed in Chap. 3, will also have the capability for "peer-to-peer" communications between satellite computers and common data buses which serve many separate manufacturing operations. There is currently a great deal of activity directed toward developing computer system networks which can tie computers and a wide variety of equipment from different manufacturers together into a common communication link. This involves software to translate different languages as well as hardware to "interface," or connect with, different types of computers and controllers.

At the same time, more functions will be integrated into the systems that run manufacturing. Common databases and distributed hierarchical systems architecture will integrate sensor-based process control, tool control, shop floor control, and administrative systems. There will be a natural tie and flow of data systems from the customer order through the various engineering and manufacturing operations, to the service of the product in the field (Fig. 5-12). Advanced data communications technologies, such as satellites and fiber optics, will help to tie such complex systems together. The computers of the future cannot exploit their capabilities in a complex manufacturing environment without a sophisticated systems architecture and advanced techniques for data communication.

5-5 MANUFACTURING TECHNOLOGIES

One of the keys to improving the productivity of manufacturing, maintaining competitive costs, and producing more complex products is the use of advanced manufacturing technologies. Manufacturing technologies are the processes, equipment, and tools which are used to manufacture products. Most of the advanced manufacturing technologies in use today and those being developed for the future depend upon, and would not even be possible without, computer control. Some of these technologies

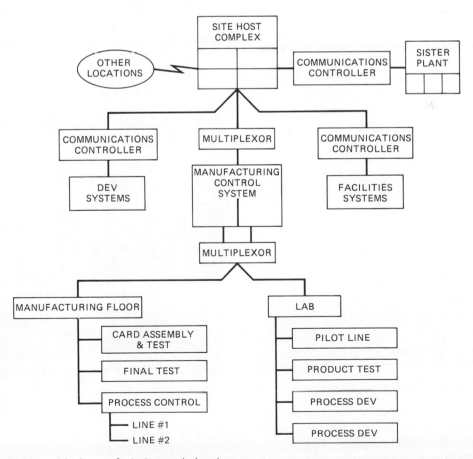

Fig. 5-11 A multilevel hierarchical manufacturing control system.

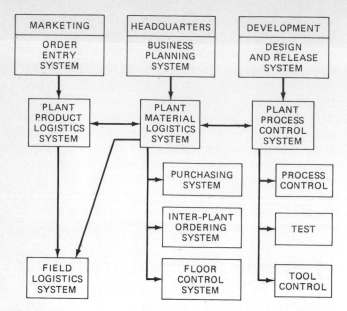

Fig. 5-12 Integrated multifunctional manufacturing system.

are generic; that is, they can be used in a variety of different applications. A number of these generic manufacturing technologies are relatively new; they are developing very rapidly and will have a significant influence on the future of manufacturing. Two areas in particular are worth noting for the purposes of identifying basic trends: automation and generic process technologies.

ROBOTICS TECHNOLOGY

The most visible area of development in the field of automation is, of course, robotics. It will continue to have an increasingly greater effect on manufacturing productivity in the future. Robotics means different things to different people. To many, robotics is still a world of humanlike machines popularized by science fiction. This, of course, is not robotics as we know it in manufacturing today, nor does it represent the trend for the foreseeable future. As we saw in Chap. 3, robotics is a natural outgrowth of numerical control (NC) technology. The major factor that makes it different from traditional fixed automation is the programmability provided by computer control. The robotics technology is composed of four basic elements:

1. A mechanical "manipulator" technology (i.e., the physical robot itself, or robot "arm")

2. Sensor technologies (which provide inputs to the robot for control similar to human senses, such as touch and vision)

3. Software languages for programming the robot's motion

4. Computers to control the entire robotic system

We will cover these technologies in depth in subsequent chapters, but for now, let us consider a few examples of how robots can be used in manufacturing.

The initial use of robots in manufacturing, and still the largest application, was for relatively simple loading and unloading tasks which are often referred to as "pick and place." Such applications are found throughout manufacturing operations where products or materials must be loaded onto or unloaded from machines. The task is usually simple and repetitive and does not require a great deal of skill by either a human or a machine. It was therefore relatively easy to place basic, unsophisticated robots in such applications.

Figure 5-13 shows a typical example of a pick-and-place robot application, which in most cases is implemented because the robot can perform without interruption or error (unlike a human operator). However, in such applications there is usually little use of the data handling and computational capabilities of computer control. This is not true of assembly applications. They are becoming the major area for the development and application of robotics in manufacturing. The very nature of most assembly tasks requires complex motions, precision, flexibility, and feedback. These are all human traits that in the past have been difficult, if not impossible, for machines to copy. Assembly operations therefore have usually been done manually, particularly for low-to-medium volume production operations that did not lend themselves to fixed automation. The development of robotics, however, has now made the automation of many of these applications possible. Robots, which come in a variety of sizes and configurations, can perform complex assembly tasks by exploiting their capabilities of programmability, sensory feedback control, and complex, precision motion control (Fig. 5-14). Substantial effort is being directed toward developing these capabilities to expand the scope of robotics applications in manufacturing. To the extent that it is successful, robots will become a basic part of all future manufacturing operations.

ELECTRON, ION, AND PHOTON-BEAM TECHNOLOGY

There are a number of emerging process technologies which can be applied to a wide variety of manufacturing

Fig. 5-13 Pick-and-place robot application: robot placing computer disks on a conveyor. *(International Business Machines Corp.)*

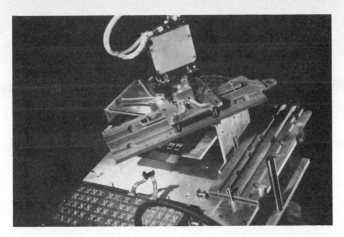

Fig. 5-14 Assembly robot application: robot assembling a computer terminal. *(International Business Machines Corp.)*

tasks; in many cases, they help to make operations feasible. One area in particular which is evolving rapidly and finding a broad range of applications in manufacturing is the technology of electron, ion, and photon beams. This is basically the control of small atomic or molecular particles by focusing their direction and motion and varying their intensity to perform highly precise tasks. This is also a technology that has been viewed by many as from the world of science fiction. But, in truth, it has come into its own as a real tool for manufacturing.

These "beam" technologies can be used in a variety of different ways, including noncontact testing, sensing, scanning, reading, and data transmission; and many more are being developed all the time. The most widespread use in manufacturing has been for machining-like applications. Because these beams can be focused to microscopic dimensions and their movement controlled for very high-speed, precision motion, they have found application in a wide variety of high-technology manufacturing tasks. Figure 5-15 shows one typical example of beam technology, which takes many forms in many dif-

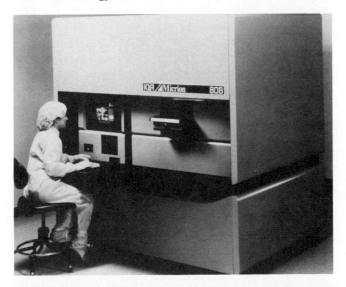

Fig. 5-15 Ion-beam system used for repairing photo masks. *(KLA Instruments Corp.)*

ferent industries. Some of the more common applications include laser machining, laser welding, ion-beam milling, ion implantation, and electron-beam and ion-beam lithography. To transform these technologies into practical manufacturing tools requires a great deal of knowledge in physics and materials science. But another essential ingredient is the computer control that makes the precise, high-speed, complex motion possible. Some of these applications, such as "direct-write" lithography in semiconductor manufacturing (i.e., without the use of conventional masks), involve a significant amount of data handling. The ability of computers to handle large amounts of data and direct the control of complex mechanisms is one of the keys to the continued development of advanced manufacturing technologies.

5-6 COMPUTER APPLICATIONS IN MANUFACTURING

A trend which is really an outgrowth of those previously discussed is the scope and nature of the application of computers in manufacturing. Here there are two major factors to consider: the presence of the computer on the factory floor and the role of the computer in the factory of the future.

COMPUTERS ON THE FACTORY FLOOR

Computers have begun to move out of the hidden domain of the computer room into the workplace on the factory floor. There they are becoming a routine aid to the manufacturing workers as well as an integral part of the manufacturing tools. As we have just discussed, advanced manufacturing technologies rely on computer control and bring with them computers of all sizes into the manufacturing process. These complex processes and tools also require operator interaction with the computer controls. This brings the manufacturing worker routinely in direct contact with computers. There will also continue to be a great deal of effort to reduce or eliminate the paperwork in the factory by relying on computer terminals for workers to record or access data. Such terminals are already becoming commonplace on the factory floor (Fig. 5-16). They can come in a wide variety of forms and functions, depending on the manufacturing process or operation involved. Sometimes a terminal includes a special data entry device such as an optical or magnetic wand which can automatically read codes on the product as it is moved through the process.

COMPUTERS IN THE FACTORIES OF THE FUTURE

There is a lot of discussion and writing on the subject of factories of the future. One thing that is common to all those visions is the role of computer-integrated manufacturing (CIM) systems, which incorporate advanced manufacturing technologies and computer control systems.

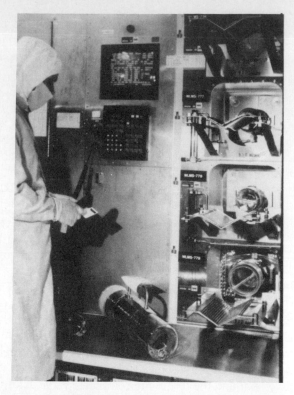

Fig. 5-16 Manufacturing terminal: operator entering data with a magnetic wand in a semiconductor manufacturing clean room. *(International Business Machines Corp.)*

That basically means that industry will continue to move toward using complex data systems that tie together the management of the entire plant. Therefore, from a systems viewpoint the factory of the future must integrate all the major functions that exist in manufacturing. This includes a multitude of technical, logistical, and administrative data systems.

To achieve this requires a clear objective and a comprehensive plan. Japan, for example, established a national goal for "peopleless factories" in order to stay competitive in the high-technology manufacturing environment of the future. Through their Ministry of International Trade and Industry (MITI), the Japanese developed a plan for such factories of the future and worked with industry to implement the concept. As illustrated in Fig. 5-17, such concepts were reduced to actual physical layouts of automated factories. Of course, some of the ingredients essential to those plans were advanced manufacturing technologies and tools and a systems architecture to manage the data that will run the factory. These are not just concepts or dreams of the future; they are actually beginning to happen. Factories of the future are already appearing, where machines do the work and very few people are directly involved with the process (Fig. 5-18). The machines are controlled by a sophisticated computer system that monitors and controls all the manufacturing operations. They use automated materials handling systems and automated tools, including a lot of robots, to do the work.

CAM concepts are now being applied in a variety of different industries. Initially, the greatest progress had been made in machining operations because of their long experience with NC technology. In the future, we will see truly automated factories in most major industries, including the automotive, aerospace, electronics, and chemical industries, which are already moving aggressively in that direction today. The individual product and manufacturing technologies involved are essential to the automated factory. However, the key ingredient that will make the concept feasible is an integrated systems approach to manufacturing.

5-7 OTHER CONSIDERATIONS

Aside from the various technologies that we have discussed, a number of other factors will influence the use of computers in manufacturing in the future. In particular, we should recognize some of the environmental and economic factors that have already affected industry.

ENVIRONMENTAL FACTORS

One thing that has certainly become apparent to the industrialized world in recent years is that the cost and availability of energy and natural resources can have a significant influence on manufacturing operations. In addition, the need to control air and water pollution has placed constraints and costs on industry which did not exist in the past. These factors have now become significant influences on the design and operation of manufacturing processes and facilities. Industry is beginning to use computers to control the machines that run physical facilities (such as the boilers and chillers that heat and cool the plants) to optimize their energy efficiency. As we saw in Chap. 4, some plants have begun to employ sophisticated computer systems which monitor and control the consumption of energy by facilities and process equipment. In large, complex process manufacturing operations, systems are also used to monitor the effluents and emissions from the factory. As processes become more complex, involving large volumes of exotic and expensive materials, such controls will become essential for both environmental and economic reasons.

ECONOMIC AND SOCIAL FACTORS

The industrial environment itself is also a major factor that will influence the use of computers in the future. Perhaps the most important trend, and the biggest challenge, is improvement in the productivity of the work force. In recent years, productivity improvement has not been progressing at historical rates, making some manufacturing operations and even some industries noncompetitive. This problem has been experienced by most industrialized nations and has contributed to the inflationary economic trend of recent years. In addition, demographic trends in most industrialized countries indicate that there

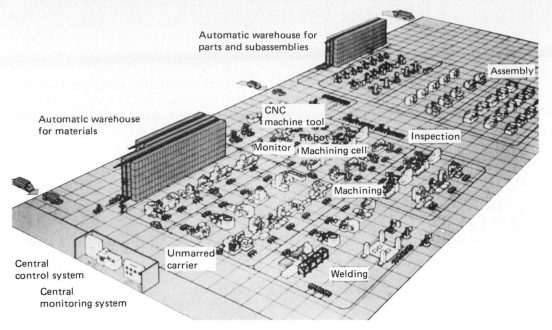

Fig. 5-17 Japanese factory-of-the-future concept: flexible manufacturing system (FMS). *(Fanuc Ltd.)*

Fig. 5-18 Automated peopleless factory in Japan. *(Fanuc Ltd.)*

could be a significant shortage of skilled labor in the future. This would further constrain their manufacturing capabilities. At the same time, there will be increasing competitive pressure from "underdeveloped" countries which have a large labor force and in some cases large deposits of natural resources. Economic studies of the factors that affect productivity have revealed that technology and capital have by far the greatest leverage (Fig. 5-19).

Fortunately, there is at least one trend that will help industry deal with the constraints and pressures of the current industrial environment that exploits this leverage. That is the continually improving productivity of data processing. As we have seen, the costs of data processing power have been decreasing while most other operational costs of manufacturing have been increasing. That factor, together with the dependency that advanced product and process technologies have on computers, will continue to motivate industry to exploit the capabilities of computers in their manufacturing operations.

5-8 SUMMARY

We have identified the major factors and key trends which will influence the role of the computer in future factories. The major influences on the use of computers in manufacturing include:

1. The technologies of the products to be manufactured (i.e., product complexity, such as advanced electronics technology)

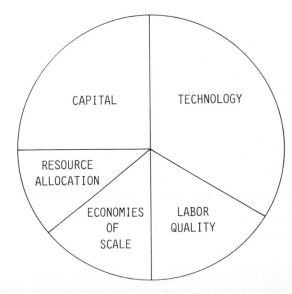

Fig. 5-19 Factors affecting productivity.

2. The capabilities of the computer (i.e., its cost, performance, and size)

3. The architecture of computer systems (e.g., networks, distributed processing, and data communications technology)

4. The technologies used in manufacturing (e.g., automation and advanced generic process technologies)

5. The applications in manufacturing (e.g., the factory of the future)

6. Other considerations, such as environmental and economic factors

The rate and form of change may actually emerge differently than current trends indicate. This could be due, for example, to significant innovations in computer technology or systems architecture that cannot be foreseen today. However, the basic factors and key trends will probably remain the same. If anything, we may be underestimating what will actually be achieved by this rapidly developing technology.

REVIEW QUESTIONS

The answer to each question can be found in the section indicated at the end of the question.

1. Identify the major factors that will influence the use of the computer in manufacturing. [5-1]

2. Explain how product technology affects the use of computers in manufacturing. [5-2]

3. What are the trends in computer capabilities that will affect the use of computers in manufacturing? [5-3]

4. Describe some of the current trends in the architecture of manufacturing systems. [5-4]

5. Explain how manufacturing technologies can influence the use of computers. [5-5]

6. Identify some examples of advanced manufacturing technologies that are dependent upon computer control. [5-5]

7. What are the functions of computer control in complex robotics applications? [5-5]

8. Describe the features of computers that make them essential in advanced process technologies. [5-5]

9. Identify some of the factors in manufacturing which drive the need for computers. [5-6]

10. Describe some of the basic concepts of the factory of the future. [5-6]

11. Explain how environmental and economic factors influence the use of computers in manufacturing. [5-7]

PART

COMPUTER TECHNOLOGIES

CHAPTER 6
COMPUTER HARDWARE

CHAPTER 7
COMPUTER SOFTWARE

CHAPTER 8
MICRO- AND MINICOMPUTERS

CHAPTER 9
ARTIFICIAL INTELLIGENCE

This second part of the book deals with one of the basic technologies involved with computer-automated manufacturing (CAM). The computer, of course, is the key tool that makes CAM possible in the first place. It was the development of the computer and electronic data processing systems that changed the way in which manufacturing operations were controlled and managed. The computer is what makes the difference between traditional hard automation or mechanization and the modern approaches to programmable automation and integrated manufacturing systems.

The objective of Part 2 is to provide a basic understanding of computer technology and its major components. It is not intended to be a thorough or complete treatment of the subject that would make the reader an expert or professional in computer science. Those already familiar with computers may find some of the material very basic and need not spend as much time on it as others might. Part 2 is designed to cover the basics of the technology and to identify the role of the computer in manufacturing applications. In addition, the state of the art of computer technology is described and the key trends that will affect CAM in the future are discussed.

Part 2 is organized into four chapters to cover each of the principal technologies involved in computers as they relate to manufacturing applications. Chapter 6 deals primarily with the hardware, the ''visible computer.'' This comprises the physical technologies, such as the electronics that make up the various types of data processing equipment. Since it is the introductory chapter to Part 2 it will also cover the basic concepts of how a computer works. Chapter 7 deals with the ''software'' that programs and controls the operation of computers. This is the ''invisible computer'' and a world of technology that is not as well known or understood by as many as is hardware. It is, however, at least as important as, and, in some cases more important than, the physical technologies to accomplish the tasks in manufacturing.

Chapter 8 focuses on the relatively new and fast-growing segment of data processing technology that is involved with small computers. This includes microprocessors, microcomputers, minicomputers, and programmable controllers (PCs). They have each found an important role in the manufacturing environment. Chapter 9 also covers a relatively new, but somewhat special, area of computer science called ''artificial intelligence'' (AI). This deals with computer capabilities that are similar to those of human intelligence, such as interpretation and reasoning. There is no way to cover this subject in depth in one chapter, and it can can easily become very complex and technical. However, because of its importance to manufacturing, it is necessary to gain a basic understanding of the technology.

COMPUTER HARDWARE

6-1 INTRODUCTION

This chapter will start this part of the book with a brief review of computer basics. It will cover what a computer does and how it does it. This will involve discussing the key elements of a computer and how they are organized to operate together. The various types of data processing equipment and computer systems will also be reviewed so that the reader has a basic understanding of what they are and what they do. The electronics technologies that are used to make computers will also be discussed. This will be done not only to identify their function, but also to help the reader appreciate their importance to the continuing advances in computer capabilities.

The emphasis in this chapter is on fundamentals. The treatment of each subject is relatively brief and is not intended to be thorough. This chapter is intended to provide enough of an understanding of computer technology, together with Chap. 7, to prepare the reader for following chapters which deal with some of the newer developments that are of particular importance to manufacturing. For those that are already familiar with the basics of computer technology, this chapter can serve as a brief review.

6-2 COMPUTER BASICS

WHAT IS A COMPUTER?

To appreciate the computer's role and potential in manufacturing, we must first understand the basics of what a computer system is and how it works. Most definitions of "computer" would include basic facts, such as that it:

Is a machine

Can process data

Can perform calculations

Can store and manipulate information

Computerized or electronic data processing is an automated approach to information handling and computation tasks that had been done manually in the past. The development of the modern computer has not only made these tasks easier to perform; it has also made possible other tasks that could not have been performed manually. As the capabilities of the computer continue to increase, we are finding more and more uses for it. This is particularly the case in helping to make manufacturing operations more productive.

As discussed briefly in Chap. 1, and illustrated in Fig. 6-1, a basic computer system is made up of:

1. A central processing unit (CPU)

2. A main memory or storage unit

3. Input devices

4. Output devices

5. Auxiliary or secondary storage devices

Together, these devices make it possible for the computer to perform three basic functions:

1. Arithmetic operations (e.g., addition, subtraction, multiplication, division)

2. Logical operations (e.g., comparison, selection, sequencing)

3. Input and output operations (e.g., printing, reading, storing)

The CPU controls and supervises the entire computer system. This is where the actual arithmetic and logical operations are performed. The CPU is composed of two sections:

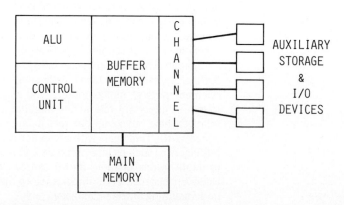

Fig. 6-1 Basic computer system.

1. The control unit. This directs and coordinates all the operations of the system that are called for by the stored program of instructions. It involves control of the input and output devices, the storage or retrieval of information from memory, and the transfer of data between memory and the arithmetic logic unit.

2. The arithmetic logic unit (ALU). This performs the arithmetic and logical operations as directed by the control unit and stored program.

The main memory or storage unit stores the program instructions and all the data being held for processing until it is released as output. There are often several levels of memory in a computer system; this is usually referred to as the "main" memory, since it is directly connected to the CPU. Auxiliary or secondary storage units provide high-capacity memory for storing large amounts of data that is not currently being processed. Typical devices of this type include magnetic tape units and magnetic disk drives.

Data is provided to the computer by input devices such as card readers, keyboards, tape units, or disk drives. Data is provided to people or to other machines by output devices such as displays or printers. The transfer of data between these I/O devices and the computer is governed by the control section of the CPU. It directs their operation by communicating with a control unit that is part of each I/O device.

HOW DOES A COMPUTER WORK?

The arithmetic and logic operations of a computer are carried out in a "binary" form. This means that data and instructions are represented by a binary code in the form of the on/off switching signals of digital electronic circuits. Before data can be processed in a computer, it must first be coded into a set of symbols which can be interpreted by the computer in the digital form in which it operates. A "bit" (BInary digiT) is the smallest unit of data that can be used by a computer. It only has two possible values, 1 or 0, representing the on or off conditions of an electronic switch. These bits are arranged in sequence into a group to form a code which represents symbols, numbers, or characters (such as the letters of the alphabet). A group of bits is called a "byte." A byte is the basic unit of information representing a single character. The basic unit of information that is processed at one time by the computer is called a "word"; it may be composed of several bytes. General purpose computers typically process 32-bit words.

The binary system is based on the number 2 raised to successive exponent powers. The first digit in a binary code is therefore 2^0 or 1. The second is 2^1, or 2. The third is 2^2, or 4, and so on. This binary code can be converted into the decimal system by entering a 0 or 1 in each of the binary digit positions. A minimum of 4 binary digits are required in order to represent any single decimal digit. Special binary coding systems were developed in order to represent larger decimal numbers as well as characters. The binary-coded decimal (BCD) system uses 7 bits to represent up to 64 characters (Fig. 6-2). The first 6 bits represent data, and the last, the "parity" bit, is for code checking. Another common coding system, the Extended Binary-Coded Decimal Interchange

Character	BCD code	EBCDIC code	ASCII code
0	000 000	1111 0000	0011 0000
1	000 001	1111 0001	0011 0001
2	000 010	1111 0010	0011 0010
3	000 011	1111 0011	0011 0011
4	000 100	1111 0100	0011 0100
5	000 101	1111 0101	0011 0101
6	000 110	1111 0110	0011 0110
7	000 111	1111 0111	0011 0111
8	001 000	1111 1000	0011 1000
9	001 001	1111 1001	0011 1001
A	010 001	1100 0001	0100 0001
B	010 010	1100 0010	0100 0010
C	010 011	1100 0011	0100 0011
D	010 100	1100 0100	0100 0100
E	010 101	1100 0101	0100 0101
F	010 110	1100 0110	0100 0110
G	010 111	1100 0111	0100 0111
H	011 000	1100 1000	0100 1000
I	011 001	1100 1001	0100 1001
J	100 001	1101 0001	0100 1010
K	100 010	1101 0010	0100 1011
L	100 011	1101 0011	0100 1100
M	100 100	1101 0100	0100 1101
N	100 101	1101 0101	0100 1110
O	100 110	1101 0110	0100 1111
P	100 111	1101 0111	0101 0000
Q	101 000	1101 1000	0101 0001
R	101 001	1101 1001	0101 0010
S	110 010	1110 0010	0101 0011
T	110 011	1110 0011	0101 0100
U	110 100	1110 0100	0101 0101
V	110 101	1110 0101	0101 0110
W	110 110	1110 0110	0101 0111
X	110 111	1110 0111	0101 1000
Y	111 000	1110 1000	0101 1001
Z	111 001	1110 1001	0101 1010
Blank	110 000	0100 0000	0010 0000
(	111 100	0100 1101	0010 1000
+	010 000	0100 1110	0010 1011
$	101 011	0101 1011	0010 0100
*	101 100	0101 1100	0010 1010
)	011 100	0101 1101	0010 1001
−	100 000	0110 0000	0010 1101
/	110 001	0110 0001	0010 1100
,	111 011	0110 1011	0010 1111
=	001 011	0111 1110	0011 1101

Fig. 6-2 Binary coding schemes. *(C. William Gear, Computer Organization and Programming, McGraw-Hill, New York, 1980)*

Code (EBCDIC), is composed of an 8-bit code plus a parity bit to define up to 256 characters. A standard code that was also developed for data communications is the American Standard Code for Information Interchange (ASCII); it is a 7-bit code that can represent 128 characters.

Each word has an address in the main memory, which the CPU uses to retrieve it. The time required to locate an address and retrieve its contents is called the "access time" of the computer; it may range from tens of nanoseconds to several microseconds, depending on the size and speed of the computer. The data and instructions to be processed are communicated to the computer in the form of these binary-coded words. The computer interprets this code as specific instructions to perform certain arithmetic or logical operations. These binary-coded instructions are called "machine language," since that is the only language the computer understands. However, it is obviously very cumbersome and difficult for humans to read or write machine language. To complicate this further, different machines use different languages. Therefore, higher-level languages have been developed to make programming easier and more efficient. These languages are converted into machine language so the computer can understand and then execute them. Programming and some of the various languages used will be covered in Chap. 7.

The CPU of the computer performs its arithmetic, logical, and control functions by using "registers." These registers are small memory devices that can receive, store, and transfer data. They are used primarily for temporarily storing data during computer operations. A register consists of binary cells that can hold the bits of data in the word of information being processed by the CPU. Several different types of registers in the CPU perform the basic functions necessary for the computer to execute a set of program instructions. An instruction word consists of an "operator," which defines the type of arithmetic or logical operation to be performed, and an "operand," which specifies the actual data to be processed. Six basic types of registers in the CPU are used to handle the data and instructions in the program sequence (Fig. 6-3):

1. **Program counter.** This holds the location or "address" of the next instruction. The CPU follows a sequence of the instructions to be performed by retrieving words from the main memory according to the contents of the program counter. The program counter automatically advances to the next instruction word address after each one is retrieved by the CPU.

2. **Memory address register.** This identifies the location or address of data that is stored in memory.

3. **Instruction register.** This holds the instructions for decoding so that they can be interpreted for the CPU.

4. **Accumulator.** This is a temporary storage register used during arithmetic or logical operations. It holds one number or subtotal, for example, while another is retrieved from memory.

5. **Status register.** This indicates the internal condition

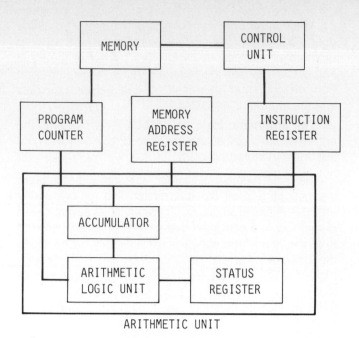

Fig. 6-3 Computer registers.

of the CPU by the use of "flags," which are 1-bit codes. Such flags can identify overflow conditions, interrupt conditions, or the results of logical decisions. This function will be covered in greater depth later in this chapter.

6. **Arithmetic logic unit.** The ALU is also a type of register. It may be a simple adder or may be capable of more complex calculations. The ALU is basically an electronic calculator.

The overall operation of the computer is controlled by an "operating system," which is a large program that manages all the computer resources (i.e., CPU, memory, and I/O). The job of the operating system is to maximize the performance and efficiency of the entire computer system. Other control programs support the internal functions of the computer rather than provide instructions for the performance of a particular application. The I/O control system, for example, works with the operating system to interpret its commands about the transfer of data between devices. The CPU must transfer instructions and data to and from memory throughout the execution of a program. Through the use of control programs and special storage devices, I/O and CPU operations can occur simultaneously. Therefore, the hardware and software of the computer work together to process data and instructions. The actual individual operations within the computer are very small, simple steps based on the fundamental on/off electronic code of the computer. Although this means that many steps are required to complete even a relatively simple operation, such as an arithmetic calculation, the computer makes up for its "simplemindedness" by using its incredibly high speed!

SYSTEM OPERATION

During the operation of a computer, data flows between the CPU, the main memory, and the various I/O devices

(Fig. 6-4). In addition, control information flows between these units to tell them what to do with the data. Control information is used to manage the overall operation of the system. In large systems which have many I/O devices, the control and data transfer process can get very complex. Several different types of schemes and devices are used to handle that process. Each I/O unit usually has its own controller so that the CPU does not have to control each individual unit directly. In addition, the computer often has separate control devices called "channels" which direct the transfer of I/O data. These channels minimize the impact of this operation on the CPU so it can concentrate primarily on computation. Channels are used to attempt to optimize the data transfer rate so that it is as efficient as possible and compatible with the internal operating speed of the computer. Channels and controllers can improve the operating efficiency by "multiplexing," which is a method of transferring data from several different sources simultaneously. This is done by filling in the waiting periods in data transmission with blocks of data from other devices.

A computer may operate in two different modes with the I/O devices. One is called "synchronous" operation. As the name implies, this means that the actions of the I/O units are controlled in a fixed sequence and timing. The alternative mode is therefore called "asynchronous" operation. It permits the I/O units to interrupt processing activity to transfer data. This is a common type of operation in a manufacturing environment, as we shall see later. We will also find that this data transfer between the many devices in a system and the CPU often is conducted through a communication link called a "data bus." These concepts will be discussed in some depth when we cover the subjects of microprocessors and computer networks.

Computer systems often have several different levels of memory in a hierarchy of storage. This is done to optimize the overall efficiency of the system by not placing the burden of all the storage tasks on the main memory. There can be "buffer" storage units in both the CPU and the I/O devices to temporarily store data during the input, computation, and output operations. These must, by necessity be high-speed but relatively small-capacity

memories. There are then higher-capacity but slower memories, including the main memory, for storing large amounts of data for longer periods during the computer's operation.

6-3 TYPES OF DATA PROCESSING EQUIPMENT

PROCESSORS

The heart of the computer system, which is often considered the computer itself, is the CPU—the central processing unit or processor. There are many different types of computers that can be used for a wide variety of purposes. They range in size, speed, and function from small portable computers for personal use to large-scale central processors used to control complex data networks. The so-called general purpose computer is the machine most commonly used for business or industrial applications. Its structure is typically like the basic one described above, and it can vary in size and speed from desk-sized "intermediate" systems to very large and fast "mainframes." Such computers are usually tied to a large number of storage and I/O devices to serve many users and perform a number of different functions. On the low end of the spectrum, aside from personal computers, there are "minicomputers," or "minis," and "microcomputers," or "micros," which typically have a simplified and specialized structure for specific or even dedicated industrial applications. These are very important to manufacturing and will be discussed in detail later. On the high end of the spectrum there are very large-scale, high-performance processors which are usually specially designed for scientific applications.

In a large manufacturing facility a hierarchy of computers are often tied together to control the production operations. At the bottom of the system hierarchy are tool controllers, which may be programmable, that are attached to each piece of manufacturing equipment that is under the control of the computer system. These controllers are then, in turn, connected to micro- or minicomputers that control a group of tools. Because these are true computers with both computational and storage capabilities, they can perform some functions by themselves, which can minimize the demands on the larger computers in the system. Large manufacturing control systems may next have a level of subhosts, which may be intermediate-size processors used to control groups of minicomputers in a major segment of the production line (Fig. 6-5). Finally, at the top of the hierarchy, is the host, or central computer complex, that manages the databases and ties all the segments of the manufacturing control system together.

The capabilities of processors vary significantly and are usually in proportion to their size and position in the hierarchy. Small controllers and microcomputers may have processing speeds (i.e., the time it takes to perform a computation or process an instruction) on the order of

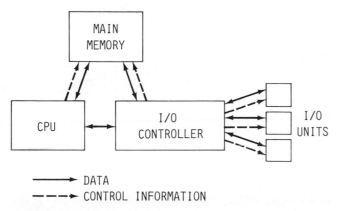

DATA
CONTROL INFORMATION

Fig. 6-4 Data flow in a computer system.

Fig. 6-5 Intermediate-sized processor. *(International Business Machines Corp.)*

thousands of instructions per second (KIPS). Their memory capacity will be relatively small and is typically less than 64K bytes. These capabilities, although quite limited, are adequate for most tool control applications. Intermediate-size processors often have speeds up to several million instructions per second (MIPS) with up to several millions of bytes, or "megabytes" (MB), of memory. The large central processors can have a wide range of performance capabilities. They usually have speeds on the order of tens of MIPS and main memories of tens of MBs. Each level of processor is selected on the basis of the performance needs of its particular application in the system.

MEMORY AND STORAGE DEVICES

A computer system will have both internal and external storage (memory). The internal memory is used to temporarily store data that is being processed, including intermediate results. There are usually several types of internal memory, including the main memory, control memory (for program control), and local memory (in the high-speed registers or buffers). The external memory is used for storing data that is about to be processed or for recording data for future use. External memory is often called either "secondary" or "auxiliary" storage. It can take several different forms, depending on the amount of data involved and the speed of access desired. Computer systems have a hierarchy of memories to provide the capability to handle all the necessary data storage tasks efficiently (Fig. 6-6). No one type of memory can satisfy all the capacity and speed requirements of the system.

Four different types of computer memory can be used for either internal or external storage:

1. Random access memory (RAM). This is a memory in which data can be stored and retrieved directly from specific addressable locations. RAMs are typically used for internal memories since they are high-speed. They are usually composed of high-density semiconductor integrated circuit (IC) devices and may have access times from tens to hundreds of nanoseconds. The technology for RAMs has been advancing very rapidly. Substantial improvements in density, speed, and cost have been made in recent years. A typical main memory is composed of many high-density RAM chips packaged on large printed circuit boards (Fig. 6-7).

2. Read only memory (ROM). This is a memory in which the data is stored permanently and can only be read out.

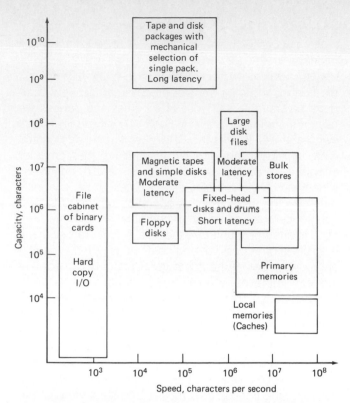

Fig. 6-6 Capacity of data storage devices. *(C. William Gear, Computer Organization and Programming, McGraw-Hill, New York, 1980)*

Fig. 6-7 Random access memory (RAM). *(International Business Machines Corp.)*

ROMs are used primarily as internal memories to store small programs that can be accessed frequently at high speeds. They are usually made of semiconductors similar to those in RAMs, but the data is entered during the manufacturing process.

3. Direct access storage device (DASD). This is a memory in which the data can be stored and retrieved randomly, but a combination of direct access and sequential searching is needed in order to reach the specific storage location. DASDs are usually external memories and are typically stored on magnetic disks. Disks can be either hard or soft (i.e., "floppy") platters, like records, which are coated with magnetic material that can be written on or read from by a magnetic head. The disks vary in size and have hundreds or even thousands of "tracks" on each side on which to store data. Large DASD systems are composed of stacks of hard disks with many read/write heads to store and retrieve large amounts of data rapidly (Fig. 6-8). The data transfer rates of such systems are often in the range of hundreds of thousands of characters per second. The speed of the DASD depends on the speed of the rotation of the disks and the speed of the movement of the head as well as, to some extent, the location of the data. The storage capacity of a DASD depends on the size of the system, the number of disks involved, and the density of storage capability on each disk. Large, high-performance DASD systems store billions of bytes (gigabytes) of data.

4. Sequential Access Storage. This is a memory in which data is stored and retrieved sequentially, not randomly. Such a memory is used for the external storage of large amounts of data as a permanent record or backup that does not have to be accessed directly by the CPU. Magnetic tape devices are the principal memories of this type and are very much like audio tapes; they can take the form of reels, cartridges, or cassettes. A tape of mylar is coated with a magnetic material and is written on or read from by a magnetic head. Since the data is stored sequentially, access time depends on the location of the data on the tape. The data is stored on tracks in high density. A high-performance tape system (usually called a "tape drive") may store thousands of bits per inch and transfer data at rates in the range of tens of thousands of characters per second (Fig. 6-9).

Fig. 6-8 Direct access storage device (DASD). *(International Business Machines Corp.)*

Fig. 6-9 Tape drive. *(International Business Machines Corp.)*

INPUT/OUTPUT EQUIPMENT

I/O equipment is composed of a wide variety of devices that are used to send data to or retrieve it from the computer. Such devices are usually connected to the computer through a control unit and an I/O channel. Instructions from the computer program being processed initiate commands that operate the I/O devices. Input devices read data from various sources and enter the data into the main memory of the computer for processing or storing. Data may come as punch cards, magnetic tape, magnetic disks, magnetic ink characters or bar codes on paper, or entry from a terminal keyboard. Output devices record data that is retrieved from the main memory onto various types of storage media. This may include the same types of media used for input (e.g., cards, tapes, disks) as well as such devices as video displays and printers.

In the previous section on memory and storage devices, we addressed two of the most common types of devices used for both input and output: magnetic tape and disk units. Some of the other types of I/O equipment and media include punch cards and punched paper tape. However, they are rarely used in manufacturing since they are bulky, have a relatively low storage density, and are easily damaged. Other devices, such as magnetic ink character readers (MICRs) and optical character readers (OCRs), are used primarily for business applications. This leaves only a few other types of I/O equipment that are commonly used in manufacturing:

1. Terminals. These are the most common devices humans use to communicate with and use the computer in an industrial environment. The many different types of terminals provide a wide range of capabilities. Most have keyboards or touch panels for data entry as well as some form of data display—either a video unit or a printer (Fig. 6-10). There are also "intelligent" terminals and workstations that include a small processor with memory for user programming and local computation. The use of such workstations can reduce the work load on the main computer system for small jobs and provide better response time for such "off-line" operations.

2. Displays. These are typically video monitors much like a television. They may only be able to display alphanumeric characters but may also handle graphics. They may be black-and-white (or one-color) or color displays. Computer displays are often used together with some other device for inputting information, such as a:

☐ Keyboard, like one found on a typewriter

☐ Light pen, which senses light from the spots illuminated on the screen (Fig. 6-11)

☐ Touch-sensitive screen, on which you can point to an object or position with your finger (Fig. 6-12)

☐ "Mouse," which you can move with your hand to direct actions on the screen (Fig. 6-13)

☐ Touch-sensitive panel or tablet, which can pick up inputs from a menu that can be pressed by the operator (Fig. 6-14)

The amount of information the screen can display will vary depending on its resolution and size (typically either 40 or 80 characters per line). The type of display device selected for a particular application will depend on the functions it will need to perform.

3. Printers. There are also many different types of computer printers, which are designed to satisfy the needs of different applications. Some are capable of very high speed printing (i.e., thousands of lines per minute) for large data output tasks, such as major batch jobs (Fig. 6-15). Others are small and relatively slow, but more than adequate for a individual user at a workstation. Some are capable of printing graphics or even color. The various types of printers can be classified in several ways: for example, by the amount of data that they print at one time

Fig. 6-11 Keyboard and light pen. *(Key tronic)*

Fig. 6-12 Touch-screen display. *(Electro Mechanical Systems, Inc.)*

Fig. 6-13 Mouse input device. *(Key tronic)*

Fig. 6-10 Computer terminal. *(International Business Machines Corp.)*

Fig. 6-14 Tablet input device. *(International Business Machines Corp.)*

Fig. 6-15 High-speed printer. *(International Business Machines Corp.)*

(i.e., "line" or "character" printers) or by the manner in which they actually print the characters (i.e., "impact" or "nonimpact" printers). Impact printers may use a print head, a print wheel, a character ball, a dot matrix, a cartridge, or a reel with embossed characters. In any case, the impact printer strikes the character onto paper through some type of inked ribbon. The nonimpact types include ink-jet, laser/xerographic, electroerosion, and thermal printers. Print speeds vary significantly from a few characters per second up to many thousands of lines per minute.

6-4 DATA COMMUNICATIONS

COMPUTER NETWORKS

The various types of data processing equipment must be tied together into a system to provide all the normal functions of a computer (i.e., input, output, storage, computation, etc.). Computer systems in large businesses and industrial operations often comprise many pieces of equipment, particularly I/O devices, which are connected to a central processor. These devices must exchange data frequently and efficiently if the computer system is to be useful. In large systems, the devices may be scattered over a number of different locations within a building or even in entirely separate facilities. To manage a large business, or one that is spread over many locations, it is often necessary to interconnect several computer systems into a network of computers that can communicate with each other. The handling and processing of data between remote locations is often referred to as "data communications," "teleprocessing," or "computer networking." Such computer networks are comprised of three major elements:

1. Computer systems. These may be central host computers that operate independently or remote processors in a distributed system which is under the control of some higher-level computer.

2. Communications facilities. These are the things that tie the equipment in the network together and provide for the exchange of data between them. They include the communication lines, special communication devices, and communication processors.

3. Terminals. These are the remote I/O devices that provide the user with access to the computer network.

Computer networks were developed to accomplish large data processing tasks economically. The first major application, which was for an airline reservation system, was developed in the early 1960s. Today there are a number of reasons for using computer networks:

1. Controlling large, complex operations that are scattered among many remote locations

2. Sharing expensive computer equipment among many users

3. Providing access to a central database for many users

4. Providing data exchange between interdependent operations

5. Providing management reports from widely scattered operations

A computer network may be used for one or more of several different types of applications, such as:

1. Data entry. The input of data to a central computer from a remote terminal.

2. Remote job entry (RJE). The input of a data processing task to be executed by a central computer from a remote terminal.

3. Record update. The changing or addition of data in a central computer file from a remote terminal.

4. Timesharing. The sharing of the resources of a central computer to simultaneously execute a number of programs that are entered from different remote terminals.

5. Process control. The acquisition of data from production equipment to monitor its performance, provide control, and generate management reports.

6. Message switching. The relaying of messages between remote terminals through the communication facilities of a computer network.

DATA TRANSMISSION

The process of transmitting data through a computer network involves several basic elements (Fig. 6-16):

1. A terminal. This provides for the input or output of data from a remote location by converting mechanical action to electrical signals and vice versa.

2. Modems (MOdulator/DEModulator devices). These convert digital signals to analog signals for transmission over communication lines.

3. A communication channel. These link the remote terminals and the central computer. It may be a wire, cable, or microwave connection, and it is usually provided by a common carrier service.

4. A communication processor. This is a special com-

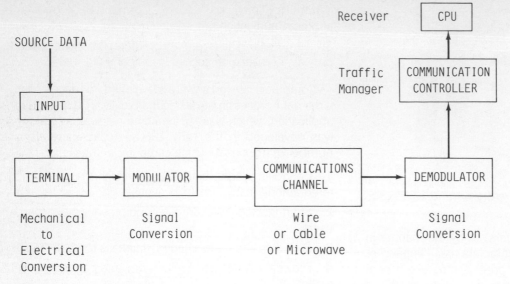

Fig. 6-16 Data transmission.

puter used to control the data transmission and processing functions in a computer communications network.

5. A computer. This is the central host processor that receives and processes the data from remote terminals.

The data that is transmitted over the communication channel must be put in a form that is compatible with the particular carrier chosen. This usually requires that the data be "serialized," or converted into a serial stream of binary digits and then "deserialized" (reconverted to machine language) when it is received. Serialization and deserialization is done by the communication processor and communication control units on terminals. The serialized binary signals are then either superimposed on or converted to an analog signal for transmission on a common carrier line such as a telephone line. This is done by a modem which converts (modulates) and reconverts (demodulates) the signal. There are several different types of modems and modulation schemes; the most familiar are amplitude modulation (AM) and frequency modulation (FM). Modems may be acoustically coupled to a telephone line or directly connected to it. Direct, hard-wired connections permit higher-speed transmission [e.g., 9600 bits per second (bps) versus 300 bps].

Several key parameters determine the performance of a communication channel:

1. Data rate. This is the speed of transmission in bits per second.

2. Bandwidth. This is the range of frequencies of the carrier signal. It can vary from the narrowband operation of a teletypewriter to the common voice-grade telephone line to a wideband digital data line, or even a direct high-speed microwave link transmitted through a satellite.

3. Direction. Data may be transmitted in one direction only ("simplex" transmission); in two directions, but only one way at a time ("half-duplex" transmission); or in both directions simultaneously ("full-duplex" transmission).

4. Error characteristics. These are factors such as interruptions in transmission, distortion of signals, and noise or interference.

5. Delay characteristics. These are differences in the timing of transmitted signals.

For equipment to send data to and receive it from each other, the communication between them must be synchronized. This is accomplished by one of two basic modes of data transmission:

1. Asynchronous transmission. This requires the use of stop and start bits before and after each character. It is the least expensive mode since it does not require a large data buffer, but it has a low utilization of the transmission line.

2. Synchronous transmission. This provides a continuous stream of characters without the need for start and stop bits. The stream of characters is divided into blocks that are transmitted at equal time intervals that are synchronized between the two pieces of equipment. This provides higher utilization of the transmission line, but it requires more expensive equipment.

Two types of communication lines are used for data transmission:

1. Switched lines. These are the dialed telephone lines available from a common carrier, which provide half-duplex transmission only.

2. Nonswitched lines. These are usually leased or private lines that are dedicated to a specific computer network and can operate in a full-duplex mode.

When data is transmitted over communication channels, it is coded using one of several standard techniques. The most common communication codes are ASCII and EBCDIC. Within these coding techniques there are also schemes to check for errors and recover them if necessary. This may be accomplished by detecting the presence or absence of a particular bit and requiring a block of data to be retransmitted.

COMMUNICATION DEVICES

In addition to the standard types of data processing equipment normally found in a computer system, computer networks require the use of several special devices which make the data communication possible:

1. Modems. As mentioned earlier, modems convert and reconvert the signals to be transmitted into a form that the communication channel or receiver can handle. There are several different types of modulation schemes, including the AM and FM techniques commonly used for radio transmission. Modems also vary in their performance, which depends in part on how they are connected to the communication channel. Some modems are "acoustically coupled" by sending audio signals through a telephone handset. This approach is relatively inexpensive, but it also offers the slowest data transmission rate (e.g., 300 bps). High-speed modems are directly connected, or hard-wired to the communication line and may operate at up to 9600 bps.

2. Multiplexors. These are special devices that can be used to reduce the cost of data transmission by combining several low-speed signals into fewer high-speed lines. Two basic techniques are used to accomplish this:

☐ "Frequency-division multiplexing" modulates the frequencies of the signal which carries the data.

☐ "Time-division multiplexing" interleaves data into one signal that must be separated by timing. This is a more complex and expensive approach, but it can provide higher transmission speeds.

3. Communication processors or controllers. These specialized computers are often stand-alone units in large computer networks. They are used to control all the communication processing functions. This includes providing the interface between the central computer and the communications network, which is sometimes called "front-end processing." The controller can also act as a multiplexor to concentrate data for high-speed transmission. In addition, it controls message switching and the data transmission traffic throughout the network.

TYPES OF NETWORKS

Computer networks have been developed in a variety of different configurations, each with its own advantages. Most can be classified into two basic types (Fig. 6-17):

1. Point-to-point. This is where each terminal or satellite computer connects directly to the host computer on its own line. The most basic form of this configuration is called a "star" network. It is a simple design that provides fast response times, but it would be expensive for large networks. In star networks, all communications must go

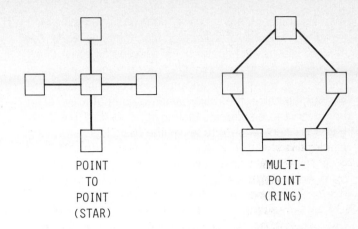

POINT TO POINT (STAR)   MULTI-POINT (RING)

Fig. 6-17 Basic computer network configurations.

through the central host computer. Therefore, if the host goes down, so does the entire network.

2. Multipoint. This is where a number of terminals or satellite computers share one line to the host computer. This requires that the communication on the line be controlled so that only one signal is transmitted at a time. Techniques such as priority interrupts, polling, and selecting are used; they will be covered in a later chapter. Multipoint networks offer lower line costs per terminal, but they do not have as fast a response time as point-to-point networks. A common form of a multipoint configuration is a "ring" network, in which communication must be passed around the ring to get to its ultimate destination. In this case, if any one device or computer goes down, the entire network will also go down.

Many variations of these basic configurations have been used to optimize different types of computer systems and applications. These will be addressed further in later chapters.

Communications in computer networks can be very complex, particularly when many I/O devices and data processing tasks are involved. The flow of data in such networks must therefore be controlled to assure efficient, error-free operation. This is accomplished by "traffic management," which is usually provided by the communication processor. It uses techniques such as communication protocols, routing strategies, control algorithms, and on-line monitoring and control, much like the modern electronic switching systems used to control telephone networks. Communication protocols involve standard line-control procedures (such as error checking and recovery, synchronization, and data transfer control) which permit different devices and computers to exchange data with each other.

6-5 COMPUTER HARDWARE TRENDS

EVOLUTION

Since the earliest practical electronic computers were developed in the late 1940s, there have obviously been many significant advances in hardware technology. In fact the evolution of computer technology is often described in terms of the number of succeeding generations of computers that have emerged over the years. Although there have been many advances in computer software and architecture, the major factor in computer evolution has been the electronics technology involved. The early systems used vacuum tubes as the basic logic switching element in the CPU. This made those computers very large, expensive, and slow by today's standards. When transistors became practical, they were used to replace the vacuum tubes in order to reduce the size and cost as well as improve the speed of computers substantially. In the late 1960s, semiconductor ICs were developed, which made it possible to replace the discrete transistor logic as well as the magnetic core memory with high-speed, high-density devices at relatively low costs. This step not only made large computers more powerful and economical; it made small computers practical for the first time.

The current generation of computers is based on an extension of this technology which is often referred to as "large-scale integration" (LSI). This has made it possible to provide very high-density and high-speed logic and memory functions in very small electronic packages. Such an advance makes it possible to further improve the cost and performance of computers; it also improves their reliability and offers opportunities for new functions and applications. Computer technology is now in the process of evolving into yet another generation. This "fifth generation" will be characterized by further advances in electronics technology and, perhaps more importantly, by some basic changes in the architecture of the computer system itself. The balance of this chapter will cover some of the trends in technology and architecture that should indicate where computer hardware is headed in the future.

COMPUTER PERFORMANCE CAPABILITIES

Advances in electronics technology are the major contributors to the continued improvements in the cost and performance of data processing equipment. Since the time of the early vacuum-tube computers, processors have become almost 1000 times faster. At the same time, the cost of computation has come down by almost a factor of 100 (Fig. 6-18). There may be no other example of a product or technology that has experienced such a dramatic improvement in productivity. It should be noted, however, that the rate of improvement in recent years is slowing down. To continue to provide faster and more powerful computers at lower costs through the use of technology alone is becoming increasingly more difficult. With the traditional approach to the design and programming of a computer, its performance is very much a function of the speed of its electronic circuitry. Although this circuitry continues to improve in cost and perfor-

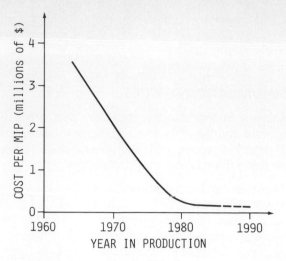

Fig. 6-18 Cost performance of data processing.

mance, as we shall see later, it is becoming extremely complicated and expensive to develop and produce. If computers are to continue to become even more powerful and economical in the future, some basic changes in their design and programming will be necessary, which will be addressed further.

The performance capabilities of computers should be measured not only in terms of their speed and cost but also in terms of the functions they provide to the user. Over the years, user requirements have become more and more demanding of the computer. This has influenced the performance requirements for computers and the types of equipment and systems configurations that are needed (Fig. 6-19). Early users of computers had to be expert programmers who typically worked directly with a large computer in a stand-alone mode. As computer systems became more powerful, they were used for large batch processing jobs. However, users eventually needed more frequent and easier access to the computer, which led to timesharing systems and higher-level languages. Today, many users and applications in industrial environments require fast response time and ease of use. To satisfy this need, large networks of terminals and intelligent workstations have become the common solution. In the future, to continue to expand their usefulness, computers will have to provide greater capabilities in areas such as:

1. "User-friendly" interfaces (e.g., higher-level "natural" languages)

2. High performance for special applications (e.g., graphics, modeling, and scientific computations)

3. Humanlike capabilities (e.g., sensing, vision, and "expert systems")

4. Smaller and less expensive individual workstations

These and other advanced capabilities will be addressed in later chapters.

Mode of operation	Single user	Batch processing	Timesharing	Networking
User needs	"An answer"	Several runs per day	Response time in seconds	Access to many systems
Typical facilities	Standalone	Mainframe and satellite I/O	Mainframe and communication controller	Local area network

Fig. 6-19 Impact of user requirements on technology evolution. *(Jean-Loup Bear, "Computer Architecture," Computer, Oct. 1984; © 1984 IEEE)*

TECHNOLOGY TRENDS

In recent years, technological advance has been extremely rapid, particularly in electronics. Perhaps the most visible trend is the ever-increasing density of semiconductor IC devices. The number of circuit elements (e.g., transistors) that can be integrated on a single semiconductor chip has been increasing at a fairly predictable exponential rate, which is approximately doubling every other year (Fig. 6-20). This has made it possible to use chips that incorporate hundreds of thousands of transistors as large building blocks for computer logic or memory. As this trend continues, chips with millions of transistors will permit us to design very powerful computer functions in relatively small and inexpensive electronic packages. This will make new computer applications feasible and will bring computer power to more people in the workplace.

The most common semiconductor technologies used today are:

"Bipolar." These are high-speed, current-driven devices

Field-effect transistors (FETs). These are high-density, low-power, voltage-driven devices

However, a number of other emerging technologies appear to offer potential advantages in density, speed, and power consumption. This may involve the use of new materials, such as gallium arsenide instead of silicon. It may also involve new semiconductor device phenomena, such as Josephson junctions, which are thin-film devices that operate at near–absolute zero temperatures. There are also "magnetic bubble" devices which can store large amounts of nonvolatile data in the form of very small magnetized areas of material. The density of IC devices will also continue to increase as a result of advanced processing techniques such as electron-beam or x-ray lithography. There are even radically different types of devices being developed that are not based on the traditional semiconductor technology, such as optical ICs that can switch at extremely high speeds, or "biochips" formed from organic molecules that may simulate the structure of the human brain. No matter which of these technologies ultimately proves to be successful, it is inevitable that devices will be available with higher performance at lower cost to build computers in the future.

Technology advances are also a significant factor in the capabilities of other types of data processing equipment in addition to the logic and memory of the CPU. Auxiliary storage devices continue to be developed with higher densities, faster access speeds, smaller size, and lower costs. Magnetic disks, for example, have been following a trend in recording density that is similar to that of semiconductors (Fig. 6-21). Advances in magnetic disk and tape technologies are making the storage of vast libraries of data (measured in billions of bytes) both feasible and economical. Alternative storage technologies are

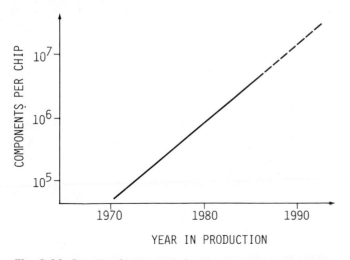

Fig. 6-20 Density of integrated circuits.

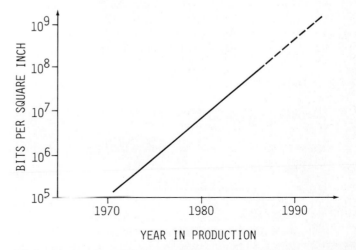

Fig. 6-21 Storage density of magnetic disks.

being developed which may have even greater storage capabilities, such as optical disks and memories.

In the area of I/O equipment, many developments will enhance their capabilities in terms of both performance and function. Display units, for example, are becoming available with higher resolution, color, touch sensitivity, and smaller size, such as the flat-panel display terminal in Fig. 6-22. This technology uses gas discharge phenomena, similar to what makes a neon sign work, rather than the common cathode-ray tube (CRT). It can achieve very high resolution images and can be packaged in a thin enclosure to save space and weight. Printers are being developed with higher speeds, finer resolution, and even color. They use advanced nonimpact technologies such as inkjets or lasers. There is even an entire area of development aimed at providing the easiest and most common form of human I/O—speech recognition and simulation.

Another area of technology which is critical to the continued advancement of computer capabilities and applications is communications. The objectives being pursued include higher-speed data communication rates, lower data transmission cost, standard interfaces, and larger networks. These will be satisfied by a variety of technology developments, such as local area networks (LANs) with standard protocols, optical-fiber links between equipment and systems, microwave satellite data communication links, and the integration of data, voice, and video transmission into a computer-based telecommunications switching system. These communications technologies will permit more users to have access to more data, more efficiently and easily than would have been possible in the past. This is essential if we are to manage complex industrial operations efficiently. The delays and expense involved in information handling can only be reduced if we replace the traditional approaches (including paper) with easy, inexpensive access to high-speed electronic data communication systems.

COMPUTER ARCHITECTURE

As the hardware technologies of computers begin to approach some physical and economic limitations (in terms of speed, power, and size), something must also be done with the basic design and organization of the computer to improve its performance. The architecture of most computers has not changed dramatically since the early machines were first developed. The so-called von Neumann machine (named after its inventor) is a computer with a program of instructions, stored in memory along with the data, which operates sequentially. This type of design was influenced greatly by the relatively high cost of hardware at the time. Therefore, its performance was optimized by getting as much out of the hardware as possible. Over the years, however, several important innovations that were compatible with this architecture provided significant enhancements to the capabilities of computers. These included:

1. Timesharing. A CPU handles data and programs from several sources at the same time by using any delays between memory access and data transmission cycles to process other available data.

2. Virtual machines. Each user has the illusion of a separate machine by using a combination of hardware and software to control the storage and processing of multiple tasks.

3. Multiprocessing. Two or more processors are used simultaneously to increase the speed of computation for a particular task.

These developments each extended the capabilities of serial processing machines, but they eventually led to the need for concurrent processing techniques to break the bottleneck created by the requirement for central control and storage. The need for more user functions and larger networks continues to grow, and alternative architectures are being pursued. Since electronics is getting less and less expensive, computer designers are finding that they can now afford to use a lot of hardware to solve the performance problems. The basic approach being pursued is to use many small CPUs operating in parallel on different parts of the data processing task. Although a number of different variations to this approach have been developed, most fall into one of three basic types of parallel architecture (Fig. 6-23):

1. Control-driven or "pipelined" machines. These use multiple processors and memories to handle multiple streams of data through a complex, centrally controlled switching network. With this approach, the instructions are executed when the control program orders them. Although both the single- and multiple-instruction versions of this architecture have the capability for much

Fig. 6-22 Flat-panel display. *(International Business Machines Corp.)*

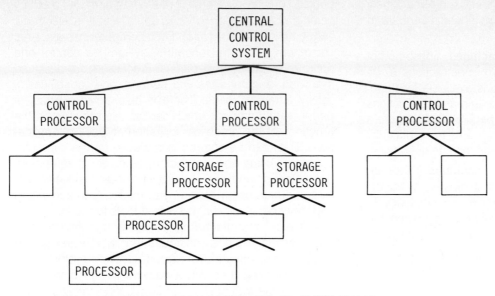

(a)   EXAMPLE OF CONTROL-DRIVEN PROCESSING OR "PIPELINING"
      (tree of many concurrent processors with central
      control system)

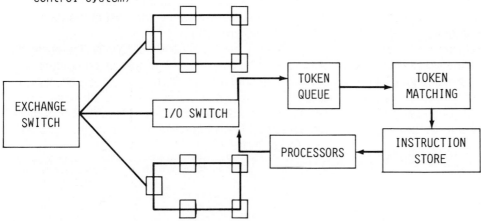

(b)   EXAMPLE OF DATA-DRIVEN PROCESSING OR DATA FLOW MACHINE
      (set of rings around which data flows and is processed)

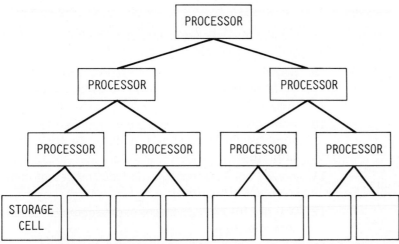

(c)   EXAMPLE OF DEMAND-DRIVEN PROCESSING OR REDUCTION MACHINE
      (tree of processors with each symbol of an instruction
      stored in a separate cell)

Fig. 6-23 Parallel processor architectures.

higher performance than conventional serial machines, they are limited by their continued reliance on centralized control.

2. Data-driven or "data flow" machines. These do not use central control, and they permit instructions to be executed whenever sufficient data is available for them. Although they improve performance in many applications by handling many instructions in parallel, they require a complex instruction control scheme and have difficulty handling complex data structures and looping operations.

3. Demand-driven or "reduction" machines. These are designed with different processors dedicated to handle specific types of operations to optimize their efficiency. In this case the instructions are executed only when the results are needed. This approach is most efficient when used together with concise mathematical languages that permit the problems to be broken down into a number of simpler tasks that can each be assigned to a specialized processor.

These new parallel processing computers all require low-cost logic and memory, complex control systems, and special programming languages that can manipulate parts of a program independently. Future computers will probably use a combination of these approaches which are tailored to the specific needs of the application involved. In general, parallel processing can significantly increase the performance of computers, but it is not as flexible in terms of the variety of tasks it can perform as the general purpose serial processing approach.

6-6 SUMMARY

A basic computer system is made up of a central processing unit (CPU), main and auxiliary memory, and inpu-

t/output (I/O) equipment. Digital computers use a binary coding system which is compatible with the electronic switching in their hardware. Each computer has a special set of programs, the operating system, to control and optimize its internal operations. Processors are designed in a wide range of sizes and performance capabilities to satisfy different needs. There is also a wide variety of peripheral equipment which can be used to support the functions of the processor for different types of applications. Data can be entered or retrieved from a computer system by using one of the many types of I/O equipment available. To handle all of the data storage tasks efficiently, computers use a hierarchy of memories, each with a capacity and speed that is tailored to its function.

A typical manufacturing computer system is composed of many pieces of equipment which are all tied together in a hierarchical structure. Such systems can also be tied together into a network in which they can communicate with each other. Data communications networks are used to accomplish large data processing tasks economically and to share data efficiently. A variety of special communications equipment can be used which will infuence the performance and cost of the data communication system.

Advances in electronics technology have increased the performance and reduced the cost of computers substantially. Although this will continue, it is becoming increasingly more difficult and expensive to make further advances in hardware technology. Much effort is therefore directed toward improving basic computer architecture and software.

REVIEW QUESTIONS

The answer to each question can be found in the section(s) indicated at the end of the question.

1. Identify the basic elements of a computer system and of a CPU. [6-2]

2. What are the basic functions of a computer? [6-2]

3. Describe the basic internal operations of the CPU. [6-2]

4. Identify some of the different types of memory used in a computer system and describe their functions. [6-2 and 6-3]

5. Describe a typical hierarchy of computers in a manufacturing system and identify their roles. [6-3]

6. What are the major types of data processing equipment? [6-3]

7. Describe some of the typical types of I/O equipment used in industrial applications. [6-3]

8. Identify the major elements of a computer network and describe their functions. [6-4]

9. Why do we use computer networks and what are some of the different ways in which they are used? [6-4]

10. What are the types of equipment required for data transmission and what functions do they perform? [6-4]

11. Identify and describe the different modes of data communication. [6-4]

12. Describe the two basic types of data communication networks. [6-4]

13. What are some of the key areas that require further development to improve the capabilities of computers? [6-5]

COMPUTER SOFTWARE

7-1 INTRODUCTION

When the early computers were first developed they could be used only by experts, and their operation was dominated by the hardware. What they did and how they operated was controlled manually by the operator. This severely limited the capability and flexibility of the computer. Therefore, along with the development of hardware technology, there was a need to continually improve the efficiency of controlling and programming the operation of computers. This is the area of software, or the "invisible" computer, which has become at least as important as the hardware in influencing computer capabilities. Although not as apparent as the physical tasks and hardware involved in manufacturing, computer software and programming have become a major factor in the success of modern production operations.

This chapter serves as the complement to Chap. 6 in establishing a basic understanding of how a computer works from a software point of view. It will not attempt to make a programmer out of the reader. However, it will cover all the key elements of programming and controlling computers to provide an appreciation for what is involved. This should permit the reader to deal with some of the more advanced concepts and applications addressed in later chapters. It also provides a foundation that could be built upon to learn how to program.

We will start with the basics of what software is and how it works. Then we will cover the two principal types of software: system software and application software. With that as a background, we can then deal with programming languages and the process of programming itself. Finally, we will review some of the recent developments and major trends in software which will influence the capabilities and use of computers in the future.

7-2 SOFTWARE BASICS

WHAT IS SOFTWARE?

"Software" is a very general term, much like "hardware," that is used to describe the programs of instructions that control the operation of computers. Computer programming was originally a manual, hardware-oriented process. The operation of the earliest machines was controlled by the operator by manually "programming" a series of switches that directed the sequence of steps to be performed by the computer's hardware, such as its registers. This tedious task was replaced by the "stored-program" concept, in which the program of instructions is read by the computer and stored in its main memory along with the data. Such programs contain the sequence of instructions that directs the computer hardware to execute specific tasks.

As we saw in prior chapters, the computer can understand instructions or manipulate data only if they are in the form of a binary code which represents the on/off switching signals of the computer's logic. This sequence of 1s and 0s must be put into a prescribed format that is based on the design and operation of the particular machine being used. The computer can then recognize and interpret the code either as an instruction for the CPU to execute, as data, or as an address (location) in its memory. The coding process is called "programming," which we will see can take many different forms.

TYPES OF SOFTWARE

There are two major types of computer software:

1. System software. These are programs that direct the internal operations of the computer while it is executing instructions from a user. They include functions such as:

☐ Translating programming languages into a form that the computer can understand

☐ Managing the use of different sections and equipment in the computer system

☐ Helping a programmer to develop or modify software

System software is generally unique to each particular type of computer.

2. Application software. These are programs that are written to perform specific tasks required by the user. They are the jobs or applications that the computer is being used for at the time. Application software is generally not unique to a particular computer. It may, however, be unique to the specific task or user involved, or it could be a common type of program that is used by many people. For example, a program which controls the movement of a robot is application software.

WHAT IS PROGRAMMING?

Programming is the process of developing a set of instructions which directs the computer to perform a desired task. This may involve the solving of a complex problem, the handling of a large amount of data, or the generation of reports or charts. Programs may be relatively short, with only a few lines of coded instructions, or extremely long and complicated, requiring thousands of lines of code. The programming process involves the analysis of the problem as well as the actual coding of the program. Although this process can be very time-consumming, depending on the task involved, much of the effort required has been automated with the use of modern computer hardware and software. Very few programmers now have to deal with the elementary binary code that is used internally by the computer. The programming task has been simplified and at the same time made more powerful through the development of more sophisticated programming "languages" that can be used instead.

TYPES OF PROGRAMMING LANGUAGES

Over the years, thousands of different coding schemes were developed for programming computers. Each was intended to improve the efficiency of either the programmer or the computer or both. Most were unique to specific computers or applications and therefore were not widely used as general purpose languages. Each has a different form and code to represent instructions. Although most programming languages could be used to perform most common tasks, they all look different when they are written. Writing programs varies in difficulty with the language used, and programs that do the same thing can perform with different degrees of efficiency.

Programming languages generally fall into one of three categories:

1. **Machine language.** This is the basic binary code that a computer understands. It is sometimes referred to as "object code." It is written with the 1s and 0s of some form of the standard binary-coded decimal (BCD) scheme to represent characters, numbers, and special codes. Each instruction contains a code for the specific operation involved (e.g., add, subtract, store) as well as an "operand" which designates data or the address for a storage location or I/O device. Machine language is unique to the design of each type of computer. A programmer must be very familiar with the inner workings of the machine in order to use machine language. To write programs for complex tasks in machine language would obviously be very time-consuming and prone to error. Most programming is therefore done in higher-level languages which are easier to use and understand. Programs written in these languages, however, must be translated into machine language before the computer can operate.

2. **Assembly language.** This is the lowest level of what are called "symbolic" languages. They were developed to reduce the effort involved in using machine language. In this scheme the operation code is represented by a symbol or word called a "mnemonic" which is easier to recognize and remember than the binary code. In addition, symbols or "labels" are used to designate memory addresses, and normal decimal notation is used for numerical values. Assembly language is machine-oriented and can be very efficient in directing the operation of a computer. However, programs written in assembly language are generally just as long and complicated as those written in machine language. Therefore, writing them can be very tedious and subject to error. A program written in assembly language is translated into machine language by an "assembler" which replaces the symbolic code with its binary equivalent.

3. **High-level languages.** These are languages that were developed to make the job of programming much easier and more efficent. They are oriented more toward the application or type of task involved than toward the machine. They are therefore sometimes referred to as "procedure" or "problem-oriented" languages. The instructions in such high-level languages use English-like statements and common mathematical symbols which make them easier to recognize and use. These statements typically generate several machine instructions; therefore, fewer are needed to write a program. Since these languages are somewhat independent of the type of computer they are used on, the programmer does not have to be very familiar with the inner workings of the computer. In order for programs written in these languages to be executed, they must first be translated into machine language. This task is usually done by a program called a "compiler."

7-3 SYSTEM SOFTWARE

OPERATING SYSTEMS

A computer's operating system is its primary interface to the user. It performs tasks for the user to make the job of using the computer simpler. It also manages all the hardware resources of the computer system during its operation. An operating system is made up of a set of special programs which perform these tasks. Different operating systems are used by different computers and for different types of applications. Some operating systems, particularly those used by large computers, can be made up of many very large and complicated programs. Most other software requires the support and services of an operating system in order to work.

To appreciate the importance of an operating system, one must consider what is involved in even the simplest of computer operations. For every instruction or basic task there are usually many individual steps that the computer is required to perform. Each character typed on a keyboard must be coded, transmitted, stored, read, and interpreted. This requires a number of electrical and mechanical operations which must be directed by the operating system, such as searching, reading, and writing on a disk. Figure 7-1 illustrates how even a relatively simple single command involves many individual operations between a number of different pieces of hardware and software. The operating system not only moves data around; it must keep track of that data, the hardware it is using, and what point in the program is executing. This can get even more complicated when several programs are being executed at the same time. It is the job of the operating system to make sure that the programs and data do not interfere with each other or get mixed up.

To accomplish these tasks, an operating system is usually made up of a hierarchy of software which deals at different levels of operations in the computer system. At the lowest level, the operations involve the internal hardware of the central processing unit (CPU)—for example, registers and memory. The system must then deal with internal instructions and procedures such as computation, subroutines, and interrupts. At higher levels there are programs to deal with tasks which require control of the timing and sequence of operations. The operating system software at the highest levels must deal with external devices and the user environment. An operating system may therefore span the entire scope of complexity and size that one can find in software. As computer systems become more powerful and complex, and as the demands of the user increase (such as for ease of use), the operating system also becomes larger and more complex.

An operating system typically performs three basic types of functions:

1. Managing tasks, such as controlling their priority, location, and status

2. Managing resources, that is, controlling and allocating the computer hardware required by each task

3. Managing data, that is, controlling I/O operations, interfaces, and files

Operating systems are usually classified by the type of environment they are designed to operate in or by the auxiliary device they are intended to support—for example, the disk operating system (DOS). There are three major types of system environments:

1. Batch mode. This was the typical mode of operation for early computers. In this case each program was executed by the computer sequentially. Therefore the operating system had to supervise the loading of each program, allocate all the resources necessary to run each program one at a time, and handle the I/O as required.

2. Multiprogramming. This mode was developed to increase the efficiency and utilization of the CPU. In this case, the operating system must control the execution of several programs at the same time. This requires the computer's resources to be shared in such a way that there is a balance in the use of I/O and CPU operations.

3. Real time. This mode is particularly important in a manufacturing process control application. In this case, the operating system is designed to respond to inputs from multiple sources very rapidly so that the user perceives no time loss.

The continued development of advanced operating systems for large computers has led to two important functions in recent years. These capabilities have made it possible to create the large networks of terminals and I/O devices serviced by a central host computer that are typical of systems in business and industry today. These functions are:

1. Timesharing. This technique permits many users to have simultaneous access to the CPU in a real-time, on-line mode. To control this situation, the operating system must assign separate parts of the memory to each user and switch the CPU operation between them by executing portions of each program. This makes it appear to each user that the CPU is dedicated to his or her program. Some timesharing systems can be used to both execute and write programs, while others are restricted to execution only or single applications.

2. Virtual storage. This technique made it possible to significantly expand the effective size of a computer's main memory without actually increasing it physically. By storing in main memory only those instructions that are needed at the specific time of execution of each portion of a program, space can be made for storing more or larger programs. The rest of the program, which is not being used at that time, is kept in auxiliary storage (e.g., on disk). The operating system must divide each program into blocks and move them between the main and "virtual" memory as they are needed by the CPU.

The operating system is the first set of software to be loaded into the computer when it is turned on and is the first to communicate with the user. As application programs are run, the operating system is still in control. When jobs are done, it loads other programs as required and "signs off" users when they are done. The operating system is therefore the heart of the system software upon which other functions can be built.

UTILITY PROGRAMS

The system programs which perform a variety of functions to help the user write and execute programs are called "utility programs." Such programs are intended to make it easier to use the system as well as easier to develop and maintain software. These typically include the following:

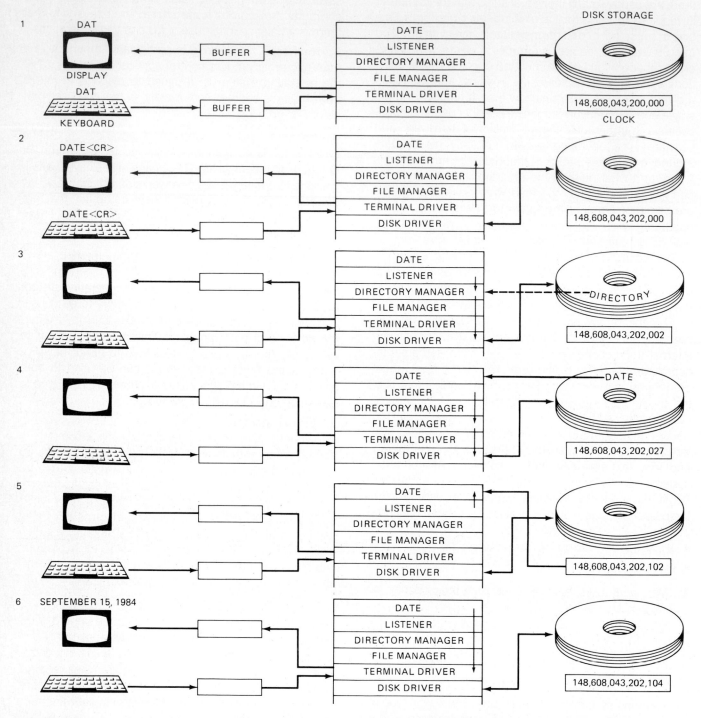

Fig. 7-1 Execution of a command. *(from "Operating Systems," Peter J. Denning and Robert L. Brown, Scientific American, Sept. 1984; © 1984 by Scientific American)*

EXECUTION OF A COMMAND sets in motion events at several levels in the hierarchy of programs that make up the operating system. The command is simply a request for the date. As each character is typed at a keyboard (*1*) it is received by a terminal-driver program, which echoes it to the display screen. When a carriage return is entered, the terminal driver passes the string of characters *d a t e* to the listener program (*2*), which interprets it as the name of a command. The listener asks the directory manager to search the directory of commands for *date*. The directory manager in turn asks the disk driver to copy the directory into a buffer in the directory manager's storage space (*3*). When the command has been found, the listener directs the file manager to load the binary code for the *date* program into memory; to do this the file manager again uses the disk driver (*4*). The listener then activates the *date* program, which reads a "clock" (*5*), a hardware device that keeps a count of the milliseconds that have passed since some fixed starting time, in this case midnight of January 1, 1980. From this number the program calculates the current date and displays it through the terminal driver as September 15, 1984 (*6*). The listener and the various drivers and managers constitute part of the "kernel" of the operating system; *date* is a utility program.

□ **I/O routines.** Programs that interpret data transfer instructions (e.g., READ and WRITE) and control the movement of data between the I/O devices and memory.

□ **Text editors.** Programs that help the user to make changes to software.

□ **Debug routines.** Programs that permit the user to try out new programs or changes interactively with the computer.

□ **Dump routines.** Programs that can print out programs or files stored in memory.

□ **Data conversion routines.** Programs that automatically convert data into the format required to communicate between the CPU and I/O devices (e.g., ASCII).

DATABASE MANAGEMENT SYSTEMS

In large industrial operations many different people may need to have access to the same data, but in different forms (such as in management reports). A central file of information that can be used for many different purposes and can be accessed by many different people is referred to as a database. The programs that are used to control such central files are called database management systems. Such systems are particularly useful when there are many large files of related data that are subject to change as well as routine reports and inquiries. The database management system must structure and store the data so that it can be easily accessed independent of the application programs or hardware used. It also usually provides data security and error recovery functions.

LANGUAGE TRANSLATORS

Since the computer can only understand binary code, all programs that are written in higher-level languages must be translated into machine language (object code) before they are executed. Therefore, the system software must include language translators for each of the languages the computer is expected to use. There are two principal types of translators for high-level languages:

1. Compilers. A compiler is a program that converts an entire program from a high-level language into machine language. Once the compiler completes the entire translation, the program can be executed. At this point it is no longer necessary to have the compiler program stored in memory. Although compiled programs are generally considered to be efficient, their performance is not as optimized as it could have been if they had been written originally in assembly or machine language.

2. Interpreters. An interpreter is a program that both translates and executes programs which are written in higher-level languages. It converts the program instructions one at a time, as they are being executed. This is particularly useful to programmers as they are writing or changing programs. They can try out a program or a portion of one interactively with the computer without the necessity of compiling, executing, and then editing the entire program. Since the program is being translated while it is being executed, the interpreter program must remain in main memory during this time. Although interpreted programs are easier for programmers to use, they are usually slower than compiled programs.

7-4 APPLICATION SOFTWARE

The software most users are familiar with are programs that perform specific functions to arrive at an intended result. Such application programs will be addressed in a number of later chapters as specific areas of manufacturing applications for computers are covered. Application programs generally fall into one of two categories:

1. Generic. These are programs which are written for common applications that many people or companies use. They are usually purchased from hardware or software developers and are available on the open market. Each user selects the program to be used after assessing its capabilities and features. There are often many competitive alternatives from which to choose. Common examples of generic programs are spreadsheets and graphics applications.

2. Custom. These are programs that are written for unique applications required by a specific user or set of users. They must be developed, tested, and maintained for that specific application, which usually requires a lot of time and effort. Such programs can therefore be very expensive and must be justified by unique capabilities that cannot be provided by generic programs. In manufacturing, it is often necessary to develop custom application programs when new processes or tools are developed. In some cases, the cost of the software development may be even greater than that of the hardware.

Many types of application software are used throughout the manufacturing industry. Some of the more common categories include:

Process control

Tool control

Production control

Design

Simulation

Accounting

Engineering

Graphics

Planning

Word processing

Data communications

Management reporting

Application programs will continue to be developed as long as there are new jobs for computers to perform or more efficent ways to perform them. The development and use of application programs involve a major part of the total resources in manufacturing operations. Individuals who are familiar with programming languages, even the simplest of the high-level ones, often find the opportunity to write some simple application programs to do a particular task that they might otherwise have to do manually.

7-5 PROGRAMMING LANGUAGES

WHY ARE THERE SO MANY DIFFERENT COMPUTER LANGUAGES?

Hundreds of different programming languages and variations to them have been developed for computers over the years. As with the development of human languages, some of this proliferation was due to people living in separate communities and environments—that is, using different types of computers for different purposes. In addition, as the art of programming advanced, languages were developed to improve the efficiency and capabilities of programming. Although many languages exist today, relatively few of them are in widespread use. In general, their differences are intended to provide different features for different applications and users. Much like human languages, each programming language has some unique characteristics in its structure and rules. This means that even though the languages may be able to perform the same tasks, they will look and operate differently. These differences are usually caused by one or more of the following factors:

1. Orientation (whether the language was designed for the machine or the user)

2. Organization and architecture of the computer hardware

3. Intended use and functions of the application

4. Skill of the programmer

5. Efficiency of the programs

6. Personal preferences, familiarity, and experience

There is no one best programming language. Different languages have definite advantages when used in specific applications and on certain types of machines. Some languages are easier to use than others but may not be as powerful or efficient. In fact, in most cases, several languages are used to actually execute a program:

One used by the user

One or more used by the programmer

One which ultimately communicates directly with the computer

FUNCTIONS OF PROGRAMMING LANGUAGES

The three principal types of programming languages are each used for their unique functions and capabilities:

1. Machine language. This is the object code, which is written in the binary form compatible with the hardware of the machine. All programs must ultimately be reduced to this form, but the actual programming is rarely done at this level. It would be an extremely tedious job and subject to errors. It would also require a great deal of skill and familiarity with the inner workings of the particular machine involved. Therefore, most programs are written in higher-level languages and then translated into machine language for execution.

2. Assembly language. Although still considered to be a low-level language, assembly language is symbolic and easier to use for writing programs. Assembly language code is similar in structure to object code (i.e., it can be translated line for line to binary form). Therefore, it still involves a lot of work to write programs. It is also unique to the machine involved. This can be an advantage for programming if you wish to optimize the performance of the program. Assembly language programs are translated into machine language by an assembler program.

3. High-level languages. These are languages that are oriented more toward the user or the procedures involved than toward the machine. They are usually easier to understand and use, since they employ more familiar codes and symbols and need fewer instructions to perform the same task as a lower-level language. High-level languages are designed for particular types of tasks or classes of users. Before the computer can execute programs in high-level languages, they must be translated into machine language by either a compiler or an interpreter program. Since high-level programs are not machine-oriented and are translated, they can usually be used on different machines.

COMMON PROGRAMMING LANGUAGES

Of the many different high-level programming languages that have been developed, only a few are commonly used for business and industrial applications. Each has certain features and capabilities that makes it unique; and each is generally restricted in its use to certain types of computers, users, or applications. Following is a brief description of some of the more common languages.

FORTRAN. This is one of the earliest and most widely used high-level language. It was intended primarily for solving engineering and scientific problems on large computers. Its name stands for FORmula TRANslation, since it uses common mathematical notation for programming equations. FORTRAN is easy to use and very efficient for mathematical computation tasks, but it is not a good choice for business applications.

COBOL. This is another early language that was developed for widespread use, in this case for large-scale data processing tasks in business-type applications. Its name stands for COmmon Business-Oriented Language. Although its statements are relatively easy to read, COBOL is not simple to program and is not suitable for complex mathematical tasks.

RPG. This is a special purpose language that was developed for generating business reports on small machines. Its name stands for Report Program Generator. It is relatively easy to use for simple tasks such as updating files and generating reports.

APL. This is a very powerful language for mathematical applications on both large and small machines. Its name stands simply for A Programming Language. APL is particularly efficient for data manipulation and matrix computations, but programmers need mathematical skills to use it.

BASIC. This has become an extremely popular language for a wide variety of applications on small computers. Its name stands for Beginner's All-purpose Symbolic Instruction Code. It is designed for ease of use and interactive operation with very simple statements. However, it is not as powerful or efficient as some of the other high-level languages.

PASCAL. This language was developed to be both easy to use and efficient on small machines. It was named after the famous French mathematician Blaise Pascal. The programming is very structured and easy to read and therefore relatively easy to learn.

LISP. This is a language that has become popular for a specific type of programming application called "artificial intelligence" (AI), which we will cover in Chap. 9. Its name stands for LISt Processing, since the programs and data are structured as lists. Although its very simple statements make it relatively easy to use, LISP is capable of very powerful functions.

These are not the only programming languages that are popular in industry today, but they are representative of the variety of types and features that exist. These languages will be modified and entirely new ones will continue to be developed, and the new languages will be more efficient, have more capabilities, and be easier to use. If one were to try to program the same problem in each of these languages, the differences between them would be apparent. Figure 7-2 illustrates what a simple program would look like if written in each of six different programming languages. The first and, of course, lowest level is the object code or machine language. Each line of binary code represents a simple step in the program that relates directly to the hardware resources of the computer. Similarly, the assembly language program uses symbolic codes to perform the task in the same manner. The higher-level languages use different techniques for accomplishing the same tasks. Their ease of use and effi-

ciency for this particular application can be seen by the number of instructions involved and the complexity of the program. In this particular case, the more powerful languages, such as APL and LISP, are obviously more efficient at data manipulation.

CHOOSING A LANGUAGE

Since so many languages are available, it may seem difficult to choose which one to use to program a particular task. A number of factors will usually influence this choice:

1. The programmer's familiarity and experience with a particular language is often the strongest factor. If you are comfortable and proficient in a particular language you would obviously prefer to use it rather than one you are less familiar with.

2. The computer being used may restrict the selection of languages that can be used to program it. Most computers can run a number of different compilers that translate the programming language used into the machine language that the computer can understand.

3. The availability of standard application programs which are written in certain languages may influence the choice. It may be more convenient to use one language over others because of its compatibility and common usage with other applications.

4. The intended application will usually have a great bearing on the choice. As discussed earlier, some languages are designed specifically for certain types of tasks where they can be very efficient. Others are more general purpose in nature and can be used for a variety of applications.

5. The technical attributes of the language are the purest form of selection criteria. These include:

☐ **Fluency.** The ease of writing and learning the language.

☐ **Legibility.** the ease of reading and understanding a program written in the language.

☐ **Maintainability.** The ease of correcting or changing the program.

☐ **Efficiency.** The performance capabilities of the language in terms of computer processing time and the use of hardware resources.

7-6 PROGRAMMING

THE PROGRAMMING PROCESS

As illustrated in Fig. 7-3, the programming process starts with an idea about the solution to a problem or task and ends with a set of coded instructions in machine-readable form. This process can be considered as three distinct stages:

MACHINE CODE				LABELS	INSTRUCTIONS		COMMENTS
00100100	01011111			SUMODDS	MOVE.L	(A7)+,A2	Pop return address from the stack into A2.
00100010	01011111				MOVE.L	(A7)+,A1	Pop address of first term into A1.
00110010	00011111				MOVE.W	(A7)+,D2	Pop n into D1.
01000010	01000010				CLR.W	D2	Assign a value of 0 to the sum in D2.
01001110	11111010	00000000	00001110		JMP	COUNT	Jump to the end of the loop to test if n=0.
00001000	00101001	00000000	00000000 00000000 00000001	LOOP	BTST	0,1(A1)	If the term addressed by A1 is even...
01100111	00000010				BEQ.S	NEXT	...then go to NEXT
11010100	01010001				ADD.W	(A1),D2	...otherwise add the term to the sum in D2.
01010100	01001001			NEXT	ADDQ.W	#2,A1	Set A1 to the address of the next term.
01010001	11001001	11111111	11110010	COUNT	DBF	D1,LOOP	Decrement D1; unless it is −1, go to LOOP.
00111110	10000001				MOVE.W	D2,−(A7)	Push the sum from D2 onto the stack.
01001110	11010010				JMP	(A2)	Go to the return address

MACHINE CODE AND ASSEMBLY CODE specify the steps of the odd-element calculation in terms of the hardware resources of the computer. The code is necessarily specific to a particular machine, in this case the Motorola 68000 microprocessor. The algorithm employed is much like that of the PASCAL procedure *SumOdds*, although it is more compact than the code that would be generated by a PASCAL compiler. Parameters are passed to the procedure on a stack and the result is returned on the stack; the address at which execution is to resume when the procedure is finished is also on the stack. The assembly-code version of the program, in which instructions take the form of "mnemonic" abbreviations, can be translated directly into the binary machine code executed by the microprocessor.

```
∇ SUM ← SUMODDS TERMS
[1]  SUM ← +/(2 | TERMS)/TERMS
∇

SUMODDS 23 34 7 9
```

TERMS	←	23	34	7	9		Initial value assignment.
(2 | TERMS)	←	1	0	1	1		Array of remainders.
(2 | TERMS)/TERMS	←	23		7	9		Compression of two arrays.
+/(2 | TERMS)/TERMS	←	23	+	7	+ 9		Reduction by addition.
SUM	←	39					Assignment of result.

APL PROGRAM calculates the sum of the odd elements in an array with a function whose operation is specified in a single line. The function has one parameter, *TERMS*, an array that "knows" how many elements it has, so that *N* need not appear in the program. An APL statement is executed from right to left except where parentheses alter the order of evaluation. In this example the expression $(2|TERMS)$ is evaluated first; it calculates the remainder left after dividing each element of *TERMS* by 2 and creates an array of the same size as *TERMS* to hold the remainders. The symbol "/" can indicate two different operations, both of which appear in the example. In the expression $(2|TERMS)/TERMS$, "/" is a compression" operator that creates a new array in which each element of *TERMS* appears only if the corresponding element of $(2|TERMS)$ is nonzero. In the symbol "+/," "/" is a "reduction" operator that reduces the array to a single number by inserting a "+" between each pair of elements.

```
program SumOddNumbers:
type TermIndex = 1 100
     TermArray = array [TermIndex] of integer;

var myTerms: TermArray:

function SumOdds(n; TermIndex; terms: TermArray): integer:

var i: TermIndex:
    sum: integer;

  begin
    sum:= 0;
    for i: = 1 to n do
      if Odd(terms[i]) then
        sum:= sum + terms [i]
    SumOdds: = sum;
  end;
begin
  myTerms[1] : = 23: myTerms[2] : = 34: myTerms[3] : = 7:myTerms[4] : = 9:
  WriteLn(SumOdds(4, myTerms))
end.
```

PASCAL PROGRAM for summing the odd numbers in an array employs a function named *SumOdds* with two parameters: an integer *n* and an array *terms*. The function consists of the statements in the panel of color; the remainder of the program sets up a particular array on which *SumOdds* operates. In PASCAL every variable must be introduced in a declaration that gives the variable's type. Some types, such as *integer*, are built into programming language; others, such as *TermIndex*, are defined by the programmer. The loop is designated by the *for...to...do...* statement and the conditional is designated by the *if...then...* statement.

```
(DEFUN SUMODDS
  (LAMBDA (TERMS)
    (COND
      ((NULL TERMS) 0)
      ((ODD (CAR TERMS))(PLUS(CAR TERMS)(SUMODDS(CDR TERMS))))
      (T (SUMODDS (CDR TERMS))))))

(SUMODDS'(23 34 7 9))

(SUMODDS '(23 34 7 9))
  = (PLUS 23 (SUMODDS '(34 7 9)))
    = (PLUS 23 (SUMODDS '(7 9)))
      = (PLUS 23 (PLUS 7 (SUMODDS '(9))))
        = (PLUS 23 (PLUS 7 (PLUS 9 (SUMODDS'( )))))
      = (PLUS 23 (PLUS 7 (PLUS 9 0)))
    = (PLUS 23 (PLUS 7 9))
  = (PLUS 23 16)
= 39
```

LISP PROGRAM calculates the odd-element sum by means of a function that calls on itself recursively. A LISP function is a list, where the first element (called the *CAR*) is the name of the function and the remainder of the list (the *CDR*) gives the parameters. *DEFUN* is a function-defining function and *LAMBDA* precedes the names of the parameters; here the only parameter is the list of numbers *TERMS*. *COND* is a conditional function that evaluates the *CAR* of the lists that form its parameters. If the result is *T*, or true, the *CDR* of the list is evaluated; otherwise *COND* goes on to the next list. Here there are three possibilities. If *TERMS* is an empty list, *NULL*, is true and *SUMODDS* returns a value of zero. If the *CAR* of *TERMS* is odd, the *CAR* is added to the running total and *SUMODDS* is called to evaluate the *CDR* of *TERMS*. If neither of these conditions is true, the *T* clause (which must be true) is reached; it simply calls *SUMODDS* with $(CDR(TERMS))$ as its parameter. Calculations are left pending during each call.

```
100 DIM T(100)
200 READ N
300 FOR I = 1 TO N
400   READ T(I)
500 NEXT I
600 GOSUB 1100
700 PRINT S
800 GOTO 2000
900 DATA 4
1000 DATA 23, 34, 7, 9

1100 REM MAKE S THE SUM OF THE ODD ELEMENTS IN ARRAY T(1..N)
1200 LET S = 0
1300 FOR I = 1 TO N
1400   IF NOT ODD(T(I)) THEN GOTO 1600
1500   LET S = S + T(I)
1600 NEXT I
1700 RETURN

2000 END
```

Fig. 7-2 Programming languages: programs written in seven different computer languages to calculate the sum of the odd elements in an array of *n* integers. *(from "Programming Languages," Lawrence G. Tesler, Scientific American, Sept. 1984; © 1984 by Scientific American)*

BASIC PROGRAM employs a subroutine to add up the odd terms in an array. The subroutine, indicated by the panel of color, has no name but must be referred to by line number; it is called by the *GOSUB 1100* statement. A BASIC subroutine also has no parameters; values are assigned to "global" variables, which the subroutine can then access. A variable does not have to be declared in BASIC unless it has subscripts, as in an array; in this example the *DIM* (for "dimension") declaration states that the array *T* can have as many as 100 elements. The *FOR...NEXT...* statement defines a loop and the *IF...THEN...* statement defines a conditional.

```
PROCEDURE DIVISION.
 EXAMPLE.
    MOVE 23 TO TERMS(1).
    MOVE 34 TO TERMS(2).
    MOVE  7 TO TERMS(3).
    MOVE  9 TO TERMS(4).
    MOVE  4 TO N.
    PERFORM SUM-ODDS.

 SUM-ODDS.
    MOVE 0 TO SUM
    PERFORM CONSIDER-ONE-TERM VARYING I FROM 1 BY 1 UNTIL I > N.
 CONSIDER-ONE TERM.
    DIVIDE 2 INTO TERMS(I) GIVING HALF-TERM REMAINDER RMDR.
    IF RMDR IS EQUAL TO 1; ADD TERMS(I) TO SUM.
```

```
DATA DIVISION
 WORKING-STORAGE SECTION;
 01 NUMERIC-VARIABLES USAGE IS COMPUTATIONAL.
    02 TERMS PICTURE 9999 OCCURS 100 TIMES INDEXED BY I.
    02 N PICTURE 999.
    02 SUM PICTURE 999999.
    02 HALF-TERM PICTURE 9999.
    02 RMDR PICTURE 9.
```

COBOL PROGRAM for the sum-of-the-odd-numbers calculation uses a procedure named *SUM–ODDS* that calls another procedure named *CONSIDER-ONE-TERM*. A COBOL procedure cannot have parameters, and so before *SUM–ODDS* is called by a *PERFORM* statement, values are assigned to *N* and to the first *N* elements of *TERMS*. The key words *PERFORM...VARYING...* define the loop and *IF...* introduces the conditional clause. In the data division the numbers 01 and 02 designate two levels in a hierarchy of data structures. *PICTURE* specifies how values are to be displayed. Only an excerpt from the complete program is shown.

Fig. 7-2 (cont'd).

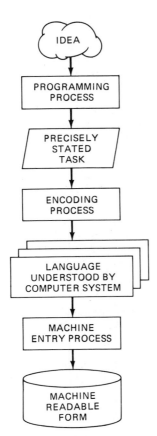

Fig. 7-3 The programming process. *(C. William Gear, Computer Organization and Programming, McGraw-Hill, New York, 1980)*

1. Programming. During this initial stage, the task is analyzed and reduced to a set of well-defined logical steps in order to reach the ultimate solution. This is the intellectual job of developing an organized approach to solving the problem. It may be relatively simple or ex-tremely complex, depending on the nature of the problem.

2. Encoding. In this stage, the program is written in the form of a code which can be used by the computer. The code-writing task can be laborious, depending on the size of the program and the difficulty of the language. Most code writing is done in high-level languages to make the job easier, faster, and less prone to errors.

3. Translation. In this final stage, the program is converted into machine-readable form (i.e., object code) which can be used directly by the computer to execute the program. This translation process may be accomplished by using an assembler, a compiler, or an interpreter.

In the past, the coding task has been the most time-consumming of the three stages. However, as computer technology and programming languages have advanced, much of the effort involved has been reduced by automated techniques and programmer aids.

PROGRAM DEVELOPMENT TECHNIQUES

Programming the solution to complex problems or large data processing tasks can involve a great deal of time, effort, and skill. Professional programmers are usually employed by industry to develop the programs necessary to perform the various tasks for which its computer systems were intended. For this process to be as efficient as possible, programmers over the years have developed a variety of techniques which help them to write programs. Once the problem is defined, the major tasks of the programmer are to:

1. Plan the solution

2. Code the program

3. Check the program

4. Document the program

The basic approach to all programming is organization and discipline. For a program to work and be effective, each step which the computer is expected to take must be spelled out in the proper sequence without error. Most programming starts with a form of "flowcharting," which lays out these steps in a visual form that permits the programmer to follow the logical sequence of steps. Figure 7-4 illustrates a flowchart of the program used in Fig. 7-2. This technique is particularly useful for short programs, but flowcharting gets cumbersome for large and complex programs. In such cases, which are typical of most industrial applications, additional programming tools and aids are needed.

One technique which is used to deal with large programming tasks is called "structured programming." It uses a very organized program structure designed to have a form which matches the meaning of the program. Programs developed with this technique are first planned at a high-level which deals with the major elements of the task. Each of these elements is then broken down into smaller subtasks for which detailed program steps are written. This organized and disciplined approach makes it easier for the programmer to complete the total job with-out getting lost in the complexity of the program. The programmer deals with one small task at a time which can be written and checked with relative ease while working within a predefined framework of the total problem solution.

An additional aid which has become essential to most programmers is a programming development system. This is a set of programs which are designed to permit the programmer to develop a program in a particular language interactively with a computer. For example, this may take the form of a multiwindow graphics package that will display both the written program and flow diagrams while it is being executed (Fig. 7-5). Other programmer aids may also be included, such as data structure and symbol references. Such computer-aided programming tools have made the programming process much more productive.

7-7 TRENDS IN COMPUTER SOFTWARE

Development activity in computer software is at least as great as that in hardware, and it is perhaps even more important. As advances in hardware technology begin to approach physical and economic limitations, more emphasis is placed on improving the performance and capabilities of computers through advances in software. In

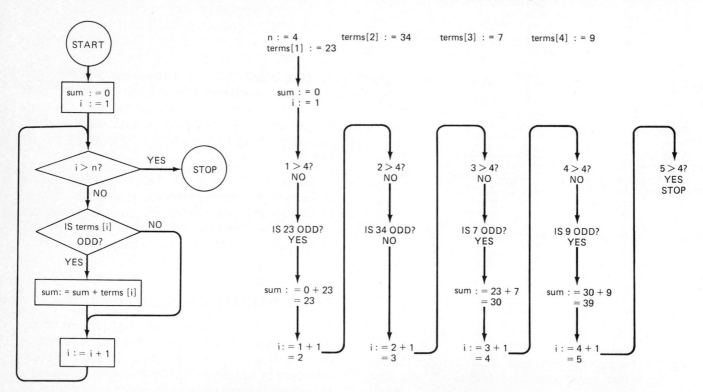

SAMPLE CALCULATION employed to illustrate the characteristics of several programming languages finds the sum of the odd elements in an array of *n* integers. The algorithm outlined in the flow chart at the left is embodied directly in the programs shown in Figure 7-2. The heart of the algorithm is a loop executed *n* times. On each passage through the loop a term of the array is examined; if it is an odd number, it is added to the running total. At the right the successive values assumed by the variables in the procedure are traced as the calculation is done for an array of four numbers. The symbol ":=" gives to the variable on the left the value computed on the right. A number in brackets, as in *terms* [1], is equivalent to a subscript: it identifies an element of the array *terms*.

Fig. 7-4 Flowchart. *(from "Programming Languages," Lawrence G. Tesler, Scientific American, Sept. 1984; © 1984 by Scientific American)*

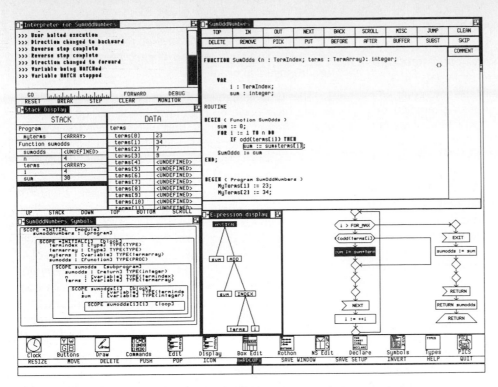

Fig. 7-5 Program development system. *(Steven P. Reiss, Brown University)*

addition, some of the major barriers to the widespread use of computers relate more to software constraints than they do to hardware considerations. For example, the largest portion of the cost of developing and installing a new industrial computer application is often due to the software involved, not the hardware. It has also become apparent that, if computers are to become more user-friendly and gain wider application by the large population of nonprofessionals in the work force, significant advances will be necessary in both system and application software. Following are some of the areas of interest and activity in software that can serve to indicate the direction in which the technology is headed.

SYSTEM SOFTWARE

Although advances in application software are usually more visible and exciting, the key to major enhancements in computer capabilities lies with the internal software that runs the system. To make computers more accessible and usable, developments are being pursued in several important areas:

1. Relational databases. Most computer applications for individuals involve database functions, such as filing, searching, sorting, reporting, and displaying data. Therefore, if computers are to continue to gain widespread use among workers who are not computer professionals, database systems that provide quick and easy access are essential. Over the years there have been a number of approaches to designing database management systems. In many cases such systems had a relatively inflexible structure that limited the ways in which data could be retrieved or changed. More recently, "relational databases" have been developed, which permit the users to manipulate data by specifying merely what they want the computer to do and not how it has to do it. Such systems use the "relationships" that exist between sets of data to retrieve information. The data is stored as 2D tables of rows and columns from which the user can make selections with a simple query language or menu-driven interface (Fig. 7-6). This flexible approach has a great advan-

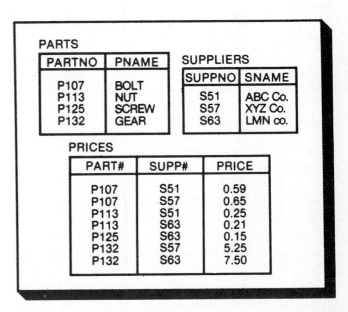

Fig. 7-6 A relational database consisting of three tables of information about parts used in manufacturing. *(International Business Machines Corp.)*

tage in applications where the database and the user needs change frequently.

2. Standard protocols. As the number of different types of computer hardware and software proliferate, it is becoming increasingly more difficult for users to communicate between systems. For large industrial applications which need to tie many people and machines together, the flexibility of open data communication is very important. To achieve this, much is being done by the computer manufacturers as well as industry committees to provide for the standardization of interfaces. This can be accomplished by using a standard operating system and language or by establishing standard data communication protocols which permit different systems to exchange data. Even if the machines involved do not use the same protocol structure, it is possible to tie them together with an intermediary system or communication "gateway" that translates the data from one structure to the other.

3. Fault tolerance. Although the reliability of computer hardware continues to improve, there are approaches in system software which can make a computer more tolerant of hardware failures. The costs involved in the failure of a large industrial computer system can be immense because of the number of applications and users that may be dependent upon its continued operation. Most large systems have diagnostic and maintenance subsystems, which are programs that monitor and detect malfunctions. In some cases, systems have built-in hardware redundancy so that if a failure occurs, the defective unit can be bypassed and repaired while the system continues to operate. The system software must also be capable of recovering from a failure by resuming the program at the point where it was interrupted.

4. Data security. Large industrial computer systems can contain a lot of valuable data that needs to be protected from accidental or intentional loss. A number of approaches that have been developed involve system software to provide such protection. The traditional simple techniques involving passwords and unlisted telephone numbers are inadequate in the age of powerful small computers that can rapidly try all the "combinations" to the "lock." One alternative approach is to secure the system software itself. This involves structuring the software in such a way as to separate the functions which must be secured from the others. A "reference monitor" is built into the system software to screen all access to the portions that are to be protected. With such an approach, different users could even have different levels of "clearance" to use the various system functions. Security audits and logs can also be built into the system software. More complex schemes involve the encryption of data or even the recognition of handwriting or voiceprints.

PROGRAMMING LANGUAGES

New computer programming languages continue to be invented and modified in an effort to improve both the ease of programming and the capabilities of software. Although this is an ongoing, evolutionary process, several recent developments are significant as indicators of future trends in software:

1. The C language. An ideal programming language should have powerful high-level functions, be usable on many different machines, and help improve the productivity of the programmer. In an attempt to achieve those objectives, the C language was developed. It was designed to be compatible with structured programming to make the program development job easier. Programs written in C are generally easier to read and maintain than programs in earlier languages like FORTRAN. At the same time, C is more powerful and flexible than more recent high-level languages such as PASCAL. Some of its functions permit faster operations with less programming code. It has also found widespread application in both small and large computers, since it was adopted as the language used for one of the new standards in operating systems—Unix. Because of these attributes, C also proved to be a useful software development tool for multiprocessing applications.

2. The ADA language. The expense of maintaining large and complex sets of software often exceeds the original costs of developing them. For this reason, there has also been an effort to create a standard language that is easy to maintain as well as use. The U.S. Department of Defense, which must use and maintain many large and complex programs over long periods of time, developed a new language based on the best features of other successful languages such as PASCAL. It was named ADA after Ada Augusta Byron Lovelace, who is believed to have been the first programmer in history and who was also the daughter of the poet Lord Byron. The language is highly structured, modular, and relatively simple. This makes it easy to use as well as easy to maintain. A program in ADA can be written in separate "modules" simultaneously by several programmers. ADA is both flexible and powerful and can provide capabilities such as interfacing with a wide variety of nonstandard I/O types, multiprocessing, and fault tolerance. It is used for program development, system software, and application programs. Although it was originally developed for military applications, ADA has also found widespread use in scientific and industrial applications such as process control.

3. Natural languages. The ultimate form for making programming languages easy to use by nonprofessionals is the natural or native languages of human speech. Many high-level languages use codes and words from natural language, but only recently has there been some success at developing natural language interfaces to computers. The major problem involved is in dealing with words or phrases that could be interpreted to have different meanings. This can be overcome by restricting the vocabulary or choices available to the user, trying out all alternatives, or resolving conflicts through dialogue with the user. Natural language interfaces have become particularly useful for database query systems to allow many users to have access to data without the help of programmers (Fig. 7-7). This capability is even being extended further with the development of voice recognition and response systems which provide direct verbal communication between the user and the computer through a natural language interface. Later chapters will address how some of these high-level "intelligent" computer functions work.

PROGRAMMING

Although the advance in the performance of computers has been very dramatic in recent years, the productivity of programmers has not improved that significantly. Programming is still a tedious art which involves a long process of definition, planning, flowcharting, testing, debugging, and so forth. At the same time, the size and

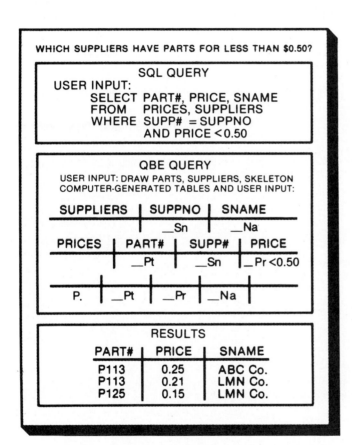

An example of using IBM's very high level database languages, SQL and QBE, to satisfy a request involving two tables from Figure 7-6. The SQL commands are expressed in a standardized block format; an example of the most common form for extracting data is:

    SELECT   some data (column names)
    FROM   some file (table names)
    WHERE   certain conditions, if any, are to be met (rows)

QBE is initiated simply by typing the table name on the display screen, and the screen returns a skeleton table with column names in it. In this example, the user builds a new table in the blank skeleton by typing "example elements" (e.g., _Pt) under existing tables and in the blank skeleton. The example elements are formed by typing an underline followed by any mnemonic the user desires. Note that "P." simply means to present the results.

Fig. 7-7 A natural language interface in a relational database application. *(International Business Machines Corp.)*

complexity of software have increased substantially which makes the programming job even more difficult and prone to errors. The solution to improving both the quality and productivity of programming is the same as it is for manufacturing—automation! The major thrust in program development is to have the programmers use some of the same tools that software has made available to the computer users. These automated tools include such functions as compilers, program development languages, and debugging programs which are commonly used in industrial applications. Of all the programming tasks, the coding effort has been the easiest to automate once the program has been completely specified. Therefore, the greatest potential for improving productivity and quality lies in some of the other tasks, such as in the design and maintenance of programs. Several key approaches are being pursued:

1. **Rapid prototyping.** This approach uses a simplified version of the program to test the software before all the functions and coding are complete. It can avert a lot of unnecessary effort by finding problems early that may have been caused by inadequate definition of the requirements for the program.

2. **Reusable software.** Effort can also be saved and errors avoided by using program modules that already exist as part of a new program. This requires that the software be standardized so that it can be easily interpreted and reused in different applications.

3. **Program development languages.** These are high-level languages that are nonprocedural in nature. That is, they permit the program developer to design a program conceptually for an application, then translate it into a procedural language for execution.

4. **Intelligent editor.** This is a tool that can generate high-level language programs from a variety of program design inputs, such as graphs, text, and program code.

7-8 SUMMARY

Software is as important as hardware to the capabilities and performance of computers. There are two principal types of software. System software directs the internal operations of the computer, while application software performs the user tasks. An operating system manages the tasks, resources, and data handling involved in executing a computer program. The development of system software features such as timesharing, virtual storage, and database management has made it possible for computers to be used by large networks of user terminals. Advanced functions such as relational databases, standard communication protocols, and fault tolerance have improved the flexibility and usability of computer systems.

Although all computers must use machine language to operate, programs are usually written in higher-level languages which are easier for humans to understand and use. There are many different programming languages and no one is best for all uses. New languages continue to be developed to provide improvements in their capabilities and ease of use. Programming tools have also been developed to make the programming job easier. Software development costs often exceed the cost of hardware in industrial applications. The key to improving the capabilities of computer systems as well as expanding their use is advances in software.

REVIEW QUESTIONS

The answer to each question can be found in the section(s) indicated at the end of the question.

1. Identify the two major types of computer software and describe what they are typically used for. [7-2]

2. What are the three principal types of programming languages and how do they differ? [7-2]

3. Describe the basic functions of an operating system. [7-3]

4. Identify the three major types of operating environments for computers and describe how they differ. [7-3]

5. What operating system functions make it possible for computers to handle large programs and many users, and how do they work? [7-3]

6. What are utility programs? Give some examples. [7-3]

7. What is a database management system and how is it used? [7-3]

8. Identify the two types of language translators and describe their differences. [7-3]

9. Identify some common programming languages and describe some of the typical types of applications for which they are used. [7-5 and 7-7]

10. What are some of the factors that influence the choice of a programming language? [7-5]

11. Describe the key steps in the programming process. [7-6]

12. What is a program development system and how is it used? [7-6]

13. What is a relational database and how does it work? [7-7]

14. Identify and describe recent advances in computer software. [7-7]

MICRO- AND MINICOMPUTERS

8-1 INTRODUCTION

As the computer and electronics technologies advanced, it became possible and economical to build small computers. This development made it practical for the first time to use computers in many applications where they would have been too expensive or too large in the past. Manufacturing, in particular, had many tasks that could benefit from the efficiency and control provided by electronic data processing. Although manufacturing was an early user of computer technology, as we saw in Part 1, there were still many jobs done manually or with traditional hard-wired control systems.

As small computers became available, the manufacturing industry found that it could afford to dedicate computers to specific tasks, such as tool control. This reduced the work load on the large central business computers and provided additional flexibility and control on the factory floor. Advances in manufacturing processes and technologies also required sophisticated controls which only computer power could provide. Many of the modern tools and much of the equipment used in manufacturing today could not operate without the small computers that are built into them.

This chapter extends the basics of computer technology that were covered in Chaps. 6 and 7 to the relatively new world of small computers. Micro- and minicomputers have become the basis for many advances in computer applications in manufacturing. In this chapter we will discuss what they are and how they work, and we will also cover, in a little more detail, some of the specifics of their design, programming, and operation. This should provide an understanding of the subject so that the reader can appreciate how minis and micros can be used when we address their applications in later chapters. It can also serve as a basis for further study, in more depth, on the subject.

8-2 MICROPROCESSORS

WHAT IS A MICROPROCESSOR?

The building block used to create most small computers and controllers is the microprocessor. It is the smallest configuration of an entire central processing unit (CPU) in a single unit, which is often an integrated circuit (IC) chip. Such microprocessor chips include an arithmetic unit, control logic, and data registers, but normally they do not include any memory or I/O control. Microprocessors were first developed in the 1970s as 4-bit processors for simple calculation or control applications. They are designed to optimize performance for specific types of applications, such as for calculators, tool controllers, and minicomputers. There is a wide range of performance capabilities and sizes for microprocessors, including from 4-bit up to 32-bit architectures. Several different semiconductor technologies and circuit families can be used for microprocessors, depending on the performance and cost objectives of the application involved.

The architecture of a typical microprocessor is made up of three sections or functional groups of circuits that are tied together by an internal bus system for data communication (Fig. 8-1):

1. **Arithmetic section.** This includes the arithmetic logic unit (ALU), which is composed of a shifter and an adder, together with the accumulator, temporary, and status registers. Their functions are as follows:

   a. The ALU performs the basic computation and logic functions (e.g., increment/decrement, AND/OR/Compare, add, subtract).

   b. The accumulator is a special register which handles the transfer of data. It holds the data to be operated on until the ALU is ready and receives the results when the operation is done.

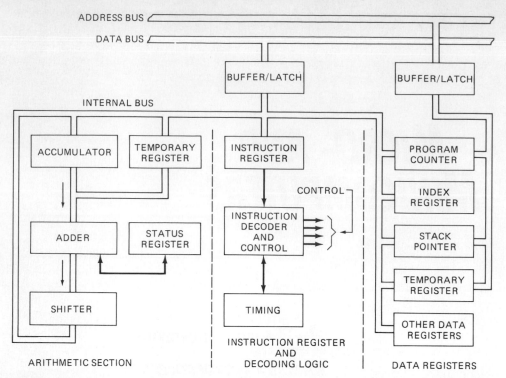

ADDRESS BUS

DATA BUS

BUFFER/LATCH

BUFFER/LATCH

INTERNAL BUS

ACCUMULATOR

TEMPORARY REGISTER

INSTRUCTION REGISTER

PROGRAM COUNTER

CONTROL

INDEX REGISTER

ADDER

STATUS REGISTER

INSTRUCTION DECODER AND CONTROL

STACK POINTER

TEMPORARY REGISTER

SHIFTER

TIMING

OTHER DATA REGISTERS

ARITHMETIC SECTION

INSTRUCTION REGISTER AND DECODING LOGIC

DATA REGISTERS

Fig. 8-1 Microprocessor architecture: functionally partitioned microprocessor with dedicated data registers. *(Edward V. Ramirez and Melvyn Weiss, Microprocessing Fundamentals—Hardware and Software, McGraw-Hill, New York, 1980)*

c. The temporary register receives data from the internal data bus and holds it for the ALU.

d. The status (or "flag") register indicates conditions after an instruction is executed (e.g., carry, parity, zero, sign).

2. Instruction register and decoding logic. This section of the microprocessor is used to store the contents of the next instruction that is to be executed by the arithmetic section. It is made up of an instruction register and decoder whose functions are as follows:

a. The instruction register holds the instructions which are coming in on the data bus for the instruction decoder.

b. The instruction decoder interprets the instruction and generates signals which execute it.

3. Data registers. This section manipulates data independent of the arithmetic unit by using a number of specialized registers such as the following:

a. The program counter holds the address of the next instruction to be executed.

b. The stack pointer is used to store other addresses.

c. The address latch selects the address to be sent from one of the registers then "latches" the address onto the address bus lines so it can be transferred.

TYPES OF MICROPROCESSORS

There are many types of microprocessors, which are designed with different features for different applications and users. Some may optimize the design for perfor-

mance while others may optimize it for low cost. Differences can often be found in the number of registers and their bit width; in the circuit families used (i.e., speed, voltages, interfaces); in the bus structure; and in the instruction sets. One special type of microprocessor, which is often used for customized, high-performance applications is called a "bit slice." This is made up of small ALUs (e.g., 4- or 8-bit) that are connected in parallel to handle larger word sizes. A bit-slice microprocessor is usually made from high-speed, bipolar microprocessor chips and controlled by "microinstructions" that are stored in a read-only memory (ROM). The most common microprocessors, however, are general purpose and can be used in many different applications. The point is that they are not all the same or even compatible. To use them, you must understand all their key features.

WHAT IS A BUS SYSTEM?

A bus is a communication link that carries data between the various sections of the microprocessor. Since they transfer an entire word of data at one time, buses come in different sizes depending on the architecture of the system (i.e., 4-bit to 32-bit). There are usually three different buses to handle each of three different types of information—control signals, data, and instructions. The functions of these buses are as follows:

1. Control bus. This transfers control signals (such as memory read or write, input or output read or write) between the microprocessor and the memory and I/O units and enables the memory and I/O to operate.

2. Data bus. This is used to transfer data to and from the microprocessor, the memory, and the I/O units. Since data must be transferred in both directions, it is a bidirectional bus, unlike the control bus, on which signals are sent in one direction.

3. Address or instruction bus. This transfers signals that select the location of instructions which are to be read from or written into memory or a register. The address data and the instruction can be transferred on the same bus by multiplexing, but this requires the use of an address latch.

8-3 MICROCOMPUTERS

WHAT IS A MICROCOMPUTER?

The term "microcomputer" is used to refer to a very small machine or set of electronic chips which contain all the basic functions of a complete computer. Microcomputers are often built into other products or pieces of equipment, such as instruments or tools, to provide computer control. However, small computer systems, such as personal computers, may also be considered to be microcomputers. The heart of the microcomputer is, of course, the microprocessor. However, it cannot function as a computer by itself. In addition to the microprocessor, which serves as the CPU for the microcomputer, there is also memory and I/O (Fig. 8-2). It is in these sections composed of registers, memories, and I/O interfaces that the

major differences exist between microcomputers and microprocessors. A microprocessor will typically use dedicated registers for storage functions, while the microcomputer uses a memory section made up of a variety of special purpose registers and memories.

MICROCOMPUTER MEMORIES

Two or more types of memory are usually used in a microcomputer:

1. Random access memory (RAM) serves as the main memory for the microcomputer. In addition to the memory array itself, the RAM also includes other support circuits, such as a clock generator for timing, address decoders for interpreting the CPU instructions, and sense amplifiers for reading and writing the signals to and from the memory cells. The RAMs used for microcomputers are usually low-cost, field-effect transitor (FET) semiconductor devices. They are also usually "dynamic" RAMs. This means that the data must be sensed and restored periodically by "refreshing" to prevent the data from being lost by gradual leaking off of the stored electrical charge. The RAM section of the microcomputer may be composed of a number of RAM chips, depending on the total size of the memory, the word size of the computer, the organization of the microprocessor, and the storage density of the RAM chips. For example, a standard 16-bit microcomputer requiring 64K bytes of RAM could use 16 memory chips, each capable of storing 64K bits of information.

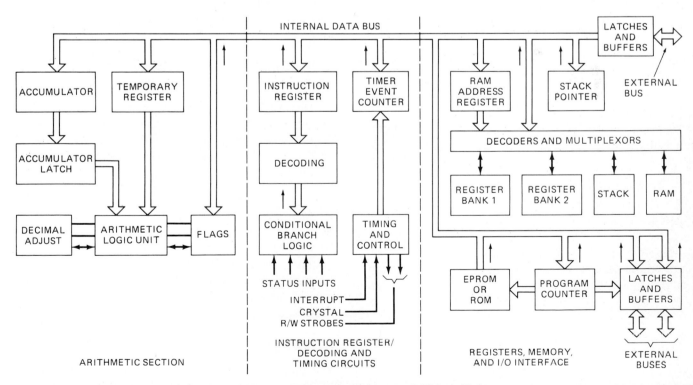

Fig. 8-2 Internal architecture of a microcomputer. *(Edward V. Ramirez and Melvyn Weiss, Microprocessing Fundamentals—Hardware and Software, McGraw-Hill, New York, 1980)*

2. Read-only memory (ROM) is used to store permanent instructions and can only be read by the CPU. Depending on the speed and storage capacity required, it may be made up of one or more semiconductor chips using either the FET or bipolar technology. In a standard ROM chip the instructions are designed into the actual circuitry and cannot be changed. Several special types of ROM, also used with microcomputers, can have the instructions stored after the chip is manufactured.

☐ Programmable read-only memory (PROM) can be programmed with electrical signals to store unique instructions for a particular microcomputer or application. Once it is programmed, however, these instructions are "burned-in" or permanent and cannot be changed.

☐ Erasable progammable read-only memory (EPROM) can be reprogrammed by erasing the stored data with either ultraviolet light or electrical signals and then writing in new instructions.

These various types of ROM are used to store special programs that are used frequently by the microcomputer in its application.

In addition to the ROM and RAM, there are a number of registers used for special purposes. Register banks can be used to switch data between registers when there are interruptions, and the stack register can store the program counter and status when an interruption or program subroutine is initiated.

8-4 I/O COMMUNICATIONS

TYPES OF I/O INTERFACES

I/O devices are connected to the microcomputer through I/O "ports" or interfaces. These are circuits which control the data that is to be transferred as well as the status of the I/O devices. They may be made up of special registers, clocks, data buses, control logic, and control lines. The I/O interface is needed to convert signals because of differences between the I/O devices and the CPU in terms of voltages, parallel versus serial data transfer, or data transfer rates. There are two general types of I/O interface:

1. Parallel. Bidirectional data transfer circuits, sometimes called "peripheral interface adapters" (PIAs), are used for direct communication between the I/O devices and the CPU. The PIAs are usually structured to be 1 byte wide, with a pair of port lines for each bit. Each port line can be individually programmed as either an input or an output by using a data direction register. The CPU can then read or send data one byte at a time through this parallel port to the I/O device.

2. Serial. When there is a need to connect a microcomputer to an I/O device or another computer that is in a remote location, telephone lines can be used. The device which permits computers to "talk" over telephone lines is called a "modem" (for MOdulator/DEMmodulator). It translates the digital signals of the computer to audio signals that can be transmitted through the telephone. The computer must have a different kind of port to connect to the modem. Data is sent one bit at a time so that only one telephone line is required. Since the microcomputer handles data in parallel form (one byte or word at a time), the data must be converted from parallel into serial form and back again. In addition, the rates at which bits and characters are transmitted may not be the same (may be "asynchronous"). The device that is used to provide this special interface is therefore called an "asynchronous communication interface adapter."

DATA TRANSFER

Once an interface is established, the transfer of data between I/O devices and the main memory can be accomplished by two different techniques:

1. Program control. This is the most basic approach to transferring data, as well as the most economical. It uses the system software and the registers in the CPU to control the interface and transfer the data from the I/O device to memory.

2. Direct memory access (DMA). This approach is intended to improve the efficiency of the data transfer process by permitting data to be read or written directly to the memory from I/O devices without passing through the CPU. To accomplish this, a special bus line is used by the I/O device to signal the CPU that it is ready to send data to memory. Then the transfer is handled by a DMA controller. The DMA scheme is often used to transfer large blocks of data to and from memory and other storage devices (e.g., disks) at high speed without wasting CPU time.

Microcomputers may also use different schemes for addressing the I/O ports. There are two main approaches.

1. Direct I/O. Microcomputers using direct I/O have separate address spaces for I/O ports. With this approach only a single-byte address is required for each input or output instruction and no memory space is used. However, control signals are required to indicate that the address on the bus is for an I/O port, not a memory location.

2. Memory-mapped I/O. In this approach the CPU handles an I/O port the same way it handles a memory location. This permits any instruction that references memory to also send data to or from I/O devices. However, each read or write instruction requires a full-word address and therefore I/O ports use up some of the memory space.

INTERRUPTS

The simplest means for a CPU to communicate with an I/O device is to read data from a port when it needs it and send data when it has data ready to send. However, microcomputers usually use one of the following two schemes for I/O communication:

1. Polled I/O. This approach uses a special line which is shared by multiple I/O devices, each of which can transmit a status signal to tell the CPU that it is ready to send data. The CPU must continuously scan or "poll" the I/O ports, looking for an I/O device that needs to be serviced.

2. Interrupt operation. This approach is faster than polled I/O and is used in applications where the I/O devices cannot wait in line for the CPU to be ready to accept data from them. This is particularly true of many manufacturing applications, such as process control, where the system must respond immediately to changes in order to maintain control of the operation. The I/O device will actually interrupt the CPU operation to transfer data. The CPU must save the contents of the program counter to know where it stopped, then branch to the interrupt routine. When the CPU is finished with the I/O data transfer it returns to its next program instruction.

Interrupts may be either internal or external to the CPU. Internal interrupts are primarily problem prevention actions, such as the detection of power failures, errors, or illegal program codes. Such internal interrupts are usually assigned a higher priority by program control than external interrupts from I/O devices. The software that handles interrupts must perform a number of tasks, including saving and restoring the contents of the status and working registers, enabling and disabling the interrupt signals, and servicing the interrupt itself.

There are several different types of interrupt schemes. With a single-line interrrupt, all the I/O devices must share one interrupt line. Therefore, the CPU must scan the I/O ports to find out which device sent the interrupt signal. A multilevel interrupt scheme, however, provides separate interrupt lines for each I/O device so that the CPU knows which device needs service. This approach is obviously faster but is more expensive.

8-5 MINICOMPUTERS

WHAT ARE MINICOMPUTERS?

As the technology advances, it is becoming more and more difficult to distinguish between a microcomputer and a minicomputer. In general, minicomputers have larger word sizes and memories as well as faster speeds. They were originally developed before the invention of the microcomputer to offer small computers that could be dedicated to specific applications where a large system was not practical or economical. This was particularly true in some real-time manufacturing applications such as tool and process control. Because of their small size and relatively low cost, minicomputers also became popular as imbedded controllers for scientific equipment and instruments.

The structure of a minicomputer is basically the same

as that of a microcomputer. It has an arithmetic logic processor, a main memory, and an I/O section, all connected by a bus system. The CPU may be a small version of the same type of logic used in large processors, or it may be a microprocessor, depending on the performance requirements it is designed to satisfy. A minicomputer can take many forms and sizes, but it is usually designed to be a self-contained package that can be built into or mounted with industrial equipment (Fig. 8-3). Minis therefore often do not look like typical computers, which have such things as keyboards, displays, and printers attached, but they operate the same way. Minis usually have one job to perform and are physically tied to it. Like microcomputers, however, they may also be tied together with other minis to a larger computer control system.

BUS CONFIGURATIONS

The I/O devices in a minicomputer system are connected to the CPU in much the same manner as in the microcomputer. However, minicomputers may have many more devices attached to them that are of different types and at a number of different locations. The interface scheme is basically the same as discussed previously and typically involves interrupts. Data communication between the CPU and I/O occurs through an I/O bus, which may be configured in several different ways (Fig. 8-4):

1. Party line bus. As in a telephone party line, several I/O devices may be connected in parallel, sharing a single line to the CPU. In this scheme, each device has a separate address, so the computer knows which one needs service when it polls for status. This is generally the least expensive and easiest configuration to control.

2. Daisy chain bus. This is a somewhat higher-speed approach without a lot of extra cost. Several I/O devices can be connected in a serial chain; each one can influence the signal from the CPU before sending the signal on to the next device. This permits the closer devices to have a higher priority in being serviced.

3. Radial bus. The highest-speed connection to memory

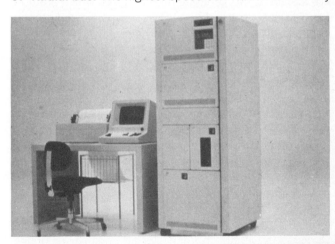

Fig. 8-3 Minicomputer. (*International Business Machines Corp.*)

PARTY LINE BUS

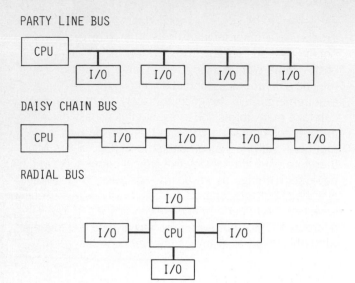

DAISY CHAIN BUS

RADIAL BUS

Fig. 8-4 I/O bus configurations.

is when a separate bus is provided for each I/O device. In this case the CPU does not need to use address lines and address codes to know the location of the device, since it has its own private line.

8-6 PROGRAMMABLE CONTROLLERS

WHAT ARE PROGRAMMABLE CONTROLLERS?

Programmable controllers (PCs) are a special sort of microcomputer. They are digital electronic devices which were developed to replace the old relay switch-type logic which had been used for many years to control machine tools. Unlike the hard-wired logic they replaced, PCs can be programmed to perform a variety of control functions and then easily reprogrammed to change those functions. They are usually small and relatively simple devices with very limited capabilities compared to normal computers. Their primary function is to perform logical operations on input signals and generate output signals which are used for machine control. Most PCs have very limited computational or storage capabilities. They are used primarily for control functions rather than the data processing or process monitoring tasks performed by micro- and minicomputers.

Although advances in technology have made it difficult to distinguish between some PCs and microcomputers, there are key differences in their design, application, and programming. PCs are designed specifically to operate in an industrial environment and perform machine control functions. Their logical operations and programming are designed to be compatible with the traditional relay logic used to control machine tools. A typical PC can be operated and programmed by a machine operator right at the workstation without the need for an engineer or professional programmer. In large manufacturing operations, a number of PCs may be tied together in a network con-

trolled by a minicomputer which monitors the performance of the production line. They can be used anywhere dedicated machine control is required as long as it involves only a limited amount of logic and storage functions. Because of their small size and dedicated role, PCs are often attached directly to the machine they control.

HOW DO PCs WORK?

A typical PC is made up of a CPU with memory, I/O interfaces, a programming device, and a power supply (Fig. 8-5). Input signals are received from switches or sensors on the machine being controlled; the inputs describe the status or performance of the machine's operation. On the basis of programs of instructions stored in its memory, the PC performs a series of sequential logic operations on these input signals. As a result, it generates output signals to control devices on the machine, such as motor starters or solenoids. Some PCs may also monitor the performance of the machine and generate data reports to a host computer.

Since PCs may be used on a wide variety of industrial equipment, they may have to handle many different types of signals, including alternating current; direct current; and binary, pulse, and analog signals. The input interface must convert the signals received from the machine to a form which can be processed by the PC. The output interface must then convert the control signals generated by the PC to the form that can be used by the machine control device. Four basic functions can be performed by a PC:

1. **Control.** All PCs generate control signals on the basis of their programmed logic. The signals may take the form of the traditional sequential on/off logic of relay control systems, or the PC may use analog control unique to the equipment operation in a particular application.

2. **Timing.** PCs must control the timing of the signals they generate in order to be compatible with the device with which they are communicating.

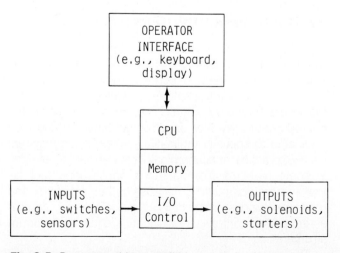

Fig. 8-5 Programmable controller.

3. Counting. PCs may be programmed to generate outputs on the basis of the completion of a certain sequence of events or number of steps in a machine's operation.

4. Computation. Although most PCs have limited computational capabilities, technology advances in microprocessors have made it feasible and economical to include some. These may include simple arithmetic operations or more complex mathematical functions, such as servo-control calculations.

HOW IS A PC PROGRAMMED?

A PC can be programmed uniquely for a specific application or machine which it is intended to control. The program may be entered through a variety of devices, such as a small keyboard, a control panel, or a computer terminal. Some PCs can even be programmed from a portable programming device that looks like a hand-held calculator (Fig. 8-6).

The most common type of PC programming, the relay ladder diagram, is a symbolic programming technique that simulates the electric circuits of switches that were used for many years to control electrical equipment. A standard relay control circuit is converted into a ladder diagram which programs the PC's logic by coding the opened and closed contacts, nodes, and branches of the circuit (Fig. 8-7). These circuits can then be easily

changed in whole or in part by reprogramming the PC. During its operation, the PC scans each network of interconnected logic elements in sequence with the input signals provided. The resulting output determines the control signals that are required.

Some PCs may also be programmed using other techniques that work more like conventional computer languages. These include Boolean logic statements to represent the relay logic, or higher-level symbolic coded languages like assembly language. The more powerful software packages available for PCs provide multiple programming functions using several levels of instruction sets. This may include English-like instructions for logic control, motion control, data communications, and math functions as well as assembly language.

PC NETWORKS

PCs were originally designed as stand-alone devices dedicated to control specific machine tools. However, as their capabilities expanded, they were found to be a convenient source of data on the factory floor. Many large manufacturing operations therefore tie PCs together into a data transfer network similar to those used for micro- and minicomputers. Such networks can be used to collect data from each machine and generate production status and performance reports for management. In addition, bus-type networks can provide for communication between PCs, to control such things as the scheduling and loading of machines in a sequential operation. Networking PCs also permits flexibility in distributing the control of a process among many devices. By breaking the process up into small steps or simple operations, the control function can be made faster and less prone to failure. When PCs are tied to a larger computer, it is also possible to load programs and reprogram the PCs directly from the host.

ADVANTAGES OF PROGRAMMABLE CONTROLLERS

Compared to the traditional hard-wired or relay logic controllers that were commonly used on machine tools and electrical equipment, PCs have several major advantages:

1. PCs are easier to program
2. PCs can be reprogrammed
3. PCs are usually smaller
4. PCs can be tied into computer systems
5. PCs are more reliable and easier to maintain

8-7 PROGRAMMING MICROS AND MINIS

OPERATING SYSTEMS

As the number and types of small computers continue to grow, the need for standardization of operating systems

Fig. 8-6 Portable programming device for a programmable controller. *(Gould Industrial Automation Systems, Andover, MA)*

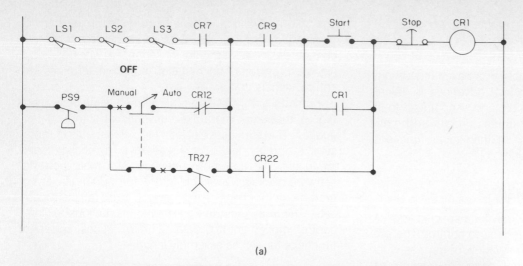

(a)

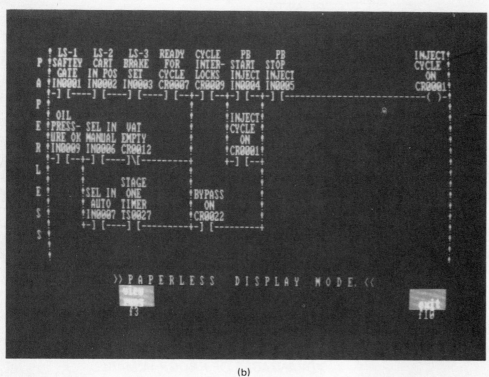

(b)

Fig. 8-7 Simplified relay ladder diagram: (*a*) relay ladder diagram converted to (*b*) a computer display for programming a controller. *(Westinghouse Electric Corp.)*

increases. If each different type of micro- or minicomputer had a unique operating system, it would be very difficult and expensive to transport application programs between systems and develop new software for broad application. A great deal of effort has therefore been spent on developing standard operating systems that can satisfy the needs of most small computers and their users. There are a number of common operating systems, particularly for personal computers, such as MS-DOS (which is a disk operating system developed by Microsoft Corporation). However, the one that appears to have the broadest application is Unix. It is a highly structured hierarchical system which can easily be expanded and modified for a wide variety of functions (Fig. 8-8). The layered approach to its software permits it to be both

efficient in its interface with the computer hardware and flexible in its interface with the user. The hierarchical structure also makes it possible to fit a relatively powerful operating system on small machines by dividing up the major functions. Only those functions which are required to service most programs (such as I/O routines) are kept in main memory. This portion of the operating system is called the "kernel." The other functions, called the "system utilities" (such as text processing or file management), are kept in auxiliary storage and retrieved when they are needed.

Unix is designed for portability between systems and uses a powerful high-level language (C). The most popular areas of application for Unix have been in text processing and program development, but many enhancements

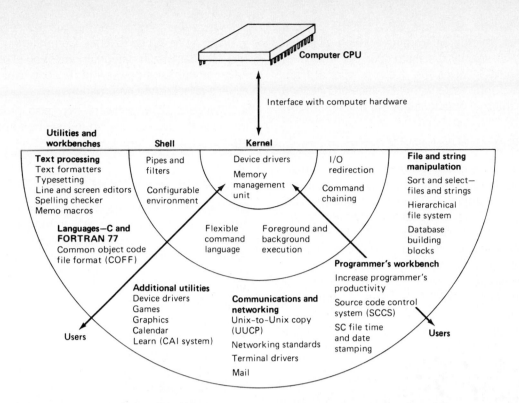

The multiuser, multitasking Unix operating system from Bell Labs is a layered system. The **kernel** interfaces directly to the computer's central processing unit (a microprocessor chip, in the case of supermicros) and is modified to run on different CPUs and to alter hardware-related operations such as memory management. Because Unix is written in C, a high-level language, it is less hardware-dependent than operating systems written in lower-level machine language. This trait makes Unix easy to transport from one system to another. Surrounding the kernel is the **shell,** which serves as a programming language and as a command language interpreter, reading lines typed by the user and interpreting them as requests to execute certain programs. Around the shell are various **utilities** and **workbenches** such as text processing and support for the C and FORTRAN 77 languages. Some parts of Unix, such as the **programmer's workbench,** which helps software developers, are not required by all users and are sometimes dropped in Unix-derived systems sold by companies that license the operating system. These firms also modify parts of Unix to meet different requirements. The kernel might be changed to run on different chips, for instance, or menus might be added to help novice users interact with the system.

Fig. 8-8 Unix operating system. *(Dwight B. Davis, "Super Micros Muscle Into Mini Markets," reprinted with permission, High Technology Magazine, Dec. 1983; Copyright © 1983 by High Technology Publishing Corp., 38 Commercial Wharf, Boston, MA 02110)*

are available that extend its capabilities to other areas, such as distributed network control and scientific computation. It is also capable of multitasking and has a large and flexible file management function. Unix is basically designed to provide for small computers most of the functions which have been available in the operating systems of large computers. The only real drawback is that some performance must be sacrificed to fit all those functions into a small computer.

ASSEMBLY LANGUAGE PROGRAMMING

When micro- and minicomputers are used in dedicated industrial applications, such as tool or process control, they are usually programmed by professional programmers. To optimize performance and minimize memory requirements (which affects the cost of the application), programmers will often write such programs directly in assembly language. This, of course, requires the programmer to be very familiar with the architecture of the machine and may necessitate a great deal of programming effort. Programming in higher-level languages would be much easier, but the resulting programs would usually require more memory and have a longer execution time.

Programmers who write programs in assembly language use an assembler that was designed for that particular machine to convert the assembly code to object code. Assembler programs often include some functions that make the programming job somewhat easier and save programming time. For example, when certain sequences of instructions need to be repeated in a program, the programmer can use a "macroassembler." This assigns a name to that particular program sequence so it can be assembled as an entire block of code. Programmers also use "pseudoinstructions" which give messages to the assembler that are not actually part of the program itself.

MICROPROGRAMMING

To improve the performance of microprocessors for special applications, it is sometimes necessary to use instruction sets different from those normally provided with standard microprocessors. This can be done by using a technique called "microprogramming," which gives the user the ability to specify the instruction set and word size. It also usually involves the use of a special high-performance microprocessor with a bit-slice architecture (Fig. 8-9). The internal operations of the microprocessor are controlled by "microinstructions" which are stored in a microprogram memory (e.g., a ROM or PROM). The performance of the microprocessor can be increased further by adding a "pipelining" register which permits the next instruction to be fetched before the current instruction cycle is completed. These techniques give the user the ability to achieve substantially greater performance from a microprocessor than would otherwise be possible with standard devices and instructions. The tradeoff, of course, is that it requires customized programming and hardware.

FIRMWARE

To increase both the performance and the reliability of small computers, some software is often provided in hardware (therefore it is sometimes referred to as "firmware") in the form of ROM. This may be done for several reasons.

1. It may be more convenient and efficient to store software that is used frequently in ROM rather than load it from a disk into main memory. This could include a general purpose operating system or a standard language interpreter.

2. Some very small computers, such as portable machines, may have a main memory of very limited size and perhaps even no disk memory. In such cases, embedding standard sets of software in ROM will save the limited memory space for application programs.

3. Since hardware is not as subject to damage or loss as

software stored on disks or in main memory, critical programs may be stored in ROM. Computers used for industrial applications, in particular, are often exposed to harsh environments and power interruptions, which will generally not affect the ROM programs.

4. Standard ROM software may also provide a stable interface for users who wish to transport application programs between different microprocessors.

SOFTWARE DEVELOPMENT SYSTEMS

Developing or modifying software can be a very time-consuming and expensive task. In the past, much of that type of work was done by professional programmers on terminals connected to large central computers. As the job of developing and maintaining application programs and custom systems spreads to the users of micro- and minicomputers, simpler and more efficient software development tools must be provided. Today, small computers can serve as engineering workstations where a user can design, code, and debug programs. A microcomputer software development system would typically include such programming tools as:

1. An assembler or compiler for translating from a high-level language

2. A text editor for making changes

3. A debugging program to help find errors and problems

4. A loader program to instruct the system to call out and execute the program

5. A library of standard program routines

6. A standard operating system to manage the execution of the program

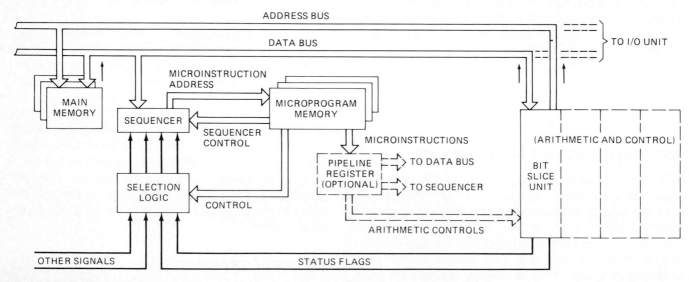

Fig. 8-9 Microprogrammable microprocessor using a bit-slice architecture (*Edward V. Ramirez and Melvyn Weiss, Microprocessing Fundamentals—Hardware and Software, McGraw-Hill, New York, 1980*)

8-8 MICRO- AND MINICOMPUTER TRENDS

HARDWARE TECHNOLOGY

As we saw in Chap. 6, computer hardware technology is advancing very rapidly. This is particularly true in the area of microelectronics, where circuit density and speed are increasing while the cost per circuit is decreasing. This has made it possible to package a great deal of computer function in a small space at relatively low cost. Micro- and minicomputers therefore continue to offer more capabilities that were previously only available on larger machines. The key to this trend is the advances in IC technology, particularly in the area of microprocessors, which are used for the basic logic functions in small machines. Since they were first developed, the performance of microprocessors has increased by more than two orders of magnitude (Fig. 8-10). This has been made possible by fabricating more circuits on each chip, increasing the switching speed of the circuits, and reducing the power consumed by each circuit. One of the major contributors to this has been the evolution in FET technology which has continued to optimize its processes and designs. It evolved from the early "P-channel" devices to "N-channel" devices, and eventually to "CMOS" (complementary metal oxide semiconductor) devices, which are widely used for microprocessors and high-density logic devices today.

As the circuit density of semiconductor logic devices increased, it became possible to provide larger computer functions on a single chip. Early microprocessors were merely small ALUs which were initially only 4 bits in length. Today, the entire architecture of a 32-bit CPU can be fabricated as a microprocessor chip (Fig. 8-11). This makes it possible for microcomputers to operate almost

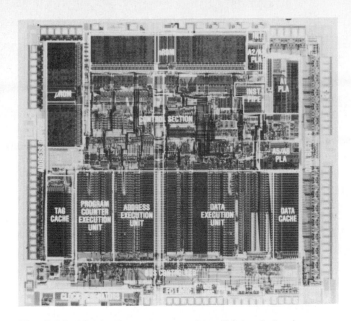

Fig. 8-11 32-bit microprocessor chip. *(Motorola Inc.)*

as efficiently as large computers. To increase their performance even more, special microprocessors have been developed, such as:

☐ Floating-point coprocessors for scientific computations

☐ High-speed bipolar bit-slice devices for custom microprocessors

☐ Microcomputer chips which incorporate memory and communication circuits in addition to a processor for parallel computer architectures

Advances in hardware technology outside the realm of semiconductors will also enhance the capabilities of small machines. In the area of I/O devices, much has been done to develop capabilities that were previously only feasible or economical with large computers. Recent advances in technology have now made it practical to provide such advanced devices as nonimpact printers, high-resolution color graphics displays, and multicolor printers for microcomputer systems. Perhaps the most significant development, however, is the advances in magnetic disk technology which have made it possible to incorporate miniature hard disk files in microcomputers (Fig. 8-12).

RELATIVE PERFORMANCE
(ratio of instructions per second)

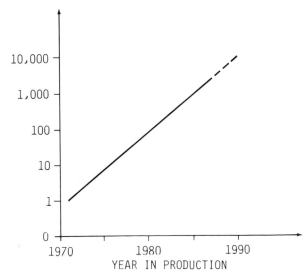

Fig. 8-10 Microprocessor performance trend.

Fig. 8-12 Miniature hard disk file. *(Hewlett-Packard)*

This can substantially increase the storage capability and operating performance of small machines so that they can handle much larger programs and databases.

SOFTWARE

The use of standard operating systems and languages is very important to the continued development and success of small computers. This permits a wide variety of programs from different sources to be compatible with many machines. Such portability of application programs and operating systems can significantly reduce the programming effort involved in supporting small computers and can make more software available to the user. An operating system with efficient program development tools will also make the programming and software maintenance job easier.

The continued success of small computers is dependent upon further developments in operating systems to provide:

1. Standardization that allows compatibility with many different machines

2. High-level languages for ease of use and programming

3. Efficient program development tools to make the programming and software maintenance jobs easier

4. Powerful capabilities to manage I/O and database functions

One major change in the approach taken to program computers will have a significant effect on the performance and cost of small machines. It has been found that instead of continuing to make the hardware and software of computers more complex in order to increase their performance, simplifying them actually yields better results. This is particularly true of applications, such as engineering workstations, that involve many complex computation tasks. The concept is known as the reduced instruction set computer (RISC). By reducing the number of instructions used in a program and increasing the speed at which the computer can execute those instructions, a simpler and more efficient computer can be produced. Most complex functions can be achieved by using a series of simple instructions, which avoids the need for the additional logic circuitry that would be required to decode and execute the more complex and less frequently used instructions. This allows the central processor to be smaller and less expensive. In addition, by using a smaller instruction set, the instruction fetch and execution cycle can be streamlined to operate more efficiently.

It is possible to execute a program faster on a lower-cost RISC machine than on a larger, more expensive conventional machine (Fig. 8-13). RISC machines, however, require special compilers to translate standard high-level languages into these simple instruction sets as well as a logic and memory architecture, which is compatible with the streamlined execution cycle.

In the area of application software, perhaps the most prominent trend for small computers is the development of very powerful and easy-to-use integrated software packages. These are sets of programs which provide many of the most common business functions that can be used by most workers without formal programming skills. Such packages typically include word processing,

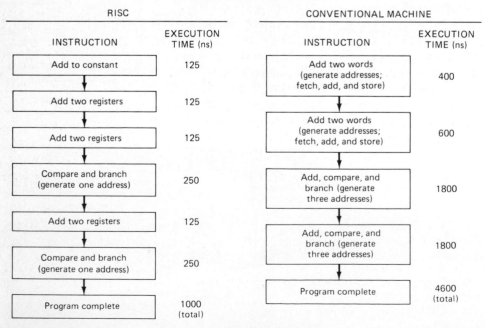

Fig. 8-13 RISC program execution. *(Stephan Ohr, "RISC Machines," reprinted with permission from Electronic Design, vol. 33, no. 1; © Hayden Publishing Co., Inc., 1985)*

spreadsheets, graphics, windowing, and database programs, which can be used in conjunction with each other. This eliminates the need for acquiring, learning, and maintaining a number of different individual programs to satisfy these needs. It also permits the user to exchange data between programs rather than reloading and reentering it for another application.

SYSTEM ARCHITECTURE

Aside from the radically different approach of the RISC machine, a number of enhancements to the architecture of small machines have been developed recently.

1. Fault tolerant designs. Since small computers are being used more often in applications where the cost of failure can be high (such as on-line transactions and real-time process control), features have been offered which can make them more tolerant of failure. The typical approach is a backup or redundancy scheme where multiple processors are used and error detection and recovery functions are built into the operating system. The relatively low cost of hardware in small machines now makes this practical for many applications that previously could not afford such protection.

2. Communications. As the use of micro- and minicomputers expands in the industrial environment, there is a growing need for tying groups of those computers together so that they can share databases and communicate with one another. To increase the communications capabilities of small computers, a great deal of effort has been spent on developing large and flexible bus structures that will permit many different types of computers, I/O devices, and applications to share a network. This requires standard communication protocols, wide bandwidths, multiple buses, and wide address capability (e.g., 32-bit). With the use of advanced hardware and software technologies, these functions, which previously were available only on large machines, are now practical on small systems.

3. Multiprocessing. Since powerful microprocessors are now feasible and economical, it has also become practical to build small computers with multiprocessing capabilities. By using several processors, the small computer can increase the efficiency of its operation significantly. Tasks such as I/O and execution can be split between them and run concurrently. This will permit small machines to handle larger amounts of data and perform more complex functions.

4. Open system architecture. To provide for the greatest compatibility of software between different types of hardware, it would be desirable for all computers to have the same basic architecture. Since the cost of software development and maintenance is often greater than the cost of the hardware involved, portability and communications compatibility are highly desirable. To achieve this, industry has attempted to develop standard approaches to system architecture which are visible in the design to all users and competitors alike. Although some of the internal functions may differ, the objective is to have standard interfaces and structure at every level of the operating system.

COMPUTERS

As a result of many of these internal advances in hardware and software technology, there have been a number of new developments in the nature of micro- and minicomputers themselves. These involve not only what they are capable of doing, but how they are used and what they look like.

1. Supermicros and superminis. By using parallel processing, high-performance microprocessors, wide address and bus architectures, and efficient operating systems, a whole new generation of small machines is being developed which can match or even exceed the performance of some traditional large machines. These so-called supermicros and superminis can be used for complex applications such as scientific computation or design. They can also be used to replace larger and more expensive processors in special dedicated applications such as process control.

2. Workstations. With such powerful capabilities now possible in small machines, there is an increasing use of desktop computers for applications that previously were only possible on large systems (Fig. 8-14). An intelligent workstation provides the user with local computational capability to reduce the dependency on remote processing by a mainframe computer. This can improve the availability and response time significantly for interactive applications such as computer-aided design (CAD). Most

Fig. 8-14 Personal computer-based graphics workstation. *(International Business Machines Corp.)*

workstations also provide communication links to a host system to gain access to central databases and store the files created by the user.

3. Portable computers. As hardware technology advances, it becomes possible to package a great deal of computer function in a small space. By using high-density microelectronics, low-profile keyboards, flat-panel displays, and low-power circuitry, fairly powerful portable computers are now feasible (Fig. 8-15). Many of these can also be connected to data communication lines for retrieving or storing information in a host computer.

4. Hardened computers. The factory environment is often too severe for delicate electronic equipment. However, as industrial processes became more dependent on computers for control, hardware developments were made to make computers more tolerant of such environments. This has been particularly successful with small computers, in which it was possible to take advantage of technological advances to tailor their design to the application (Fig. 8-16). Some of the features of such computers are:

Wide range of operating temperatures because of low-power, temperature-tolerant circuitry

Sealed enclosures to prevent exposure to contaminants and corrosives

Low-power operation to avoid the need for cooling

High immunity to electrical noise from industrial equipment

Rugged construction to prevent damage from vibration and shock

High reliability and ease of repair because of high-density electronics with error detection and diagnostic capabilities

As the hardware and software technology of micro- and minicomputers continues to advance, they will offer even more improvements in cost, performance, and function. The small computer will become a basic tool for all manufacturing workers both on and off the factory floor.

Fig. 8-15 Portable computer. *(International Business Machines Corp.)*

Fig. 8-16 Industrial computer in a factory environment. *(International Business Machines Corp.)*

8-9 SUMMARY

Small computers were made practical and economical by the development of the microprocessor. This is the basic logic building block for micro- and minicomputers. The microprocessor, together with memory and I/O, can perform all the basic functions of a computer. It can communicate with I/O devices and other computers through a variety of interface and bus schemes, depending on the performance and complexity involved. Programmable controllers (PCs) are a special type of microcomputer. They usually have more limited functions and are dedicated to a particular application. PCs were developed specifically for industrial environments and have significant advantages over the hard-wired and relay logic controllers they replaced.

The development of small computers expanded the possibilities for computer applications on the manufacturing floor. They may be dedicated to specific tasks or tied together into a larger manufacturing control or management system. By using standard operating systems and high-level languages, micros and minis can offer flexibility, expandability, and portability. Micros are often programmed at a relatively low level to optimize their performance in a particular application. However, they can also be easy to use in general purpose applications with high-level languages.

The power and performance of micro- and minicomputers are increasing rapidly with the advances in hardware and software technology. Today the small computers can have many of the same capabilities that were previously only possible on large machines. This permits them to handle larger programs and provide more flexibility at the user level. Many advanced functions are being incorporated into small computers, including wide addressing, reduced instruction sets, parallelism, and fault tolerance. With such capabilities, micro- and minicomputers will have key roles in the future of computer-automated manufacturing (CAM).

REVIEW QUESTIONS

The answer to each question can be found in the section(s) indicated at the end of the question.

1. What are microcomputers and minicomputers and how are they different? [8-3 and 8-5]

2. Describe a microprocessor. [8-2]

3. Define a bus system and describe its purpose. What are some types of buses? [8-2]

4. Describe the functions of RAM and ROM. [8-3]

5. Describe how a microcomputer interfaces and communicates with I/O devices. [8-4]

6. How do PCs differ from microcomputers? [8-6]

7. Identify some of the principal functions of a programmable controller. [8-6]

8. How are programmable controllers programmed? [8-6]

9. Identify some of the advantages that programmable controllers have over hard-wired or relay logic controllers. [8-6]

10. What are some of the different ways in which microcomputers can be programmed? [8-7]

11. Describe the advantages of standard operating systems and programming languages. [8-7]

12. What is a software development system and what does it typically include? [8-7]

13. Identify some of the major contributors to the improvements in the capabilities of microcomputers. [8-8]

14. What is a RISC machine? [8-8]

15. Describe some of the advanced forms and functions of micro- and minicomputers. [8-8]

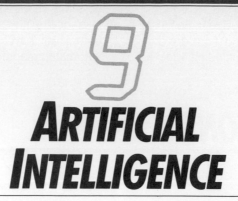

ARTIFICIAL INTELLIGENCE

9-1 INTRODUCTION

Artificial intelligence (AI), is a new and rapidly developing field of computer science. It deals with computers performing humanlike functions such as reasoning and interpretation. AI may also take the form of simulating the human senses of vision, hearing, touch, and even smell. These senses usually serve as a source of information for the human brain to use in making observations, drawing conclusions, and formulating decisions. Computer systems are being developed which operate with the same basic approach to problem solving, rather than the fixed, sequential, programmed logic of conventional computers.

AI is no longer just a field of academic research or science fiction imagination. Significant advances have been made in both hardware and software to make these capabilities possible. AI is an important, emerging area of technology with great potential in a wide variety of applications. This is particularly true for the manufacturing industry. "Intelligent systems" may be used for routine decision-making tasks (such as process control) and for sensing in manufacturing operations (such as robotic assembly). AI is starting to be used in real applications, and its use and capabilities are being extended rapidly. It will be a key factor in significantly expanding the use of computers in the future by providing greater capabilities and making them much easier to use.

"Artificial intelligence" is a general name that has been applied to a broad field of developments in computer science. It is an extension of the basic hardware and software technologies that were covered in Chaps. 6 through 8. This chapter will cover the basics of AI. However, AI is too complex and changing a field to address in depth here. The objective of this chapter is therefore to provide enough information to permit an understanding of what AI is, how it works, and where it can be used.

9-2 BASICS OF ARTIFICIAL INTELLIGENCE

WHAT IS AI?

"Artificial intelligence" is a term that is usually applied to functions performed by machines that would normally require some properties of human intelligence. This typically involves either humanlike reasoning or senses. AI is not just a hardware or software technology. It involves an entire process of data collection, computer logic, and data processing. Development efforts in this area require advances not only in computer science but also in the basic understanding of human senses and the brain. AI machines do not operate like conventional computer systems, which follow instructions step by step. They use association, reasoning, and decision-making processes much like the human brain would to solve problems. Some can also "learn" from their experience and communicate in natural language.

Such capabilities are often taken for granted by humans. However, they all require a great deal of knowledge and reasoning, which conventional computers do not possess. To simulate these capabilities in a machine, new approaches to computer hardware and software had to be developed. Special programming languages were developed that manipulate symbols rather than numbers. New computer architectures were devised to handle large databases and parallel processing. This new-generation computer system differs from contemporary computers primarily in its ability to draw conclusions and make decisions based on inference and deduction from a base of knowledge. Because of their ability to handle data with some level of meaning and understanding, such machines can be used as intelligent, problem solvers rather than merely as computation devices that follow strict instructions.

AI can simulate many human capabilities, but it cannot (yet) create. This requires imagination and intuition, not just logic. However, AI is being used successfully for a number of industrial applications, primarily in four areas:

1. **Expert systems.** These are systems that provide decision-making capabilities for specific applications that require expert knowledge.

2. **Natural language.** This may take a variety of forms wherever a machine understands and interprets a natural language (e.g., English).

3. **Artificial senses.** These are special systems to simulate one or more of the human sensing capabilities, that is, vision, touch, hearing (or speech), and smell.

4. Robotics. This is an extension of mechanical robot technology that uses adaptive control, sensing, and learning capabilities to perform physical manipulation tasks.

WHY DO WE HAVE AI?

Although AI research has been conducted for many years, it is only recently that it has become a popular field of serious interest to industry. This is due primarily to two factors:

1. The need for AI technology. As industrial operations become more complex and the costs of doing business increase, problems develop which can potentially be solved by some of the capabilities of AI such as:

☐ Limited availability and high cost of skills

☐ Inefficiencies and costs associated with routine decision making and controls

☐ Large databases which humans and conventional computer systems cannot handle

☐ Change and variability in products and processes

☐ Uncertainty and lack of a single solution

☐ Expanded use of computers by nonprofessionals

2. The feasibility and practicality of AI technology. Recent advances in both hardware and software technology have made it possible and economical to provide effective AI systems. The key influences have been:

☐ High-density, high-performance electronics at relatively low cost

☐ Powerful high-level programming languages and operating systems

☐ New system architectures capable of highly parallel data processing

☐ Increased understanding of the functions of the human brain

AI was not just invented. It has evolved from computer science as a field whose time has come for practical application. When both the need and the capability exist, technologies are usually implemented successfully.

HOW DOES AI WORK?

The AI process is made up of three basic steps:

1. Data collection. This includes the acquisition of information from knowledge sources (e.g., experts) to establish a knowledge database as well as the access to input data for the task at hand.

2. Computer logic. Unlike conventional computer logic, AI machines use facts and rules from the data collected to draw inferences and conclusions and reach solutions to problems.

3. Data processing. To manipulate the large amounts of data typically involved in AI tasks requires high-performance parallel processing techniques with a very efficient I/O data transfer system.

Implementing the AI process requires the use of some of the most advanced concepts in two different fields of science:

1. Computer science. AI implementation is drawn particularly from the areas of "intelligent" machines, parallel architectures, and logic programming.

2. Cognitive science. This is the science of knowledge and perception. From this field we are learning how to simulate the way in which the human brain processes information to associate facts and deduce conclusions.

AI therefore requires a new generation of computer systems. Such systems are typically comprised of three major elements (Fig. 9-1):

1. The hardware system (intelligent machine). The computer must manage the knowledge database, perform the logical processing, and provide interfaces to the user and other machines.

2. The software system. AI software must represent knowledge in symbolic form, provide access to a relational database, and use a logical technique for solving problems without step-by-step instructions.

3. The external interface. AI systems must be both easy to use and efficient in interactions between the user and the machine. This requires a high-level query language, interfaces to various natural sources of data (e.g., vision and speech), and a logic programming language.

A variety of approaches has been used for the hardware and software of AI systems; some will be covered in more depth in later sections of this chapter. The basic operation of most systems, however, is similar. A user makes an inquiry or states a problem to be solved. In some cases the user may provide additional data in support of the problem statement. The system then uses facts and rules from its knowledge base to conduct a logical deductive process on the problem and search for a conclusion. Since such conclusions are based on a reasoning process, AI systems can make mistakes! Conventional computers only make mistakes when there are errors in the data or program. But even if the data is correct, AI systems may reach the wrong conclusion because of the number of possible alternative solutions involved. The step-by-step, algorithmic approach to problem solving by conventional computers is usually the most efficient when only one specific solution exists. AI systems, however, can deal with uncertainty and very large amounts of data, which permits them to solve problems that conventional computers cannot.

HOW ARE AI SYSTEMS USED?

AI is beginning to be used in a wide variety of industrial applications. Situations which require handling large amounts of data, dealing with uncertainty, or natural interfaces are all candidates for using AI. Following are several brief examples of how and where they can be used:

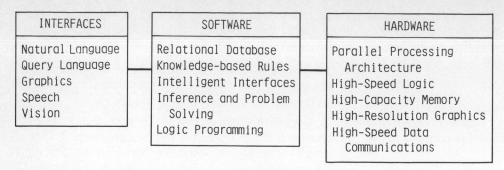

Fig. 9-1 Basic elements of an artificial intelligence system.

1. Expert systems use specialized knowledge to solve problems where human skills or capabilities are limited. Such situations may involve large amounts of data which are too large or subject to error for human experts to handle. They may also act as a substitute for a human expert when one is not readily available or is too costly to consult. Early applications for such systems included analysis of seismic data for oil well drilling and medical diagnosis. These same capabilities can be extended to engineering and management tasks in the manufacturing environment, such as:

Product design

Process planning

Production scheduling

Process control

Line modeling and simulation

Diagnosis of equipment malfunctions

2. Information retrieval systems can also benefit from the use of AI technology by making large databases accessible to nontechnical users through natural language interfaces. Some AI systems can interpret natural language or even speech by using a knowledge base of grammer, definitions, or sounds. This permits them to translate user inquiries or requests into commands that the computer can understand. Such systems can be used for libraries, management reports, financial transactions, production tracking, and a wide variety of similar information retrieval applications. The key to their success in such applications is ease of use and the ability to deal with large and changing databases.

3. The simulation of physical human capabilities is perhaps the most popular AI application. This, of course, includes industrial robotics, where adaptive control systems can make robots more flexible. Sensory capabilities such as vision and touch can significantly expand the scope and nature of applications for robots and other automated equipment. The ability to recognize or sense the presence of objects avoids the necessity of elaborate programming and special fixturing. This can make the automation of many more manufacturing operations practical and economical.

4. AI systems can also be used for computer-aided instruction (CAI). By combining some of the same capabilities used for expert systems and information retrieval, AI can be used to teach skills. It does this by permitting the user to try out alternatives and providing explanations to answers. This can allow routine subjects as well as specialized skills to be taught more efficiently. In a rapidly changing industrial environment with a limited availability of skilled workers, training is an important and continuous activity.

5. The efficiency of computer programming and operation can also be improved by using the capabilities of AI. For example, AI techniques can be used to manage some of the internal tasks of an operating system, saving some of the software development effort which is normally required. They can also be used to automate the coding and translation of routine programs. Since the costs of software development and maintenance often exceed the costs of computer hardware and manufacturing tools, such automation offers substantial potential benefits.

9-3 EXPERT SYSTEMS

BASIC ELEMENTS OF AN EXPERT SYSTEM

An expert system is a form of AI that uses a base of knowledge to solve problems that would otherwise require some specialized human expertise. Such a system is put together by first obtaining knowledge from experts and storing it in the form of rules. These rules are then used to guide a reasoning process. This reasoning process—called "heuristic" or "empirical," which means based on experience—is applied to a problem statement and data to reach a conclusion or solution. Knowledge-based expert systems use a number of advanced techniques which can be thought of as a hierarchy of building blocks (Fig. 9-2):

1. Logical process. This is the base or foundation of the expert system. It consists of the logical tools which perform the problem-solving tasks. Included here are:

☐ Symbolic programming, which manipulates symbols that represent objects and relationships

☐ Propositional or predicate calculus, which is a form of logic that uses rules to deduce conclusions

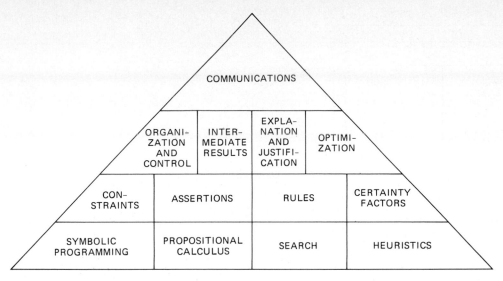

Fig. 9-2 Techniques used in knowledge systems. *(Frederick Hayes-Roth, "The Knowledge-Based Expert System: A Tutorial," Computer, Sept. 1984; © 1984 IEEE)*

□ Search methods, used to seek solutions from alternative approaches

□ Heuristics, using rules derived from experience to guide a reasoning process

2. Knowledge representation. Knowledge is stored in forms that can be used in the problem-solving process. These forms include constraints, assertions, rules, and certainty factors (i.e., confidence). Such structured information can then guide the system to a solution by eliminating alternatives and associating facts.

3. System operation. Although expert systems may be designed to operate in a variety of different ways, they usually have several common characteristics. They organize and control their problem-solving activities. Many will offer intermediate results and explanations of their conclusions to the user. This permits interaction and the possibility to change the line of reasoning being used. Some systems can even "learn" from their own experience and optimize their performance.

4. Communications. To be useful, an expert system must do a great deal of communicating. It must have the hardware and software required to communicate efficiently with not only the user, but also experts, knowledge engineers, databases, and other computers.

To put all these techniques to work solving problems requires the implementation of a system which is made up of three major elements (Fig. 9-3). Each plays an important role but represents a distinctly different environment. Together, the elements of the system span the scope of activities from development tools to knowledge acquisition and maintenance to actual operations.

1. System development. This involves the use of AI techniques to obtain and code information into a knowledge database and provide an interface to the user.

2. Knowledge system. This is the actual hardware and software of the system itself. It is comprised of the knowledge database, the logical processor ("inference engine"), and the operating system and programming languages.

3. System operation. The operational phase of an expert system involves a wide variety of traditional data processing activities, including data acquisition, data transfer, and communication.

HOW DOES AN EXPERT SYSTEM WORK?

An expert system uses information that has been acquired from experience to formulate and test solutions to a problem. That expert information or knowledge is represented by rules which influence the logical reasoning process of the system. Three principal approaches are used to represent knowledge and execute the logic of an expert system.

1. "IF-THEN" rules. The most common technique uses simple two-part rules which identify conditions or patterns together with actions or consequences that should follow. For example, "IF the temperature is 100C AND the pressure is one atmosphere, THEN water will boil." The logical process evaluates the problem statement and data by searching through the rules available for a solution that works. This may be accomplished by using several techniques (Fig. 9-4). "Forward chaining" starts with the initial data and tries to find a rule whose consequence (i.e., THEN statement) matches. Alternatively, "backward chaining" selects a rule first and tries to match its consequence to the initial data. Some systems combine these techniques to improve their efficiency in seeking a solution.

2. Semantic networks. This is a technique which represents relationships among objects and provides a means for more efficient logic processing. Semantic networks

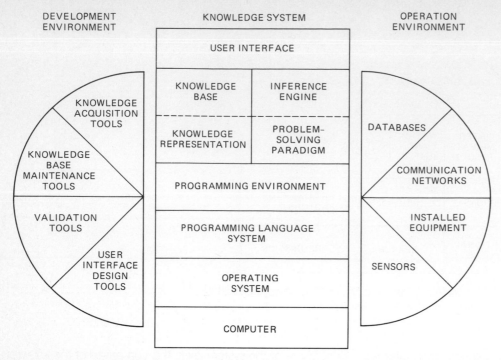

DEVELOPMENT ENVIRONMENT   KNOWLEDGE SYSTEM   OPERATION ENVIRONMENT

Fig. 9-3 Major components of a knowledge system. Knowledge is acquired with system development, then put into operation to solve problems. *(Frederick Hayes-Roth, "The Knowledge-Based Expert System: A Tutorial," Computer, Sept. 1984; © 1984 IEEE)*

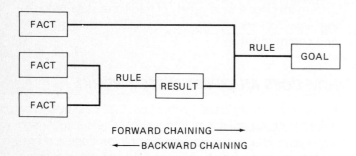

FORWARD CHAINING ⟶
⟵ BACKWARD CHAINING

Fig. 9-4 Forward and backward chaining. *(International Business Machines Corp.)*

use information stored in memory which is linked to other information (Fig. 9-5). For example, a typical semantic network would link one concept or object to another by a classification relationship, such as "A computer IS A machine." Another variation of this technique uses "frames," which are packages of knowledge that describe both fixed and variable attributes of objects. The fixed part identifies the permanent attributes, while the variable part identifies those attributes which may change or vary between objects in the same class. For example, a computer can be classified as a machine, but it can have many different features and functions. Computers may vary in their hardware and software as well as their architecture and technology. Once a framework is built, it can be enhanced by acquiring additional information about objects which may help the system reach a conclusion faster.

3. Formal logic. This is the traditional form of mathematical logic which uses propositional or predicate calculus. It is sometimes also referred to as "first-order logic,"

which uses mathematical symbolism to represent propositions (i.e., assertions) and relationships (or truth functions) between objects. For example, the fact that all metals have melting points and gold is a metal could be represented in formal logic statements as:

ALL X, MELTING POINT(X) IF METAL, METAL(Gold)

The function, MELTING POINT(X), therefore has the value "true" when X is a metal, and as a consequence gold also has a melting point. The formal logic technique is more precise than the others but has been more complicated to implement for expert systems.

A number of variations to these three basic approaches to problem solving may be used, depending on the nature of the problem, the amount of information available, and the type of system being used. For example, the computer can be programmed to ask for confidence levels when obtaining information from the user. These can then be applied statistically in the search for a solution. There is even an entire field of study called "fuzzy logic," in which a line of reasoning is developed on the basis of uncertain or partial evidence. For example, the term "usually" defines a relationship which is not precise but may be helpful in solving a problem. This, of course, is a common approach in human reasoning, but computer problem solving has traditionally been limited to the formal logic of mathematics. Although the results of such fuzzy reasoning are not as accurate as those derived from formal logic, a wider variety of more complex problems can be dealt with using this approach.

When the information base is large and the problem is

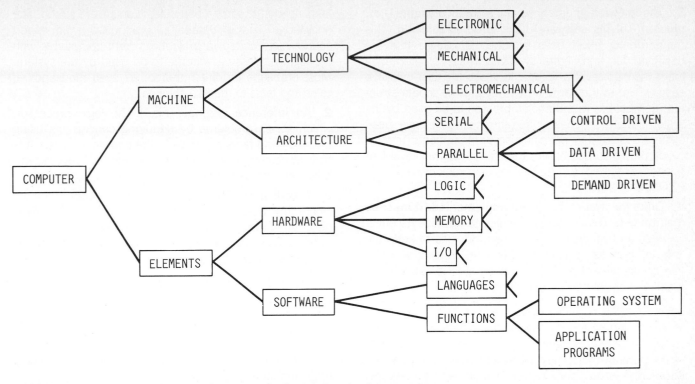

Fig. 9-5 A simplified semantic network representing knowledge about a computer.

complex, it is desirable to break it down into smaller, simpler problems which can be solved in parallel. An expert system can also "learn" from its experience to be more efficient in solving a particular type of problem. The knowledge it has about a problem can be used to suggest shortcuts to a solution. It can also try a "brainstorming" approach, in which it tries out many different ideas and retains only those that prove to be the most successful at solving the problem. Problems with limited uncertainty and few solutions can often be solved most efficiently by analogy or formal logic. Whichever approach is used, the basic objective of an expert system is to efficiently represent and process facts and ideas on a computer to solve a problem.

KNOWLEDGE ENGINEERING

One of the key elements of expert systems is the field of knowledge engineering. This includes the development of problem-solving techniques and the acquisition and maintenance of knowledge bases. The overall task involves several major stages of activity (Fig. 9-6):

1. Knowledge must be obtained from expert sources. This includes the basic principles involved as well as relevent experience. Depending on the uniqueness and complexity of the problem, this can be a substantial and ongoing effort. If the problem is a classical one that has been solved many times (e.g., an engineering design problem), most of the necessary knowledge may be available in textbooks. If it is new or somewhat subjective (e.g., failure analysis), it may have to rely mostly on the experience of experts.

2. A knowledge system must be designed and organized. This includes identifying the characteristics of the problem, designing a structure or system architecture to organize the knowledge, and selecting a technique to represent and use the knowledge for solving the problem. A substantial effort may be needed to develop a new system. It would then also have to be tested to verify that it works.

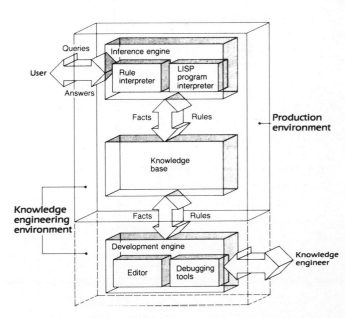

Fig. 9-6 Building an expert system. *(Paul Kinnucan, "Software Tools Speed Expert System Development," reprinted with permission, High Technology Magazine, March 1985; Copyright © 1985 by High Technology Publishing Corp., 38 Commercial Wharf, Boston, MA 02110)*

3. The knowledge database and inference engine (i.e., problem-solving software) have to be established. This programming effort uses symbolic languages that are designed for knowledge systems. These programming tools use high-level languages to make the user interface easy, while, at the same time, they provide a data structure compatible with the logic process to be used by the system. Many of these programming tools have been custom designed for specific classes of problems. In most cases they also use one of the standard lower-level languages designed for AI system applications, such as PROLOG or LISP.

4. Once the system has been implemented, it must be maintained. This involves changes to improve its effectiveness and efficiency as well as additions to keep the knowledge base up to date. This is an ongoing task that can be significant if the application area is subject to a great deal of change and new knowledge.

All four stages in the development and implementation of a knowledge system require the efforts of specialists called "knowledge engineers." These are not just professional programmers; they are experts in their own right in knowledge acquisition, representation, and processing. Once a system has been developed, however, it should not require a knowledge engineer to use it. The objective of an expert system is to provide the end user with easy access to expert knowledge and efficient problem-solving tools.

USING AN EXPERT SYSTEM

An operational expert system is made up of three major components (Fig. 9-7):

1. A knowledge base. This is where all the rules and facts needed to solve a particular problem are stored. The knowledge in this database is acquired from expert sources and coded in a symbolic form compatible with the logic processing scheme to be used by the system. Most problems require hundreds or even thousands of rules and facts to reach a solution.

2. An inference engine. This is the logic processor, which solves problems by deducing conclusions using the rules and facts available in the knowledge base. It is basically a computer that is programmed to process symbols that represent objects and their relationships.

3. A workspace. This is the part of the system's memory where the description of the problem is stored. That description may be developed from information obtained from the user or inferred by the system from the information in the knowledge base.

Expert systems may also include several other features which enhance their capabilities, such as:

☐ A natural language interface for communicating with the user

☐ A knowledge acquisition system to obtain additional rules and facts

☐ A system to provide explanations of the reasoning process to the user

In the operation of an expert system, the computer uses the facts obtained about the problem from the user together with its knowledge base about the subject and then applies its problem-solving techniques to find a solution. The computer reasons by processing symbols that represent objects. The objects may be concepts, rules, processes, or physical objects. Most of the computer processing activity involves matching objects or joining several together to establish relationships that can be

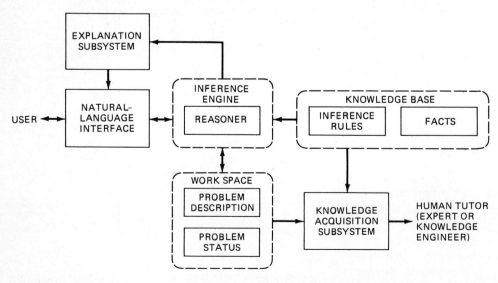

Fig. 9-7 Major components of an expert system of today—arrows indicate the flow of knowledge during operation. *(Paul Kinnucan, "Computers That Think Like Experts," reprinted with permission, High Technology Magazines, Jan. 1984; Copyright © 1984 by High Technology Publishing Corp., 38 Commercial Wharf, Boston, MA 02110)*

tested against rules (e.g., a computer IS A machine). To make the job of searching and modifying the knowledge base easier, information is usually organized into groups that relate to specific subjects (e.g., machines). In addition, to avoid the necessity of entering duplicate information, common properties can be described for a class of objects (e.g., size and shape). When the system tries to match or join objects, it determines their properties by looking them up in the knowledge base. If that information does not exist there, the system may infer by using its logic processing techniques or it may be able to obtain it directly through the use of some sensor input (e.g., monitoring process parameters). Since erroneous information could lead to incorrect conclusions, controls or restrictions are often applied to data in the knowledge base. For example, this may be an either/or condition or a range of acceptable or possible values.

An expert system is usually intended to be used interactively so that it can operate as a consultant with the user. To solve the problem the system may ask the user for additional information and confidence levels that can be applied to it. The user may also interrupt the system for explanations of the reasoning or questions being used. This user interaction is usually provided by a natural language interface and flexible graphics system (Fig. 9-8). The graphics display may provide "windows" which permit the user to see several different types of information simultaneously, such as the questions, answers, explanations, and even graphical illustrations.

When an expert system is initially developed, it usually does not match the performance of real human experts.

However, as the knowledge base is expanded and the system's efficiency is improved, it may eventually exceed it! Using an expert system can offer many advantages:

1. It can act as a substitute for a scarce or costly expert skill

2. Its problem-solving capabilities are superior to those of conventional computer systems

3. It can work continuously and does not need to be replaced

4. It is easy to replicate

5. It is easy to maintain and expand

6. It can be improved and changed

With the present state of the art, however, expert systems have a number of disadvantages which must also be considered:

1. Expert systems are expensive to develop

2. They are more limited than humans in their ability to store and manipulate abstract information

3. They rely primarily on the use of heuristic reasoning based on "rules of thumb" rather than first-order logic principles

4. They have limited ability to interact with humans

5. The expert knowledge they use may not be well developed or organized

6. They are not able to handle new situations

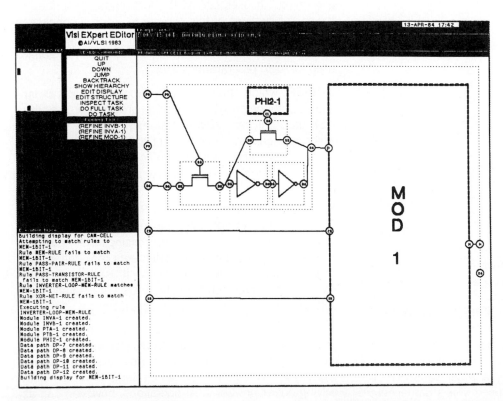

Fig. 9-8 Expert system graphics interface. *(Tom M. Mitchell, Rutgers University)*

TYPES OF AI SOFTWARE

AI systems are mostly software—not hardware. Although we will see in Sec. 9-5 that advanced approaches to computer hardware are also required to make AI systems practical and efficient, the key to their capabilities and performance is in advanced software. AI software is made up of high-level languages and programming tools which are designed to focus on the problem-solving tasks and relieve the burden of coding and translation in lower-level languages. There are typically five different types of software in an AI system:

1. System software. This is the internal operating system software which manages the computer resources, controls operations, and provides some utility functions. It is the lowest level of software and must ultimately translate all instructions into machine language. In specially designed AI computers this software may be unique, but many small systems operate under standard operating system environments such as UNIX.

2. Problem-solving software. This is the software used by the inference engine to perform the actual AI logic functions. It uses a special logic programming language which is focused on problem solving rather than procedures. This is called the "kernel" language of the system and is the lowest level that a programmer would normally have to deal with. Unlike the machine and assembly languages of normal computers, these kernel languages are very high-level languages.

3. Knowledge base software. This is the software used to manage the knowledge database. It uses a high-level inquiry language to provide access to the relational database.

4. Intelligent interface software. This is the software that provides the interface to the end user or application. It may include natural language, speech, vision, or graphics.

5. Development software. This includes the software tools for the system programmer, application, developer, or knowledge engineer. A variety of intelligent programs are being used for these tasks to reduce the time and effort involved in software development. Tools may include editors, debuggers, compilers, and automatic programming techniques.

The heart of the AI system software, of course, is the problem-solving inference engine. Rather than develop new software for each application, many systems use software "shells," which provide all the intelligent logic functions without the knowledge base. In most cases these shells were derived from expert systems that had previously been developed for unique applications, but were later found to be useful for a wider variety of AI tasks. Some of the most widely used shells came from early expert systems that were developed in the late 1960s and early 1970s for medical and scientific diagnosis applications. These were then stripped of their unique knowledge databases, leaving the logic software that can be used to interpret and draw conclusions from knowledge. Such shells are designed to be transportable for use on many different computers, including micros and minis. Their capabilities, however, are limited and may not be adequate for large, complex, and unique applications.

AI SOFTWARE LANGUAGES

In AI systems, knowledge is represented in a symbolic form which allows it to be stored and processed efficiently by the inference engine. The languages used are therefore often referred to as "knowledge representation languages" (KRLs). This representation of knowledge information usually takes two forms:

1. Objects. An object is a group of attributes that describes something. Objects can also be associated with other objects by using symbolic references or links in memory.

2. Rules. A rule is a collection of statements or facts that defines relationships between objects.

AI languages are nonprocedural in nature; they focus more on what task has to be done than on the step-by-step process of how to do it. These languages are sometimes referred to as "object-oriented," since the data they deal with is organized into objects which can be addressed individually or related to one another. Programming in the kernel languages of AI is called "logic programming." It uses one of the techniques described in the earlier sections of this chapter to reach a conclusion by matching the objects and rules with the goals or objectives of the problem. The two most common languages used for logic programming are:

1. LISP, which stands for LISt Processing. In this language, both the programs and data are structured as lists. LISP uses only one kind of statement, a "function call," which can call on another function or repeat itself. This permits LISP to be used for an efficient form of "functional programming" and recursive problem-solving techniques.

2. PROLOG, which stands for PROgramming in LOGic. This derivative of the LISP language is based on the symbolic logic of inference processing. The language consists only of declaration statements that define relationships between objects and data. It uses the IF-THEN rules and backward chaining technique of resolution logic to draw inferences and reach conclusions.

Versions of these and other languages are sometimes used in AI systems. METAPROLOG, for example, operates in a PROLOG environment to provide a higher level of logic, which directs the overall strategy of the problem-solving operation. Some AI operations, however, can actually be done more efficiently with procedural languages.

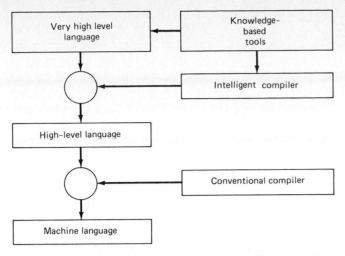

Fig. 9-9 The use of an intelligent compiler in software development. *(Max Schindler, "Expert Systems," reprinted with permission from Electronic Design, vol. 33 no. 1; © Hayden Publishing Co., Inc. 1985)*

Those operations might be handled as subroutines in a program using a compatible, high-level language such as PASCAL.

SOFTWARE DEVELOPMENT

Developing software for AI systems requires the same types of programming tools used in conventional programming. These include text editors, debugging programs, graphics aids, and program monitors like those you might find in any programming environment. To make the job of developing AI software simpler and more efficient, however, special programming tools and languages have been developed. These use very high level languages to define the program logic, which, in turn, can be efficiently translated into a common language (e.g., LISP or PROLOG) for lower-level operating system functions. This is accomplished by an intelligent compiler and a partially automated programming process which performs the coding in the common language and compiles it into machine language (Fig. 9-9).

The most time-consuming and difficult part of developing AI systems is the acquisition and coding of knowledge information. By using a software shell, a programmer or knowledge engineer can create the knowledge base for the particular applicaton involved. The use of the special logic definition languages and software development tools can simplify this process and minimize the effort required to implement an AI application. Windowing graphics packages, in particular, have proved to be very useful in establishing and maintaining a knowledge base as well as in debugging the problem-solving logic (Fig. 9-10).

9-5 COMPUTER HARDWARE FOR AI SYSTEMS

For the most efficient handling of AI applications, special computers are used which are designed for logic programming and parallel processing. The architecture of these machines and even the logic technologies used in them may be designed around a particular language structure (e.g., LISP machines and chips). Some of the advanced computer concepts discussed in Chaps. 6 and 7 are especially well suited for AI systems. The non–Von Neumann parallel processing architectures, for example, are very efficient for managing relational databases and performing logic functions with large amounts of data. Since AI programs use very high level languages with limited functions, RISC architecture is also a useful approach for improving operational efficiency. Unlike con-

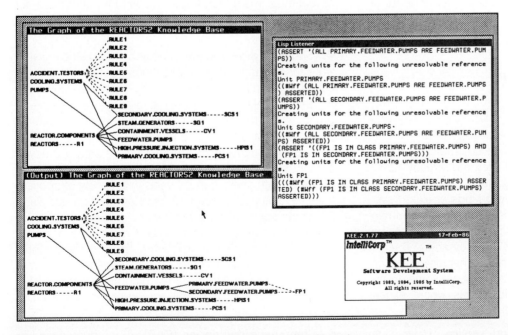

Fig. 9-10 System development with a software shell. *(Intellicorp)*

ventional computers whose performance is measured in terms of millions of instructions per second (MIPS), AI computers are measured in terms of logic inferences per second (LIPS).

The characteristics required of these AI computers include the ability to:

1. Perform very high speed inference operations
2. Utilize parallelism in the structure of the hardware and software
3. Support very large databases of knowledge
4. Perform very fast associative retrievals
5. Use a natural language interface with humans

To provide all these capabilities, AI systems are usually comprised of three main subsystems:

1. The relational database, which provides the knowledge needed to solve problems
2. The inference engine, which uses the knowledge and logic processing to solve problems
3. The interfaces between the user and the machine, which provide for interaction during programming and problem solving

There are many approaches to the design of such machines. One is illustrated in Fig. 9-11. The machine is made up of a number of processors with different specialized functions to optimize system performance and efficiency. Some systems use special auxiliary processors to handle incoming data and prepare it for processing by the AI system. This is particularly useful in real-time process control applications. Hierarchical memories and I/O control processors are also used to handle the transfer of large amounts of data. Today, to provide all those functions in a high-performance machine requires a relatively large and expensive computer system. In the future, advances in technology should make small but powerful AI machines practical.

9-6 AI APPLICATIONS IN MANUFACTURING

EXPERT SYSTEMS

Although knowledge-based expert systems are still in their infancy and were first applied commercially in the fields of medicine and science, they are already being used widely in the manufacturing industry. At the current stage of development, two different kinds of expert systems are being used. Some actually perform like human experts to solve complex problems. Others, however, are designed to perform simpler tasks. They actually serve not as substitutes for experts, but rather as reference books, manuals, or instructors. Following are some examples of manufacturing application areas where expert systems have been implemented sucessfully:

1. **Equipment maintenance and service.** Diagnosing and repairing problems in the operation of complex equipment requires a great deal of skill and experience. The engineers and technicians who service such equipment must be highly trained and continously kept up to date to perform effectively. As the amount and variety of equipment increase and equipment design becomes more complex, it becomes more difficult to obtain and train maintenance and service personnel. Expert systems have the potential to simplify the job of troubleshooting problems by drawing upon the knowledge of the equipment designers as well as the experience of expert maintenance personnel. This approach can be applied to a wide variety of equipment which is either used as an end product by a customer in the field or installed in a manufacturing operation. A typical example is troubleshooting electronic instruments (illustrated in Fig. 9-12).

2. **Production ordering and scheduling.** The task of interpreting customer orders into detailed schedules for production parts can involve a lot of data and many transactions. In most companies that manufacture many complex products, this is a time-consuming and error-prone job. Expert systems can use the knowledge of bills of materials and ordering rules to generate production schedules efficiently and accurately. One of the earliest applications of this type was successfully used to configure customer orders for computer equipment. This approach could also be applied to the traditional production control job of scheduling and tracking a product through a manufacturing line.

3. **Modeling and simulation.** Computers can be used to avoid mistakes in the layout and scheduling of production lines. By modeling the line and simulating the flow of product through the manufacturing process, potential problems can be detected and alternative schemes tried without having to actually run a real production line. Expert systems with graphics aids can be used to design or optimize manufacturing lines without the need for professionals in simulation or production engineering (Fig. 9-13).

4. **Software development.** The job of developing software for manufacturing applications is often more time-consuming and expensive than the hardware involved. Automating either the handling of physical objects in production or the data that controls the operation requires a great deal of software. To minimize the time and effort involved in programming, more efficient software development tools must be provided. As we have already seen, expert-system shells can be used in a variety of applications, avoiding the need to develop system software. Automated programming techniques using very high level definition languages and intelligent compilers can also reduce and simplify the programming effort significantly. In addition, AI-based compilers can generate more efficient assembly code than standard compilers, resulting in greater productivity from the computer resources being used.

5. **Computer-aided design (CAD).** As products become

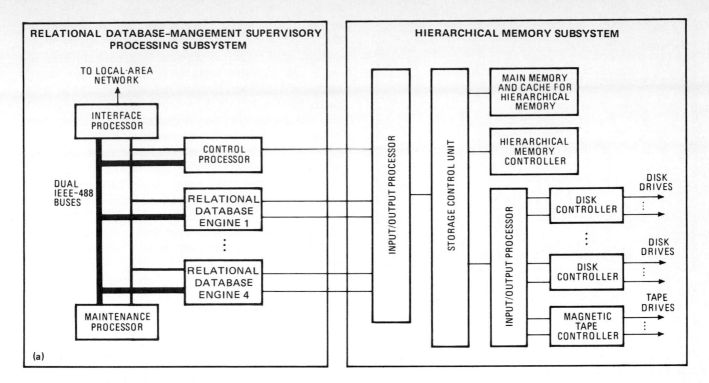

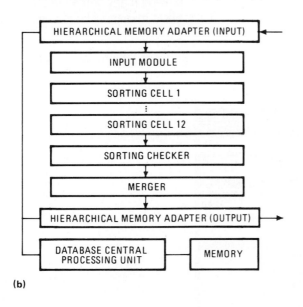
more complex, the design process must be automated to minimize the time, cost, and errors involved. Expert systems offer the potential to provide additional tools to the designer beyond the highly structured design systems that existed previously. The use of logic programming and knowledge-based rules can add automated reasoning and deduction to a complex design problem. This could permit designers to interact with the design process, increase flexibility in design constraints, try out alternatives, and optimize the design.

6. Test. The testing operations in manufacturing become larger and more complicated as the product complexity increases. To minimize the time and effort involved, and in some cases to even make testing feasible, automated, computer-controlled test systems have been developed. There are opportunities, however, to improve the efficiency of this process even further by using AI techniques. Expert systems, using a knowledge base of design and test information, can improve several stages of the testing process, by such methods as:

☐ Providing design aids to ensure testability

☐ Automating the generation of test programs

☐ Building self-testing functions into products

☐ Providing intelligent aids for diagnosing equipment malfunctions

7. Computer-aided instruction (CAI). Training skilled workers is a major activity for most large manufacturing companies since the products and processes are so complex and change so frequently. For many years comput-

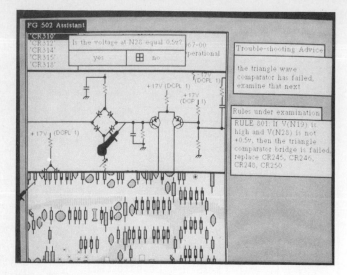

Fig. 9-12 Expert system as an intelligent assistant. *(Tektronix Inc.)*

require sophisticated controls to maintain production throughput and product quality. Large manufacturing operations, particularly in process-intensive industries, already use computer systems to control the production equipment. Skilled operators are still required in most cases, however, to run complicated tools. Expert systems offer the capability to aid operators and production management in monitoring the operation of processes and equipment to detect variations and potential problems. An intelligent control system can collect thousands of real-time process measurements and can use a knowledge base of rules and expert operator experience to advise or alert human operators (Fig. 9-14). Such systems will make it easier for operators to control a process and in some cases may even make a very complex and sensitive process practical.

ers have been used to supplement classroom instruction with structured tutorials and testing aids. Expert systems, however, have the ability to significantly enhance the capabilities of CAI. In additon to providing a more comprehensive source of knowledge, expert systems offer the ability to simulate and practice learning situations as well as a natural language interface for ease of use.

8. Process control. Complex manufacturing processes

DATABASE MANAGEMENT

Large manufacturing operations are driven by large amounts of data. To run a factory efficiently requires that the data be readily accessible to those who need it to perform their jobs. AI techniques, and relational databases in particular, can provide many of the features and functions that are desirable in managing large databases. These include:

1. Natural language interfaces which permit access by nontechnical users

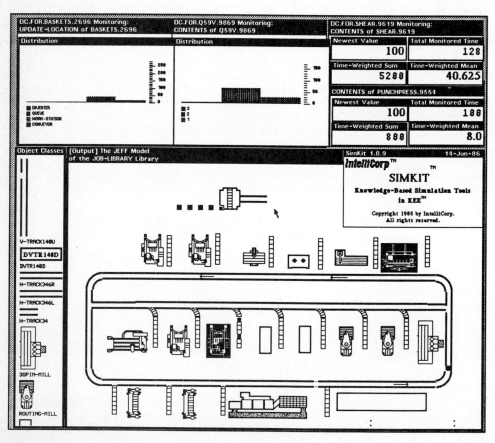

Fig. 9-13 Expert system as a line simulator. *(Intellicorp)*

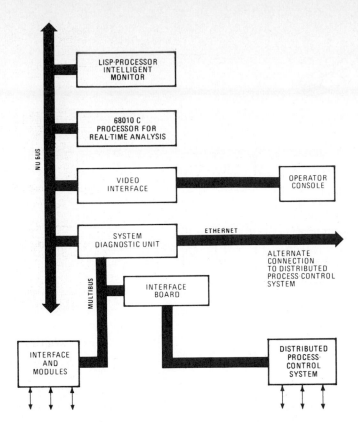

Fig. 9-14 An intelligent process control system which can monitor many points in the process and reprogram its monitoring patterns. *(J. Robert Lineback, "LISP Machine Provides a Shell for Industrial AI Applications in One of the First Expert Systems to Go to Work," reprinted from Electronics Week, Aug. 27, 1984; Copyright © 1984, McGraw-Hill Inc., all rights reserved)*

2. The ability to support frequent changes in the database and in user needs

3. Expert aids for designing and building a database application

4. Efficient management of computer resources in the acquisition, preparation, manipulation, and storage of manufacturing data

There are many common applications for database management systems in manufacturing today which should only expand in the future as intelligent systems become more readily available. Some applications are:

Production schedules and status

Process descriptions and routings

Technical references (e.g., manuals and procedures)

Management reports

VISION SYSTEMS

The use of AI technology has made machine vision a practical tool for manufacturing. AI computers and software can be combined with television cameras and other optical sensors to permit machines to perform tasks that would otherwise require human skills (Fig. 9-15). Vision systems analyze and interpret 2D camera images to infer 3D objects. This process consists of several steps:

1. Image formation. The camera's image of an object is broken down electronically into a matrix of very small picture elements ("pixels"). Each pixel is graded and coded by its light intensity and position to form an electronic "image" in the computer memory. Some systems are more sensitive and precise in their formation of images because they store in memory more data about the number of picture elements or shades of light intensity.

2. Image analysis. The stored image is then described by the computer in terms that can be related and compared to known objects. Several techniques can be used to accomplish this. The most common is "edge detection," which segments the image into regions and defines lines to represent the edges of the object.

3. Image interpretation. The object which has been described in outline form in the computer's memory is compared to the symbolic models of other objects in the knowledge base. Several approaches can be used for this step. Objects are usually recognized either by their form in comparison to a template image or by their features (e.g., area, length).

Most vision systems today have limited capabilities and perform relatively simple tasks. Objects are usually restricted in their variety, prevented from overlapping or intermingling, and handled in fixed positions. Although such restrictions often make the application practical and economical, they do not permit the flexibility that would be required in most manufacturing operations. To provide that flexibility requires the acquisition and processing of large amounts of data. Some systems, for example, use multiple cameras or range-finding instruments to add depth perception. Different techniques are also used for increasing the detail of image detection, from the simple black or white to shades of gray or 3D shapes.

A vision system should be tailored to the needs of the application. If the detection and interpretation task is simple, then a sophisticated system is obviously not required. The most common applications for machine vision in manufacturing today include:

1. Inspection tasks which have been automated to check the quality of products against their requirements. This job can often be done more efficiently and accurately by computer than by human visual inspection.

2. The identification and sorting of parts used in production, which also lends itself to automation. How the parts are stored and handled will determine the level of sophistication required of the vision system used.

3. Robot guidance systems, which have become the most popular application for machine vision. With vision capability, robots can be much more flexible and can handle more complicated and precise tasks than robots requiring fixed programs. If a robot is given the ability to search for and detect objects, it can be freed of many of

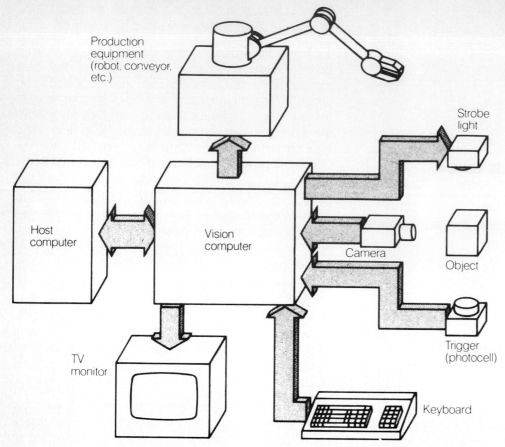

Fig. 9-15 A typical vision system: the photocell alerts the vision computer when an object enters the camera's field of view. The strobe light enables the system to "freeze" rapidly moving objects for analysis. The keyboard and video monitor are used primarily for system programming and debugging. *(Paul Kinnucan, "Machines That See," reprinted with permission, High Technology Magazine, April 1983; Copyright © 1983 by High Technology Publishing Corp., 38 Commercial Wharf, Boston, MA 02110)*

the traditional restrictions found in manufacturing applications, including special fixturing and programming.

9-7 TRENDS IN ARTIFICIAL INTELLIGENCE

AI technology is still in its early stages of development and application. As computer hardware and software technologies advance, many of their new capabilities can be applied to improving the performance and lowering the cost of AI systems. Manufacturing is already a major user of AI techniques, but AI has the potential to significantly expand the scope of its use in the future. Following are highlights of some of the key trends in AI which can influence its application in manufacturing.

EXPERT SYSTEMS

Knowledge-based expert systems are just beginning to find successful commercial applications. As the basic technology advances, the scope of expert systems, capabilties and uses will expand dramatically (Fig. 9-16). The following will be some of the key contributors to these expanded capabilities:

1. "Deep knowledge" systems will be able to reason on the basis of the fundamental laws of nature and the basic scientific theories that are used by experts. This will permit them to improve their own performance and deal with new situations, just as human experts do to obtain their experience.

2. Expert systems will become more efficient in acquiring their knowledge. Techniques will be developed to obtain knowledge directly from human experts and other expert sources without the aid of knowledge engineers. This may be accomplished by natural language interviewing and reading.

3. The scope of knowledge used by expert systems will be expanded beyond the specific applications involved in order to deal more effectively with complex and changing problems. This will include "world knowledge" of the general subject matter.

4. Expert systems will also expand their knowledge bases by learning from their own experience. They will not only be able to use the rules and facts supplied to them by others, but will also deduce new information to guide their reasoning on the basis of results of their own problem-solving experiences.

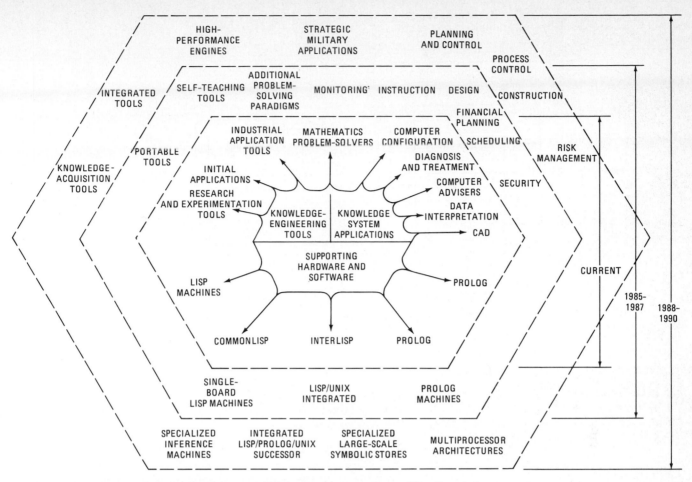

Fig. 9-16 Commercialization of knowledge systems. *(Frederick Hayes-Roth, "The Knowledge-Based Expert System: A Tutorial," Computer, Oct. 1984; © 1984 IEEE)*

5. Beyond just applying rules to facts, expert systems will deal with more abstract reasoning concepts, such as analogies and metaphors. This will permit them to handle problems that are less structured and have more subtle implications.

6. The architecture of expert systems will expand to include additional functions (Fig. 9-17), such as:

☐ Learning subsystems which permit the expert system to acquire knowledge automatically by learning from its own experience

☐ Conceptual networks (taxonomies) which allow the system to make decisions based on the meaning of words

☐ Behavioral models which use simulations to predict the behavior of other systems

☐ Direct access to external data sources from sensors or other systems

HARDWARE

One of the key elements required to make AI systems more powerful and economical is advances in computer hardware technology. Some of the developments that can be expected include:

1. Highly parallel processors which can handle large amounts of data and pursue multiple paths to solve problems

2. Highly efficient machines dedicated to specific functions, such as relational databases or logic programming

3. Powerful personal computers for AI applications, which will include 32-bit processors; high resolution, bit-mapped displays; windowing; mouse pointers; and natural language interfaces

SOFTWARE

The software used by AI systems is perhaps even more important than the hardware. Advances will be required in the following areas:

1. Intelligent software development tools which permit automatic translation of a problem statement into a machine language program

2. Natural language interfaces that are transportable, to permit the system to use databases that exist on other machines

3. Reduced costs for the development of AI systems by using intelligent software development tools and flexible AI shells

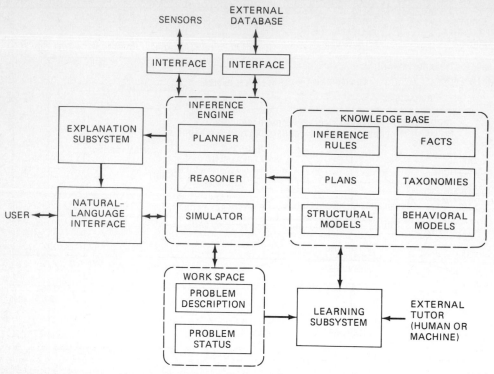

Fig. 9-17 Architecture of future expert systems—arrows indicate flow of knowledge during operation. *(Paul Kinnucan, "Computers That Think Like Experts," High Technology Magazine, Jan. 1984; Copyright © 1984 by High Technology Publishing Corp., 38 Commercial Wharf, Boston, MA 02110)*

APPLICATIONS

Once the technologies have been developed, the challenge remains to use them productively. Some of the most important trends in the application of AI in manufacturing are:

1. The programming of machine vision systems directly from computer-aided design systems.

2. Embedding intelligent processors in the tools used in the manufacturing process to control their operation.

3. The integration and automation of product design and manufacturing process planning. As illustrated in Fig. 9-18, the decision processing capability of expert systems can be used to tie the geometric modeling of design systems

to the automatic generation of NC data which controls tools on the factory floor.

These trends offer the potential to eliminate many of the inefficiencies and costs involved in traditional manufacturing operations. Some of the more obvious benefits include:

Less paperwork

Less time from product development to manufacturing

Fewer errors

Smaller losses in the process

Shorter production cycle time

Less labor required

Less management and administrative support required

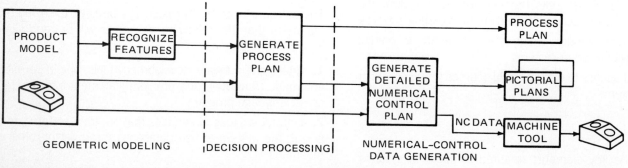

Fig. 9-18 Integrated product design and process planning system. *(J. Robert Lineback, "AI Transforms CAD/CAM to SIM," reprinted from Electronics Week, Dec. 17, 1984; Copyright © 1984, McGraw-Hill Inc., all rights reserved)*

9-8 SUMMARY

Artificial intelligence (AI) is the application of computer systems to the performance of functions that would normally require some properties of human intelligence. AI evolved from advanced concepts in hardware and software that permit computers to simulate human reasoning and sensing. AI systems are made up of intelligent hardware, software, and interfaces. They are used for expert systems, for database managers, for artificial sensors (such as vision systems), and for guiding robots.

Expert systems use rules and facts from a knowledge base to reach conclusions in solving problems. They can serve as a substitute for scarce or expensive skills, or they can provide intelligent aids to complex tasks. Intelligent database systems provide easy access to large amounts of data using natural language interfaces and relational processing techniques. Sensory systems, such as machine vision, use AI hardware and software to interpret and translate signals from external sources. Robotics combines the technologies of mechanical manipulator systems with computer control and AI sensors to provide flexible tools for manufacturing.

AI systems use nonprocedural languages that deal with objects and their relationships rather than with quantities and computations. They need special computers whose designs are based on parallel processing and logic programming. AI is beginning to be used for a variety of manufacturing applications. As computer hardware and software technology advances, the scope and complexity of AI applications will increase.

REVIEW QUESTIONS

The answer to each question can be found in the section(s) indicated at the end of the question.

1. What is artificial intelligence and how does it differ from conventional computer systems? [9-2]

2. Why do we need AI and how is it used in industrial applications? [9-2]

3. Describe the major elements of an AI system. [9-2]

4. What is an expert system and how can it be used? [9-2 and 9-3]

5. Describe how an expert system works. [9-3]

6. What is a knowledge base and how is it created? [9-3]

7. Identify some advantages and disadvantages of using expert systems. [9-3]

8. What types of software are used in AI systems? [9-4]

9. Explain how AI languages differ from conventional computer languages. [9-4]

10. Describe some of the characteristics of the computer hardware required for AI systems. [9-5]

11. Identify some examples of applications for AI systems in manufacturing. [9-6]

12. Describe the basic elements of a vision system and how it works. [9-6]

13. Identify some of the advances in computer hardware and software that will influence the capabilities and applications of AI systems in the future. [9-7]

PART THREE

COMPUTER-AUTOMATED ENGINEERING

CHAPTER 10
COMPUTER GRAPHICS TECHNOLOGY

CHAPTER 11
COMPUTER-AUTOMATED DESIGN

CHAPTER 12
COMPUTER TOOLS FOR ENGINEERING ANALYSIS

In computer-automated manufacturing (CAM), the manufacturing process starts with the design of the product. For a product to be manufacturable and economical in an automated environment, it must be thoroughly engineered during the design process. Computers have proved to be an invaluable engineering tool for designing products and developing manufacturing processes. Part 3 deals with the technologies involved and how they are applied to engineering tasks. The scope of this subject is often referred to as computer-aided engineering. However, because of the rate at which the technology in this area is advancing, and because the emphasis of this book is on computer automation, we choose to refer to it as *computer-automated engineering (CAE)*.

Chapter 10 addresses computer graphics technology, which includes the hardware and software that provide the engineering functions. Graphics is a very important and rapidly developing field within computer technology that is being used in a wide variety of applications both in and out of manufacturing. Chapter 11 covers the use of the computer as a graphics design tool in computer-aided design (CAD). CAD systems are used to design mechanical and electrical parts, assemblies, and entire products. Chapter 12 deals with the use of the computer for engineering analysis. Included here are tasks such as the analysis of stresses and motion, which permits a designer to evaluate the physical properties and behavior of objects without actually having to build them or test them. Such engineering tasks usually involve a great deal of data handling and computation as well as graphical representations. These can be automated through the use of advanced computer graphics technology.

Part 3 will attempt to familiarize the reader with the basics of the computer technologies and applications involved. It will describe how graphics systems work and how they can be used for common engineering tasks that are usually involved in the design process. The emphasis will be on what can be done with graphics systems in a manuafacturing environment rather than on providing a detailed tutorial on how to operate specific types of hardware and software.

COMPUTER GRAPHICS TECHNOLOGY

10-1 INTRODUCTION

Computer graphics technology includes the hardware and software involved in generating graphic representations of objects and data on a display device. When used for a graphics application, the computer provides the user with several different functions. It performs computations, stores data, and serves as a design tool. The basic technologies involved are within the scope of the computer hardware and software covered in Part 2, but they are specialized for graphics functions.

Although some of the early computers used cathode-ray tube (CRT) devices to display data, it was not until the mid-1960s that systems were developed which could generate graphics displays interactively with the user. This "interactive computer graphics" technology has evolved since then as a powerful tool for design and engineering tasks. Initially, these systems only had the capability to make relatively simple black-and-white line drawings using a large central computer. Today, individual engineering workstations can generate full-color 3D images that can be animated, changed, and mathematically analyzed. Computer graphics has become an essential tool for manufacturing. It is used to develop the initial design concepts for products, perform engineering analyses, generate formal product design documentation, simulate manufacturing operations, and provide manufacturing process data.

This chapter will cover the major elements of the technology that makes such functions possible. It should provide a basic understanding of the hardware and software involved and how they work in graphics applications. One chapter, however, is not enough to teach the detailed inner workings of graphics hardware or the specific techniques of graphics programming and system operation. It should be adequate, however, to help the reader understand what can be done and what is necessary to provide the design and engineering functions that will be addressed in Chaps. 11 and 12. The chapter is organized such that it first introduces the reader to the basics of the technology and then it covers in more detail the specifics of the hardware and software involved. Since 3D graphics has become a very important function

for this technology, Sec. 10-5 is devoted to describing the techniques used for geometric modeling. Next, the scope of applications that one may find for computer graphics in manufacturing is presented. Finally, some trends are addressed which should influence the capabilities and applications for the technology in the future.

10-2 BASICS OF COMPUTER GRAPHICS

WHAT IS A COMPUTER GRAPHICS SYSTEM?

Computer graphics systems have become the principal user interface to computer-aided engineering (CAE) systems which drive automated design and manufacturing processes (Fig. 10-1). The hardware and software of a graphics system permit the user to interact with the computer that maintains the central database which serves both the computer-aided design (CAD) and computer-automated manufacturing (CAM) operations. The graphics system is the basic tool that is used to create, analyze, and change designs and processes; it replaces the drawing board and slide rule of the past.

A computer graphics system usually takes the form of a terminal or workstation like the one illustrated in Fig. 10-2. Although the most prominent component of the system is the display, a basic graphics system is made up of four major hardware elements (Fig. 10-3):

1. Processors. Every graphics system must have a computer to support it for computational and storage tasks. This computer may be either central or local. It must also have a special processor, usually built into the workstation itself, to perform unique graphics functions.

2. Display. This is typically a CRT device which can display both graphics and alphanumeric data. Such displays may have a wide variety of features and capabilities for graphics applications.

3. Input devices. The most common type of input device, found on almost all graphics systems, is a keyboard. However, there are also many other special devices which aid the user's interaction with a graphics system.

4. Output devices. Hard copy of what is displayed on a

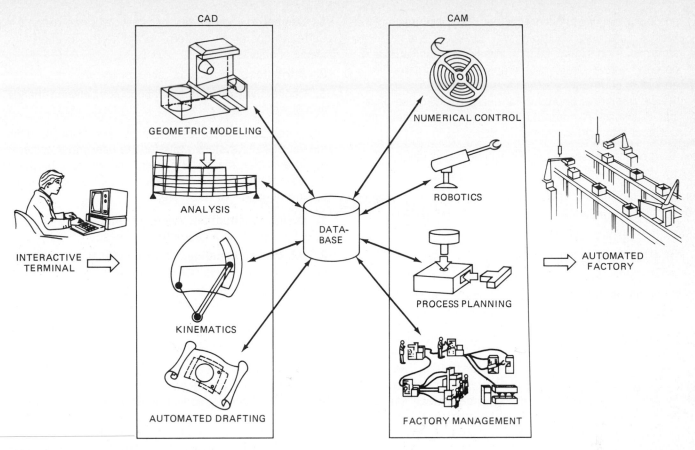

CAD

GEOMETRIC MODELING

ANALYSIS

KINEMATICS

AUTOMATED DRAFTING

INTERACTIVE TERMINAL

DATA-BASE

CAM

NUMERICAL CONTROL

ROBOTICS

PROCESS PLANNING

FACTORY MANAGEMENT

AUTOMATED FACTORY

Fig. 10-1 Computer-aided engineering system. (*reprinted from John K. Krouse, What Every Engineer Should Know About Computer-Aided Design and Computer-Aided Manufacturing, by courtesy of Marcel Dekker Inc., New York, 1982*)

graphics terminal can be provided by a variety of different kinds of printers and plotters which are specially designed to handle graphics.

A computer graphics system is interactive. This means that the user works together with the hardware and software to create, analyze, or change the graphics data. The user provides input, and the computer processes the data and displays the output. A computer graphics system optimizes the roles of the computer and the user. The user is the designer or engineer who thinks, develops concepts, and makes decisions. The computer assists the user in nonconceptual tasks where it can be much more efficient than humans: in computation, storage, and displaying of information.

WHY DO WE HAVE COMPUTER GRAPHICS SYSTEMS?

Graphics is an efficient means of communicating large amounts of complex data and the results of computations to a user. Pictures contain more information than alphanumeric characters. We can see a lot more in a picture of

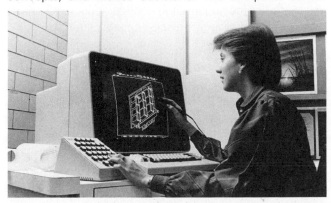

Fig. 10-2 Computer graphics workstation. (*International Business Machines Corp.*)

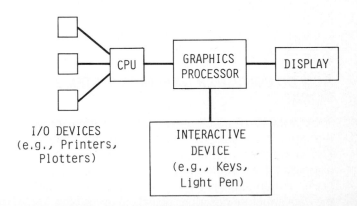

I/O DEVICES (e.g., Printers, Plotters)

CPU

GRAPHICS PROCESSOR

DISPLAY

INTERACTIVE DEVICE (e.g., Keys, Light Pen)

Fig. 10-3 Basic computer graphics system.

113

an object than we can learn by reading a description of it. Pictures are also a simpler way to organize data in a summary form so that it can be easily communicated and comprehended. Graphics is simply a more human-oriented way to communicate than text.

The traditional approach for designing is to use physical drawings. A computer graphics system offers the designer many advantages:

1. It can increase the productivity of the designer
2. It is more efficient than using paper
3. Designs can be made faster
4. Designs are easier to change
5. The results have better quality
6. It can provide different views
7. It offers added dimensions such as color and depth
8. It makes drawings easier to store and retrieve

A graphics system takes advantage of the characteristics and strengths of the computer to enhance the capabilities of the user. Its speed and accuracy relieves the user of routine, repetitive, and complex tasks. The computational power of a computer increases the logical and analytical ability of the user. Its large storage capacity serves as an extension of the user's memory.

Computer graphics systems are particularly well suited to help us respond to the dynamic nature of the manufacturing environment. A graphics system can give us easy access to a large amount of information stored in a central database. That information can be complex and can change continually. Most manufacturing operations are geared for fast turnaround, which requires efficient communication and rapid feedback for control.

In addition to the needs and benefits for the manufacturing user, we have computer graphics systems because the technology has progressed to the point where it can provide the functions and features required. The performance of data processing systems in general has increased substantially in terms of speed and capacity. At the same time, the cost of computation and storage has been reduced significantly. Advances in computer hardware have made small, powerful workstations feasible. Complex software has also been developed to permit the efficient generation of high-resolution pictures. It is the result of all these factors that has made computer graphics systems a useful and practical tool in today's manufacturing environment.

HOW DO COMPUTER GRAPHICS SYSTEMS WORK?

The hardware of a computer graphics system provides the input, computation, storage, and output functions. The software includes an operating system (which is usually called the "graphics system"), an application program, and an application database. In graphics systems the data structure describes objects in terms of:

1. **Geometric coordinates** (typically in terms of the XY Cartesian coordinate system) to define the location and shape of objects
2. **Attributes** such as color, style, or texture
3. **Connectivity,** to identify relationships between shapes and objects

An interactive graphics system is "event-driven." This means that the system waits for the user input or command and then reacts. The interactive process between the user and the system in creating designs can typically be described in four basic steps:

1. **Design construction,** where a computer model of the object is created in a geometric database
2. **Information handling,** where computational tasks are performed (e.g., measurements, dimensioning, or views)
3. **Modification,** where changes, updates, transformations, or moves are made to the design
4. **Analysis,** including engineering analyses for stress, motion, and physical properties

To build a graphics model, data is first entered into the system by using a variety of input devices, such as a standard keyboard, special function keys, a light pen, or a cursor (e.g., a mouse). The computer processes this data by using analytical geometry to represent shapes in terms of mathematical expressions. These computation and storage tasks are performed by the central processing unit (CPU). The graphics processor, however, is a special set of hardware and software which is used to create and display the geometric model of the object. It converts the digital signals from the CPU into graphics commands for function generators that produce standard shapes (such as circles, lines, and symbols). These function generators send signals to the display device (such as a CRT) which writes the shapes on the screen. Hard copy of the design displayed can be obtained from output devices such as graphics printers and plotters.

TYPES OF GRAPHICS SYSTEMS

There are a number of different types of computer graphics systems, differentiated in terms of the functions they can perform, the content and configuration of their hardware, and their graphics display capabilities.

1. **Functions or applications.** There are basically three types, as follows, in order of increasing function:

☐ **Automated drafting systems.** This was the earliest application for computer graphics. It was derived from numerical control (NC) technology, which was used for plotting and drafting machines. Typical applications include the automation of mechanical, electrical, and architectural drawings.

☐ **Design systems.** These are the basic computer-aided design systems used to create, change, and store complex designs. This can also include mechanical, electrical, and architectural design systems.

☐ **Engineering systems.** This is the more powerful class of computer-aided engineering systems, which include engineering analysis, modeling, and simulation capabilities, in addition to the basic design functions.

2. Hardware configurations. There are three basic types:

☐ A terminal connected to a central mainframe computer

☐ A graphics controller or subsystem supporting a terminal

☐ A stand-alone workstation supported by its own minicomputer

3. Graphics display capabilities. These capabilities vary in complexity to achieve more realism in terms of:

☐ **Dimensions.** 2D shows a flat image; 21/2D includes some depth perspective, as in isometric drawings; and 3D creates volume models of objects.

☐ **Color.** Displays may be either monochrome (e.g., black and white) or color (with varying numbers of shades available).

☐ **Resolution.** The ability to display fine detail.

HOW COMPUTER GRAPHICS SYSTEMS ARE USED

Computer graphics systems can perform many useful functions, including:

1. Replacing paper as the medium used for drawing
2. Presenting large amounts of information rapidly
3. Communicating visually instead of with text
4. Enhancing the interpretation of data
5. Displaying the results of computations as they occur
6. Making and observing changes
7. Simulating and verifying processes
8. Representing theoretical results graphically

These functions can be applied to many activities throughout the manufacturing process, from the initial design of the product to the control of the tools and processes which produce it (Fig. 10-4). Typical graphics applications include:

1. Creating conceptual designs
2. Generating drawings and documentation
3. Modeling designs of products and processes
4. Changing or modifying designs
5. Conducting engineering analyses
6. Reviewing and evaluating designs
7. Establishing a manufacturing database
8. Classifying and coding parts
9. Planning processes

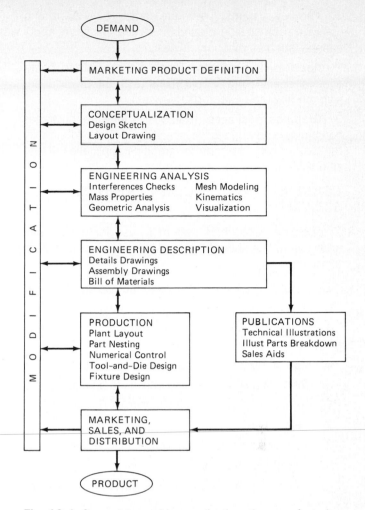

Fig. 10-4 Computer graphics applications in manufacturing. *(reprinted from Daniel L. Ryan, Computer Aided Graphics and Design, by courtesy of Marcel Dekker Inc., New York, 1979)*

10. Controlling tools and processes
11. Generating management reports

Using a graphics system to create or change a design involves several basic operations, including:

1. Drawing an object
2. Changing an object
3. Attaching objects to each other
4. Translating objects (e.g., rotation or zoom)
5. Magnifying or scaling objects
6. Clipping objects (to fit the image to the screen)
7. Animating objects (i.e., making them appear to move)

Drawing an object itself can involve a number of operations for the graphics system, such as creating lines, symbols, letters, and geometric shapes.

To provide these functions, a wide variety of graphics systems have been developed over the years. Some are unique proprietary systems which are used only by the companies that developed them. Others are available commercially for common use. Graphics systems usually take one of three forms:

1. Turnkey systems. These are complete hardware and software packages developed for general classes of applications (such as mechanical design).

2. Application products. These are specialized packages for specific user applications (e.g., the physical design of integrated circuit (IC) devices).

3. Graphics packages. These are software packages containing subroutines for specific functions to simplify the task of programming for graphics applications (e.g., standard function generators).

CONSIDERATIONS IN USING GRAPHICS SYSTEMS

In selecting a graphics system to use for a particular application, one should first consider the requirements in terms of:

1. Resolution (degree of detail in the displayed image)

2. Accuracy and repeatability of computations

3. Response time from the input to the displayed output

4. Rate of interaction (frequency of user inputs)

5. Complexity of images and objects to be displayed

6. Realism (e.g., color, solid images)

7. Ease of use

8. Human factors (e.g., convenience, comfort)

9. Cost (payback)

10. Robustness (i.e., ability to sustain some mistakes by the user or malfunctions in the system)

All systems have some limitations. These should be understood before selecting one for a particular application. The limitations are often considered in terms such as:

1. Size of the output (either the display screen or printer paper)

2. Number of colors or shades available

3. Dimensions of the image (2D, 2½D, 3D)

4. Flicker in the displayed image (due to the frequency with which it is regenerated)

5. System performance (e.g., processing speed, memory size)

6. Portability (i.e., the compatability of the software with different types of hardware)

7. Personnel skills required to use the system

Despite the limitations of any particular graphics system, most offer a substantial number of advantages to the user, including:

1. Reducing errors and improving accuracy in drafting and analysis activities

2. Increasing the rate at which data can be communicated and comprehended (since images are more efficient than text)

3. Increasing analytical capability (i.e., engineering computations and analyses)

4. Increasing the productivity of engineers and designers

5. Increasing the creativity of designers

6. Providing the ability to respond to change easily and rapidly

7. Avoiding the necessity of fabricating and testing prototypes of new designs

8. Creating a database for manufacturing

9. Reducing the cost and time for developing and manufacturing a new product

10. Improving the quality, availability, and cost of providing management information (e.g., business graphics)

10-3 GRAPHICS HARDWARE

DISPLAY TECHNOLOGIES

The most common type of display device used for computer graphics systems is a CRT. Although CRTs look like television monitors, most graphics displays have special electronic circuits for creating the images on the screen from computer data. The capabilities of the display depend on the type of CRT used. Two basic approaches are used to generate images on a CRT (Fig. 10-5):

1. Stroke writing. The image is constructed by drawing line segments (vectors) in much the same way as a human would draw.

2. Raster scan. The image is constructed by scanning the electron beam horizontally across the screen to create an array of dots or picture elements (pixels) just like in a conventional television tube.

Two approaches are also used to maintain the image on the screen once it has been created:

1. Storage tube. This retains the image on the display screen continuously. It does this by using an extremely long-persistence phosphor and a storage grid that is continuously flooded with electrons to maintain the original image drawn by the electron gun.

2. Refresh tube. This regenerates the image at a high frequency to overcome the fading of the original image drawn on the screen. It requires a special display processor and buffer memory to store and regenerate the original image. Although this approach is more complicated, it also provides the capability to easily change and move images during the refresh cycle.

The approaches used to generate and maintain images have been combined to develop three basic types of CRT graphic display devices:

1. Direct view storage tube. This uses a storage tube and the stroke-writing approach. It can create high-reso-

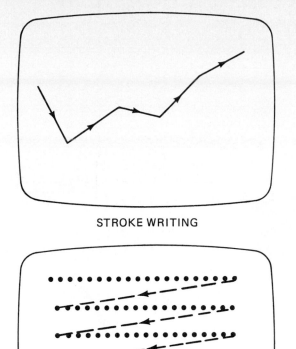

STROKE WRITING

RASTER SCAN

Fig. 10-5 Approaches to generating images on a CRT display.

the fixed number of points (pixels) it generates. Since each pixel must be stored in memory, high resolution requires a large buffer memory. This can range from a low-resolution tube with an array of 256 × 56 (which would require approximately 65K bytes of memory) up to high-resolution tubes of 1024 × 1024 or more (which would require more than 1 million bytes of memory).

Even with relatively large arrays, however, the raster scan approach cannot create perfectly straight lines like stroke writing can. The image is subject to a staircaselike effect, which can be minimized but not eliminated (Fig. 10-6). Color capability can add clarity to complex images by separating features as well as making them appear more realistic (Fig. 10-7). However, this is also more expensive since it requires three electron guns (like in a color television) and additional memory to store the color or shade information for each pixel. Some advanced displays can provide as many as 256 shades or color hues.

No one type of display is perfect. Each has its own advantages and disadvantages, which should be considered when selecting a display for a particular application. For example, for applications that require color, solids, animation, and interactive design, the raster scan CRT has clear advantages over the other two technologies. Some graphics systems even use two different types of displays, such as a storage tube for text or reference drawings and a raster scan for interactive design (Fig. 10-8).

Fig. 10-6 Staircase effect on raster scan displays. *(Richard Fichera, "Rendering Adds Realism to Graphics," reprinted from Electronics Week, Oct. 22, 1984; Copyright © 1984, McGraw-Hill Inc., all rights reserved)*

lution, flicker-free images at a relatively low cost. However, since the entire image must be completely erased and regenerated in order to change it, the direct view storage tube cannot provide dynamic viewing or frequent interaction. Its use of stroke writing also limits its ability to represent solids by filling in areas of the image with lines. In addition, its low brightness normally requires dim lighting.

2. Vector refresh CRT. Also known as the "random scan CRT," this is the oldest type of device used for computer graphics. It employs a refresh tube and stroke writing so that it draws an image on the screen, but then regenerates it at a high frequency to sustain it so that it appears not to flicker (typically between 40 and 60 times per second). This results in a very high-resolution image that can also be very interactive for changing and dynamic viewing. However, it is the most expensive approach and has limited color capability.

3. Raster scan CRT. This is becoming a very popular type of graphics display because of its versatility. It combines raster scanning and refresh technologies to provide capabilities that can include high resolution, color, dynamic viewing, high interaction, and solid images. In its basic form it is like a television without the receiver circuitry. The resolution of a raster scan CRT is limited by

Fig. 10-7 Realistic color graphics display. *(International Business Machines Corp.)*

Fig. 10-8 Dual display system for simultaneous viewing of text and graphics. *(International Business Machines Corp.)*

INPUT DEVICES

Three major categories of devices are used for interactive input to a graphics system. These devices may be used to either create, point to, or move objects.

1. Keyboard devices. This, of course, is the most common type of input device that allows a user to interact with a computer system. Graphics systems, however, use a variety of different kinds of keyboard devices, and a system can often use more than one.

☐ **Standard keyboard.** Incorporates the normal alphanumeric keys for entering nongraphic data, as well as direction keys for moving the cursor.

☐ **Function keys.** Separate keyboard for special graphics commands which are usually identified on an overlay (Fig. 10-9).

☐ **Tablets.** Touch-sensitive panels with blocks for special graphics functions (Fig. 10-10).

2. Picks. These are devices used to point at objects or positions on the screen. The most common is the light

Fig. 10-9 Function keyboard with light pen. *(International Business Machines Corp.)*

Fig. 10-10 Tablet input device. *(Calcomp)*

pen, which is very easy to use. It detects light on the screen by sensing a peak in brightness when the electron beam excites the phosphor on the CRT. This causes the light pen to generate a signal which identifies its location on the screen. It operates only when switched on by pressing it against the screen surface. Since its operation is based on a scanning electron beam, the light pen can only be used with a refreshed CRT (Fig. 10-9). With the proper application programming, light pens can also be used to draw on the screen.

3. Analog devices. This category includes a variety of devices that can be used to measure, point to, or code coordinate positions by converting analog signals (e.g., speed, direction, distance, rotation) to digital signals. They are often used to trace drawings so that the designs can be entered into the graphics system. There are two basic approaches to analog input devices:

☐ **Flat-surface devices.** These are digitizer tables that trace drawings. They can use a number of different techniques, such as electromagnetic, touch sensitive, sonic, incremental, and light detector techniques. Each differs in speed, accuracy, and cost. The electromagnetic digitizer is the most common and precise (Fig. 10-11).

☐ **Positioning tools.** These are devices that the operator can move, which in turn signal the cursor on the screen to move in the same direction and speed. They include the mouse, joystick, trackball, and dial (Fig. 10-12). Their principal advantages are ease of use and speed.

Fig. 10-11 Digitizer tablet input device. *(Calcomp)*

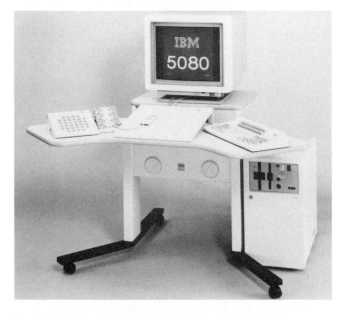

Fig. 10-12 Graphics system with positioning input devices, including mouse and dials. *(International Business Machines Corp.)*

OUTPUT EQUIPMENT

A variety of different types of computer output equipment are used to provide hard copy of the graphics information created on a display. This information may be output when it is created or when it is retrieved from storage. The output may be generated directly from the user's terminal or workstation while it is on-line, or it may be generated off-line from a remote terminal.

There are three basic classes of graphics output equipment, each of which includes several different types. The selection of a particular type of output equipment depends on user requirements, such as speed, quality, size, and cost.

1. Plotters. This is the most common class of output equipment used for graphics applications, particularly for drafting systems. There are two basic types of plotters:

☐ **Pen plotters.** The pen plotter—an outgrowth of NC technology—uses stepping motors to control the motion of the pen relative to the drawing board. It traces the image as an outline drawing stored in the computer. Because of this, pen plotters cannot handle a lot of intricate detail or solid-color areas efficiently. Although they are also relatively slow, they can be inexpensive and can be equipped to use multiple pens for color plotting. Pen plotters are most often used in a "flatbed" form to plot horizontally on single sheets of paper (or another drawing material), just like a drafting board. The drum- or "beltbed"-type plotter can also handle continuous feeding of paper.

☐ **Electrostatic plotters.** These are faster and can handle more intricate designs. An electrostatic charge is used to produce an image on paper, and then a toner is applied to the paper to create the printed image. This is accomplished either by using a matrix of needles to "write" the electrostatic image or by using a photoconductor which is written on by a light from a CRT or laser, as in copying machines (Fig. 10-13).

2. Printers. This is a common class of output equipment used for all types of computer applications which primarily involve alphanumeric data, not graphics. Recent advances in printer technology, however, have made some printers suitable for graphics applications. There are two basic types of printers:

☐ **Impact printers.** Although there are many types used for general purpose computer applications, only dot matrix printers can provide the versatility of printing variable patterns from computer signals. They use a small array of wires which can be individually actuated to print a pattern through a ribbon. These can be relatively inexpensive and even provide color. However, they are also relatively slow and limited in size.

☐ **Nonimpact printers.** These use a variety of new print-

Fig. 10-13 Electrostatic plotter. *(International Business Machines Corp.)*

ing technologies that permit individual picture elements to be printed from computer signals. They can provide fast, high-quality outputs. These include thermal and xerographic printers, which are often used to provide convenient but not necessarily high-quality hard copy during the design process. Perhaps the most versatile type, however, is the ink-jet printer, which can also print in color (Fig. 10-14).

3. Photorecorders. This is a special type of output equipment which uses photographic techniques to record graphic images. They are used when high-density storage or projection are desired. The recorded image size can vary and can include microfilm or standard 35-mm colorslides.

SYSTEM ARCHITECTURE

The configuration of a typical computer graphics system consists of the following basic hardware (Fig. 10-15):

Display terminal or workstation

Input devices (e.g., digitizer, light pen)

Output devices (e.g., printer, plotter)

Processor

A graphics workstation may be an intelligent terminal or minicomputer to provide local computer power to the

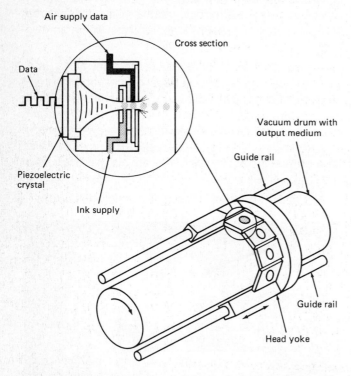

Ink–jet color printer: the digital–image data are transmitted as electronic pulses to the ink jet's piezoelectric crystal, causing it to constrict. Small drops of colored ink are ejected and accelerated by moving air toward the paper as it revolves on a drum beneath the carriage carrying the ink jets. The carriage contains a jet for each color and advances one raster line for each drum revolution so that the full color image is completed with one traverse of the carriage.

Fig. 10-14 Ink-jet printer. (*Harry S. Watkins and John S. Moore, "A Survey of Color Graphics Printing," IEEE Spectrum, July 1984; © 1984 IEEE*)

Fig. 10-15 Computer graphics system. (*International Business Machines Corp.*)

user and fast response time for interactive graphics. Its job is to interface with the processor, translate computer commands into graphics functions, and generate the graphics image. Included in the processing hardware within the workstation may be special graphics functions, such as:

☐ **Function generators.** These are special circuits designed to generate signals to direct the display to produce standard images such as lines or circles. By imbedding these functions in hardware, the amount of data and software required to draw simple images is reduced.

☐ **Graphics processors.** These are microprocessors which help speed up the generation of complex graphics images by sharing the processing load with the main CPU. They may perform some of the special geometric computations directly or convert the data and graphics commands from the CPU into signals that drive the display.

☐ **Buffer memory.** This is a local memory to store graphics data while it is being used by the display. It is more readily accessible and may even be faster than the main memory of the CPU.

The hardware of a computer graphics system is tied together by data buses, much like those discussed in Part 2. Their design depends on the size and complexity of the system involved. The data bus which ties the input and output devices to the processor is sometimes called a "graphics pipeline."

10-4 GRAPHICS SOFTWARE

TYPES OF SOFTWARE

A complete graphics system requires four different types of software (Fig. 10-16):

1. Operating system. As with any computer system, operating system software is required to manage the transfer of data and the computer resources.

2. Graphics system. This is the software that creates images from geometric data. It is internal to the computer system and does not interact directly with the user.

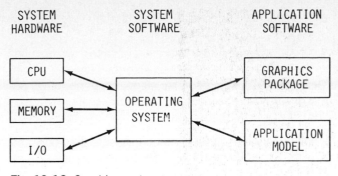

Fig. 10-16 Graphics system.

3. Application program. This is the software that interacts with the user. It converts user input to geometric data and provides commands for the construction and manipulation of the geometric model.

4. Database. As with other data-intensive computer applications, software is required to manage the database in which the geometric information is stored.

To implement a graphics system, each of these types of software may be developed or purchased by the user individually, or as part of a package.

FUNCTIONS OF GRAPHICS SOFTWARE

Some of the major functions performed by the graphics software are:

1. Converting user commands and data inputs into structured geometric data

2. Generating primitive graphic elements from the data, which are used to build a geometric model

3. Converting user editing commands into modifications to the model

4. Describing the geometric model to the graphics system in the form of commands for displaying the image

5. Converting user commands for changing the displayed image into geometric transformations (e.g., scaling, zoom, rotation)

6. Controlling the display (e.g., updating the image, windowing, clipping the image)

7. Providing utility functions (e.g., measurements)

Most of these graphics software functions are accomplished by algorithms that have been written in a programming language that can manipulate graphics data which is in the form of mathematical equations.

GRAPHICS DATA

To build a geometric model requires four basic kinds of data:

1. Geometric coordinates (i.e., location and shape)

2. Attributes (e.g., color)

3. Connectivity relationships

4. Nongeometric relationships (e.g., text)

The input description may be made in several different ways. For example, a circle can be defined by specifying its center point and radius, the endpoints of its diameter, or points on its circumference. The data describing these graphic elements is converted into mathematical equations by using analytical geometry so that they can be processed by a computer. Graphics data is represented in different forms during the process of building and displaying a geometric model. It is first organized into a description of the object in the application program and then structured into a data file that represents the object as instructions for graphics primitives that can be understood by the display controller.

The graphics database created by the application program describes objects in terms of what are called "world coordinates." This means that their location is identified in terms of measurements that relate to the "real world" of the object. To display an image of this object, however, requires that the description be transformed into "device coordinates" compatible with the display system. In addition, in many cases the view selected to be displayed hides portions of the object from the viewer. This requires a mathematical process called "clipping" which eliminates those parts of the image that fall outside the viewing window.

The graphics database also contains the relationships between objects (e.g., intersections, edges, corners). It may also store standard shapes and symbols that may need to be replicated. The database may be organized in a variety of ways to permit the user to optimize the efficiency of access and storage space for a particular application. For example, graphics data can be stored by its coordinates, by graphics elements, or by the total geometry of the object. It may also be organized into a hierarchy of objects which permits the user to break down a complex object into levels of simpler and perhaps replicated parts. This structure lends itself well to the use of a relational database for efficient retrieval, storage, and modification of large amounts of data. It may also be desirable to maintain separate storage for graphics and nongraphics data.

VIEWING THE MODEL

Viewing a 3D model of an object on a 2D surface (i.e., the display screen) requires that the graphics system transform the data into a geometric projection. Depending on the capabilities of the system, this may be a simple parallel projection or a more realistic projection which provides perspective from the user's viewpoint. This process requires the user to specify the volume to be viewed as well as a projection plane and viewport (viewing surface). The objects in the 3D model are then clipped against the specified view volume and projected onto the projection plane. The resulting image is then mapped into the viewport for display. This process involves a series of operations by several different sets of hardware and software in the graphics system: a compiler structures the data, a processor transforms the data, a file stores the data and updates, and a controller drives the display.

GRAPHICS PROGRAMMING

Graphics programs were originally written in assembly language to optimize the efficiency of the processor. Today, however, most graphics systems are programmed in standard high-level languages such as FORTRAN, BASIC, or PASCAL. There are also special graphics command languages for user input which involve three types of commands:

1. Command selection. These are the special graphics operations which are usually selected by using function keys (e.g., creating a line, defining a point, or rotating an element).

2. Positioning. These commands are used to define the location and size of the object being created.

3. Pointing. These commands are used to identify objects which are to be revised or moved.

Graphics programs are often written with logical or virtual input devices. These are the software counterparts to real physical input devices such as a light pen or keyboard. This technique is used to provide software which is independent of the hardware it is used on, and it can simplify the programming job. Some examples of names assigned in programs to refer to such logical input devices are:

PICK (for pointing devices)

LOCATOR (for positioning devices)

KEY (for alphanumeric keyboard input)

To use such a standard graphics program on a specific set of hardware requires that specialized programs called "device drivers" be written for each unique type of input device to enable it to transfer data to its logical input device counterpart.

INTERACTIVE TECHNIQUES

The key to interactive computer graphics is that the user can make inputs and changes to the geometric model and the image can be displayed as frequently as desired. The computer and the user work together interactively to build and display the model. This is accomplished by using a variety of input devices, such as keys, buttons, light pens, and dials, to invoke graphics functions. There are three general types of interactive commands:

1. Menu commands. These are standard graphics functions such as adding, deleting, moving elements. They may be selected by using a program function keyboard or by using a light pen and a display menu.

2. Macro commands. These are single commands

which are used to represent a standard sequence of operations, such as dimensioning.

3. Viewing commands. These are used to view the model interactively during the process of building the model. They include defining graphic elements, performing geometric transformations (e.g., rotation, scaling, translation), and transforming coordinates (i.e., changing views).

Many different techniques can be used to create and edit a geometric model in an interactive graphics system. Some typical examples include:

Defining graphic elements (lines, points, circles, etc.)

Generating shapes (e.g., curves, surfaces)

Moving objects

Deleting objects or elements

Duplicating objects or elements

Rotating objects

Scaling objects

Trimming elements

Several convenient techniques are used for the construction of relatively simple objects:

☐ **Sketching,** which permits the user to draw freehand on the screen using a light pen

☐ **Dragging,** which permits the user to "pick up" an object on the screen with a light pen and move it to another position.

☐ **"Rubber-banding,"** which permits the user to pick a point in a line with the light pen and stretch or bend the line to change its shape, just as one can with a rubber band

These will be covered further in Chap. 11.

GRAPHICS SOFTWARE STANDARDS

Software standards are developed to permit programs to be compatible and transportable to a variety of different hardware systems. Graphics software, in particular, since it is often very complex, is usually written by combining a number of existing subroutines that may differ from one computer to another. This would ordinarily require that they be rewritten in order to be transportable between machines. Software standards can provide the ability to use applications programs which are independent of the graphics hardware used.

As the application of computer graphics has grown in recent years, there has been a great deal of effort among users and suppliers of graphics technology to develop standards for industry. A number of different standards projects address both the format of data and the interfaces between hardware and software. Some of the major standards being developed and used by industry today include:

1. IGES (Initial Graphics Exchange Specification). This was developed by the American National Standards Institute (ANSI) for the exchange of graphics databases among computer-aided design systems.

2. NAPLPS (North American Presentation Level Protocol Syntax). This was also developed by ANSI to provide a standard data structure for graphics and text.

3. GKS (Graphics Kernel System). This was developed by the International Standards Organization (ISO) to define a software boundary or interface between applications programs and graphics support software.

4. SIGGRAPH (Special Interest Group in Computer Graphics). This is an ongoing project of the Association for Computing Machinery (ACM) to develop a standard system architecture and software structure for computer graphics in order to provide maximum flexibility for the user. Its approach is to develop modular software packages that can be added together for more graphics functions. It is intended to provide a framework in which specific packages of standard graphics software can operate.

The general view of a graphics system which uses such software standards is illustrated in Fig. 10-17. Device drivers provide the unique software interface to individual pieces of equipment. The graphics kernel system provides the device-independent interface to the application program.

GRAPHICS APPLICATION PROGRAMS

Graphics software packages usually provide three types of functions for the user:

1. Geometric construction. This includes the operations to construct 2D objects such as points, lines, circles, and cones; and 3D objects, such as surfaces and solids. It provides the input functions (e.g., keys, locators, buttons) and the graphic output primitives.

2. Display management. This includes operations such as scaling, windows, and zooming. It should also provide display control functions such as clipping and segmentation for efficient viewing and updating.

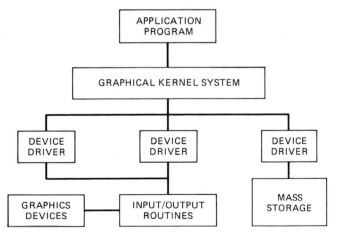

Fig. 10-17 The use of graphics software standards. *(Stephen Evanczuk, "Graphics Standard Gets Boost in U.S.," reprinted from Electronics, Aug. 11, 1983; Copyright © 1983, McGraw-Hill Inc; all rights reserved)*

3. Data management. This includes providing capabilities for filing, retrieving, and maintaining central data libraries. It involves all the data transfer operations during the building, modifying, viewing, and storing of the graphics model.

Graphics applications are extremely data processing-intensive. That is, a great deal of computation, data transfer, input and output, and storage is involved. Only a relatively small portion of the computer power in a graphics system is used for the actual display of images.

Some of the functional requirements of a good graphics application program include:

☐ Interactive techniques and languages

☐ Geometric transformations

☐ Image enhancement techniques (e.g., hidden line removal)

☐ Manipulation of objects

☐ Maintenance and extendability of the software

In addition, a number of other considerations are important in designing or selecting graphics software:

☐ Human factors, such as simplicity, ease of use, error recovery, and tolerance of misuse (robustness)

☐ Performance in terms of functional capabilities and speed

☐ Transportability between different hardware systems

☐ Costs of obtaining and using

10-5 GEOMETRIC MODELING

BASICS

A geometric model is a mathematical representation of the geometry of objects. Using mathematical models to represent real-world objects permits computers to display and manipulate images as data. Such models can be used not only to create images for display but also to represent the structure and behavior of objects for engineering analyses (e.g., physical properties). In an engineering design application, geometric models can be used for several purposes:

☐ Creating and modifying engineering drawings

☐ Creating a model for engineering analyses, such as finite element analysis (FEA)

☐ Generating NC data for production tools

A computer graphics system provides the facilities for creating, modifying, and storing geometric models of objects. The user constructs the geometric model by providing input which describes the shape and properties of the object. The computer converts this data into mathematical and symbolic representations that can be stored, retrieved, and modified. To completely describe an object requires several different types of input data, including:

☐ Data elements and their relationships (e.g., coordinates of points, vertices, edges, and shapes)

☐ Geometry of the object (i.e., spatial layout and shape)

☐ Connectivity (i.e., topology or relative location of elements)

☐ Application data (e.g., physical properties, descriptions, text)

Geometric models may be created in three different dimensional representations, depending on the capabilities of the graphics system being used:

1. 2D. XY-coordinate drawings for flat objects.

2. 2½D. Depth perspective with a constant cross section (i.e., no sidewall features).

3. 3D. Detailed geometric representation in all three dimensions.

In order to minimize the demands on computer resources and programming efforts, the computer models of solid objects are simpler than real visual or photographic images. A variety of different techniques that may be employed to represent solid objects will be addressed further in this section. Some models only represent the surfaces of objects, while others represent the entire solid.

To create and view geometric models on a computer graphics system involves several data transformation operations. Primitive graphic elements (i.e., standard 2D or 3D shapes) which are stored in the graphics system library must be converted to the user's coordinate system when they are incorporated into a model of an object. Displaying an image of the object on a screen requires a transformation of the model to a 2D plane with the perspective and view specified by the user. This involves computational techniques for projecting and clipping the image. Providing realistic views requires a combination of perspective projection, shading or coloring, and the removal of hidden lines or surfaces.

WIRE-FRAME MODELS

The most common and simplest approach to creating a geometric model of a 3D object is the wire-frame representation (Fig. 10-18). It is made up of a collection of lines and arcs that are connected to represent the boundary edges of an object. To create the model, the user specifies points and lines in space, using commands on a function keyboard. Standard shapes can be generated simply by inputting a minimum amount of data (e.g., a circle can be defined by a center point and a radius). The points and lines in one view can automatically be project-

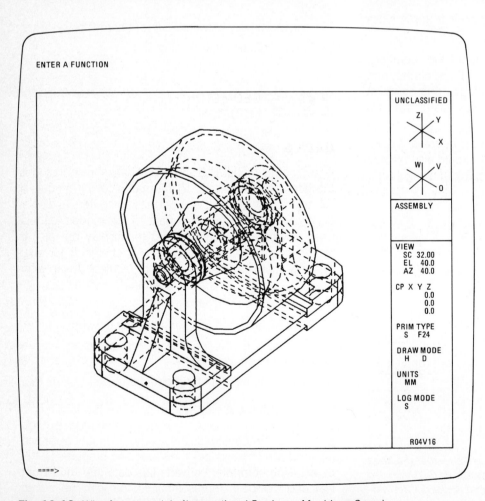

Fig. 10-18 Wire-frame model. *(International Business Machines Corp.)*

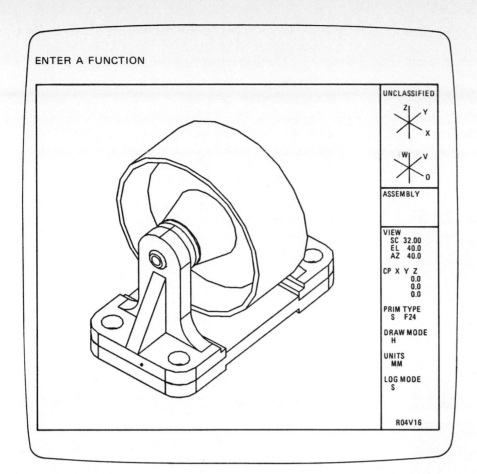

ENTER A FUNCTION

UNCLASSIFIED

Z Y
X

W V
O

ASSEMBLY

VIEW
SC 32.00
EL 40.0
AZ 40.0

CP X Y Z
0.0
0.0
0.0

PRIM TYPE
S F24

DRAW MODE
H

UNITS
MM

LOG MODE
S

R04V16

Fig. 10-19 Wire-frame model with hidden lines removed. *(International Business Machines Corp.)*

ed into another view. Once details or features on an object have been created, they can be automatically duplicated and placed in their proper positions.

The wire-frame model is relatively easy to create and does not demand a lot of computer resources. However, it is not a complete representation of a 3D object since it does not provide the true relationship between edges and surfaces. It also does not differentiate between the outside and inside of an object. This can cause the displayed images to be confusing and ambiguous, particularly for complex objects. The wire-frame model is also unable to represent sculptured 3D surfaces accurately and permits impossible and artificial objects and views to be generated. Wire-frame images can be clarified by eliminating the hidden lines (Fig. 10-19), but this requires a lot of additional computer effort.

SURFACE MODELS

To provide a clearer image of 3D objects, models were developed that would describe an object by representing the shape of its outside surfaces (Fig. 10-20). These surface models represent the shell of an object, but not the internal details or mass properties of a solid. Several techniques are used to generate surface models:

1. Ruled surface. Surfaces are generated by moving a

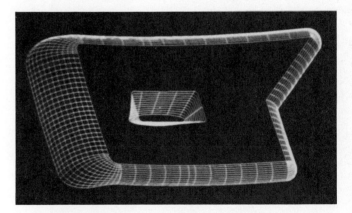

Fig. 10-20 Surface model. *(McDonnell-Douglas Manufacturing Industry Systems Co.)*

straight line through space with its endpoints tracing curves.

2. Surface revolution. Surfaces are created by rotating a curve through a circle.

3. Sweep surface. Surfaces are generated by moving one curve through another.

4. Sculptered surfaces. Complex curved surfaces can be created by connecting planar surface shapes into a quiltlike pattern to approximate the surface.

125

To enhance the clarity of surface-model images, a number of techniques have been developed to remove hidden surfaces and create the illusion of depth and curvature (Fig. 10-21):

1. **Geometric relationships.** The graphics program tests the data in the geometric model for the relative positions of surfaces to determine visibility and depth.

2. **Pixel contents.** This technique uses a special memory called a "depth buffer" to store information on the relative depth position of each pixel in the image. The graphics program then compares the points to determine which are closer and therefore visible.

3. **Illumination.** The surfaces of the object are shaded or colored to simulate curvature.

4. **Ray tracing.** The graphics program traces a theoretical ray of light from the viewer's eye to each pixel in the image. It then performs computations to find intersections between rays and the surface of the object to define reflections, transparencies, and surface texture.

Although all of these techniques provide more realistic-looking images, they require considerably more computer resources and still do not result in a true model of a solid object.

SOLID MODELS

Solid models are the only true representatiion of solid objects. They include all the information necessary to define both the surface and interior of an object as well as its mass properties. Solid models evolved from the wire-frame and surface-modeling techniques to overcome some of their deficiencies. Because they require a lot of computer resources, they only became feasible and practical as the cost and performance of computer graphics technology improved. Solid models provide unique features such as:

☐ Cross-sectioned and exploded views of objects and assemblies (Fig. 10-22)

☐ Realistic 3D views can be provided by removing hidden lines and surfaces and using color shading

☐ Mating sections of complex objects and assemblies can be checked for continuity, fit, and clearances

A model of an object can be manipulated, examined, and changed in great detail, much like a real prototype or physical mock-up.

A solid model can be broken down into its component parts, reassembled, cross-sectioned, simulated in motion, analyzed, and modified. These capabilities allow a solid model to be used for many applications and shared by several users. It can be used to help to generate drawings, conduct engineering analyses, develop manufacturing process plans, and program tool control commands.

A number of different approaches to solid modeling have been developed. Two major techniques are in use today. Neither is necessarily better than the other, since they both have advantages and disadvantages.

1. **Constructive solid geometry (CSG).** This is basically a "building-block" approach which creates a 3D model of an object out of simple solid shapes or primitives, such as blocks, spheres, cones, and cylinders. The creation of the model involves several basic steps (Fig. 10-23).

☐ Primitive solid shapes are created or selected from a stored library.

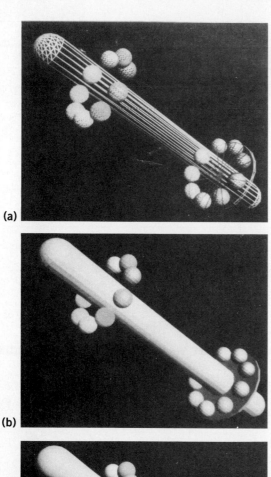

(a)

(b)

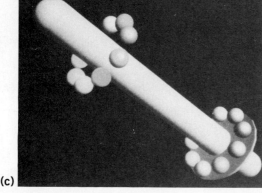

(c)

Fig. 10-21 Techniques for enhancing the image of surface models: (*a*) with wire-frame rendering there is ambiguity about the location of the balls at the bottom of the rod; (*b*) the depth-buffering technique improves the image; (*c*) brightness interpolation adds considerably more realism to the display. *(Richard Fichera, "Rendering Adds Realism to Graphics," reprinted from Electronics Week, Oct. 22, 1984; Copyright © 1984, McGraw-Hill Inc., all rights reserved)*

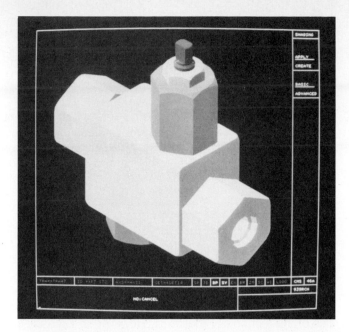

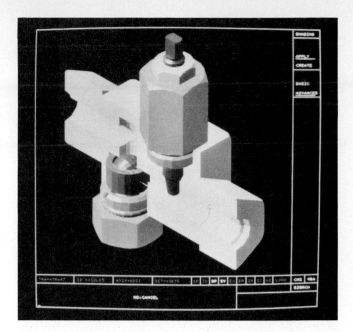

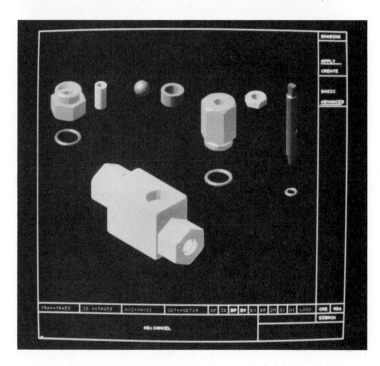

Fig. 10-22 Creating cross-sectioned and exploded views from solid models. *(International Business Machines Corp.)*

☐ The user specifies the dimensions, locations, and orientations of the primitives.

☐ The primitives are combined by standard mathematical operations for adding, subtracting, and intersecting sets of geometric data.

☐ The graphics system replaces the primitives with the resulting combined solid shape.

In this approach, the model is stored as a list of solid primitive shapes (which in turn are represented by mathematical equations and parameters) and the operations which combine them. It is a relatively easy approach to defining a geometric model and it uses the computer memory efficiently. However, because it builds the model out of standard shapes, CSG does not lend itself well to

unusual and sculptured surfaces. In addition, it would be difficult to translate the model into a line drawing or wireframe display.

2. Boundary representation (B-rep). This type of model defines the surfaces which bound a solid object. However, unlike surface models, the data structure in a boundary representation model contains enough information to determine where any point lies on the inside as well as the surface of the object. The model is stored as a list of surface shapes (faces, edges, and vertices) and their connectivity relationships (Fig. 10-24). Two different techniques can be used to create a boundary representation model:

☐ **Polygon mesh.** A set of connected polgon-shaped

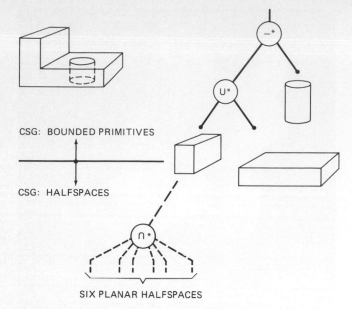

CSG: BOUNDED PRIMITIVES

CSG: HALFSPACES

SIX PLANAR HALFSPACES

Fig. 10-23 Creating a solid model with constructive solid geometry. *(A. A. G. Requicha and H. B. Voelcker, "Solid Modeling: A Historical Summary and Contemporary Assessment," IEEE Computer Graphics and Applications, March 1982; © 1982 IEEE)*

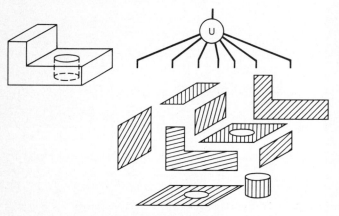

Fig. 10-24 Creating a solid model by boundary representation. *(A. A. G. Requicha and H. B. Voelcker, "Solid Modeling: A Historical Summary and Contemporary Assessment," IEEE Computer Graphics and Applications, March 1982; © 1982 IEEE)*

planar surfaces define an approximate boundary of the solid object.

☐ **Parametric bicubic patches.** Coordinates of points on the curved surfaces are defined with 3D equations. (This is a more accurate, but also more complex method.)

The images generated by a boundary representation can be enhanced by creating more planar faces and using algorithms to create smoothing effects for curvature. Since the B-rep model is not limited by standard primitive shapes, it allows unusual shapes and sculptured surfaces to be modeled. In addition, since it includes a model of the surface of the object, it can be used to generate wire-frame drawings or displays. It cannot, however, be converted automatically to a CSG representation.

Although it uses a lot of computer resources, boundary representation is not as accurate as a CSG model.

In addition to these two common techniques to create solid models, there are several others that have been developed which have unique capabilities or advantages in certain applications. Some break down the solid object into standard-sized "cells." Others use sweeping techniques to "extrude" a solid object from a 2D shape. There are also hybrid techniques which combine some of the features and functions of the others to take advantage of their unique capabilities.

10-6 APPLICATIONS OF COMPUTER GRAPHICS

Computer graphics technology has been found to be a tool for improving productivity in a wide variety of applications throughout industry. Some are obviously more demanding than others which may not use all the capabilities of the technology that exist. However, in each case, most of the basic functions of the hardware and software addressed in this chapter are used.

AUTOMATED DRAFTING

Automated drafting is one of the earliest and least demanding of the application areas, but it is still very widely used. Automated drafting systems have been developed for electrical, mechanical, and architectural drawings. They provide for the creation of views, scaling, dimensioning, and transformations (e.g., rotation and isometrics). Automated drafting systems are faster and more accurate than manual drafting, and productivity improvements of anywhere from 5:1 to 15:1 have been reported.

DESIGN

Computer-aided design systems have been developed for electrical, electronic, mechanical, and architectural applications. The design process in all of these areas can be very data-intensive and time-consuming. Some modern, complex products, would be extremely difficult, if not impossible, to design manually. Automated design systems not only speed up the design process; they also provide for design reviews and checking. Such design systems will be covered extensively in Chap. 11.

ENGINEERING ANALYSIS

A number of engineering activities are involved between the initial design of a product and its ultimate manufacture. Computer graphics techniques have been developed which use the database established by an automated design system to perform these engineering analyses more efficiently and accurately than was ever possible before. These activities include the analysis of stresses, motion, and physical properties; simulation of operation; process planning; and NC programming. They will each be addressed in detail in Chap. 12.

INFORMATION DISPLAYS

Since graphics displays can communicate information more efficiently than hard copy text, they are being used more frequently in industry for a variety of information display tasks. These include information retrieval systems, telecommunications, management information systems, and manufacturing control systems. For example, graphics displays are being used on the factory floor for process monitoring and control (Fig. 10-25). They can be very easy for process operators to use, and at the same time they can communicate important information in the form of graphics and text.

BUSINESS GRAPHICS

The power of computer graphics is also finding many applications in administrative and management offices, where they help run the business. Since graphics is more efficient than text in communicating information, graphics can be used to summarize complex data and focus on the most important information. A business graphics system can make information more visible and accessible to managers. It can also provide lower-cost and higher-quality charts than the traditional manual techniques.

APPLICATION CONSIDERATIONS

When deciding to use or select a computer graphics system, a number of factors should be considered:

☐ Graphics applications require more computer power for data processing tasks than for generating pictures on a display. It is these computational and data handling tasks that will determine the computer requirements for a given application.

☐ The characteristics of the intended application should be understood. For example, for a design system one needs to know the nature of the designs and the design process as well as the types and styles of drawings that are involved.

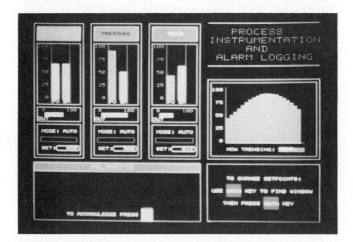

Fig. 10-25 Graphics display of process control application. *(Industrial Data Terminals Corp.)*

☐ To select the proper system, one needs to understand the characteristics of the hardware and software, including their functions and performance, flexibility and portability, user-friendliness and support.

☐ A company preparing to use the system must consider personnel implications, including skills requirements, training, working conditions, and human factors.

☐ Finally, there may not even be an opportunity to obtain a system unless it is economically justified. It should improve productivity and have a good return on investment (ROI).

10-7 TRENDS IN COMPUTER GRAPHICS

DISPLAY TECHNOLOGY

The most active and visible part of computer graphics technology is, of course, the display unit. Therefore, improvements in its cost, performance, and function are very important to users. Some of the desired objectives are:

☐ Higher-resolution images for more detail and realism

☐ Higher-quality text displays for easier reading

☐ More subtle shading and coloring for greater realism

☐ Larger screens for greater content and ease of viewing

☐ Faster response for updating, scrolling, windowing, and animation

☐ Smaller, lighter, more rugged displays for portability

There are efforts to achieve these objectives with both conventional and new display technologies:

1. CRTs. Although CRT technology has been around for many years, much can still be done to improve its capabilities:

Increasing the number of pixels to several million is feasible and substantially improves resolution

Using a "bit-mapped" scheme, where each pixel can be individually addressed (as opposed to the conventional approach of grouped arrays of characters), can also provide higher resolution

Black-on-white displays can be easier to read than conventional single-color (monochrome) displays

Simplified designs for color monitors can reduce costs

Flat CRTs have been developed for portable applications

2. Alternative display technologies. Several other display technologies have been developed which have some basic advantages over conventional CRTs:

They can be made as large flat screens

Each pixel is individually addressable by a matrix of wires rather than a scanning electron gun

They more easily produce high-resolution, high-contrast images

They are generally lighter, more rugged, and consume less power than a CRT

Three major alternative technologies have been considered for computer graphics applications:

☐ **Plasma,** which uses electrical energy to illuminate a gas sealed inside a glass panel (like a neon sign); see Fig. 10-26

☐ **Liquid-crystal displays (LCDs),** which use an electric field to switch pixels from a polarization that transmits light to one that blocks light (as in liquid-crystal watches)

☐ **Electroluminescence,** which creates images from phosphor pixels that give off light when voltages are applied

GRAPHICS CHIPS

Many of the improvements in the performance of graphics systems, for both the processor and the display, will come from IC technology. Many graphics functions that had traditionally been performed by software can be executed much faster with hardware. In addition, special purpose chips can be used to relieve the CPU of some of the graphics processing functions. The graphics functions which lend themselves best to using special chips are:

1. Controlling the image on the screen (e.g., windows, zooming, generating primitive shapes)

2. Performing computations (e.g., transformations, clipping)

3. Improving the efficiency of data transfer from memory to the processor and display controller (e.g., video random access memory, or video RAM)

SYSTEM ARCHITECTURE

To optimize the performance of systems for graphics applications, some companies are developing unique architectures rather than adapting general purpose hardware and software. Some of the key elements of advanced graphics systems are:

☐ "Geometry engines," which perform complex geometric computations using floating-point arithmetic and parallel processing techniques

☐ Standard interfaces for data transfer and device-independent applications programs

☐ Engineering workstations that include high-performance processors (32-bit), large memories (1 MB or more), and high-performance color graphics displays

☐ Distributed networks that tie many workstations to a shared database with high-speed communication links

GEOMETRIC MODELING

Modeling techniques continue to be improved to provide more function and greater efficiency. Some examples of such improvements are:

☐ Simplified solid models that can be run on personal computers

☐ Automatic generation of finite element meshes and NC programs

☐ Special purpose hardware to perform image enhancement functions (e.g., ray casting)

☐ Combining capabilities of different representation techniques (e.g., sculptured surfaces with solid models)

GRAPHICS APPLICATIONS

As the capabilities of computer graphics technology increase and its cost decreases, its application will continue to expand. Some of the trends in manufacturing include:

☐ Shop floor graphics terminals, which must be rugged displays for process monitoring and control, with color graphics (Fig. 10-27)

☐ Dynamic simulation for animation of manufacturing operations (e.g., robots and materials handling systems)

Fig. 10-26 Flat gas panel display with high-resolution graphics. *(International Business Machines Corp.)*

Fig. 10-27 Shop floor color graphics terminal. *(Industrial Data Terminals Corp.)*

☐ Integrated CAD/CAM systems that have a common geometric database for development and manufacturing

☐ Automated geometric models in which the computer designs, analyzes, simulates, tests, modifies, and generates manufacturing data, but in which minimal human interaction or even visible graphics is required

10-8 SUMMARY

Computer graphics technology evolved from the basic hardware and software capabilities of computers to perform design and engineering tasks. Interactive graphics systems permit the user to work with the hardware and software to create, analyze, and change graphic representations of objects. Interactive graphics systems can significantly increase the productivity of engineers and speed up the process of designing a product.

A wide variety of graphics systems are available, in terms of the functions and performance capabilities that can be tailored to the needs of the application in which they will be used. This can range from relatively simple drafting systems to complex, high-performance design and engineering systems. There are also many types of graphics hardware to choose from, each with its own advantages and disadvantages. Input, output, and display devices are available with features to satisfy any graphics application (e.g., speed, accuracy, and color).

Graphics software provides an interface that enables the user to construct a geometric model and display and manipulate images of it. This software structures the data, transforms it, performs mathematical computations on it, and issues commands for generating graphics images. The user interacts with the system by using high-level commands and special input function devices to construct the models and display the images. Standards have been developed by industry to simplify the programming job and make graphics software more portable between systems.

Graphics applications involve a lot of computations and data handling. This requires more computer power than just the actual generation of graphics displays. Creating a geometric model of an object permits computers to analyze its structure and behavior as well as create and manipulate images. A number of different techniques are used to represent objects with geometric models. Only solid models are a true representation of an object. They are complete and unambiguous, and they provide details on internal as well as external structure. A variety of techniques can be used to enhance the images generated by models to improve their realism (e.g., hidden line removal, shading, depth perspective).

Computer graphics is used in many industrial applications to improve productivity in the office, in the laboratory, and on the factory floor. As the technology continues to advance, these applications will expand even further. It is one of the key tools which makes computer-automated manufacturing (CAM) possible.

REVIEW QUESTIONS

The answer to each question can be found in the section(s) indicated at the end of the question.

1. What is computer graphics and how is it used in manufacturing? [10-1]

2. Identify the basic elements of a computer graphics system and what they do. [10-2]

3. List some of the advantages of using computer graphics systems. [10-2]

4. Describe the basic steps in designing on an interactive computer graphics system. [10-2]

5. What are some of the considerations in selecting a graphics system? [10-2 and 10-6]

6. Identify the types of graphics displays available and describe their major differences. [10-3]

7. Describe the major types of input and output devices used in graphics systems. [10-3]

8. What types of software are required for a graphics system and what are their functions? [10-4]

9. Identify the types of data required to create a geometric model. [10-4]

10. Describe the basic process by which user-input data results in images on a display screen. [10-4]

11. Identify some of the user operations and techniques involved in an interactive graphics system. [10-4]

12. What is the purpose of graphics software standards? [10-4]

13. Define a geometric model and describe how it can be used. [10-5]

14. Identify the major types of geometric models and describe the major differences between them. [10-5]

15. What are some of the techniques that can be used to make the images generated by geometric models more realistic? [10-5]

16. Describe some of the techniques used to create solid models. [10-5]

17. List some of the major applications for computer graphics in industry. [10-6]

18. Identify some of the key trends in computer graphics technology that will influence its capabilities and applications in manufacturing. [10-7]

COMPUTER-AUTOMATED DESIGN

11-1 INTRODUCTION

Computer-automated design (CAD) is an outgrowth of numerical control (NC) and computer graphics technologies. The earliest CAD systems were developed in the 1960s, primarily for mechanical design applications. They were so widely used as an automated drafting tool that CAD was often referred to as computer-aided drafting. Many CAD systems continue to be used today as drafting systems which provide hard copy output for documentation. Physical drawings and printouts are still the common way in which information is extracted and communicated from a computer system. Advances in interactive computer graphics technology, however, have made the automation of the design process and the elimination of paper possible. As products became more complex and subject to change, the role of the computer as a design tool increased substantially. What started as a design aid has become an automated system for design. Computer-aided drafting led to computer-aided design and ultimately to computer-*automated* design.

There are several basic reasons for using CAD systems:

To increase the productivity and creativity of designers

To improve the quality of the design

To improve the efficiency of communications (in terms of documentation, accessibility, speed, and standardization)

To create a database for engineering and manufacturing

To make the design of large, complex products and systems practical and economical

A CAD system uses the strengths of the computer together with those of humans to solve problems. Such systems are used for a variety of design-related tasks, including:

The creation, review, and evaluation of designs

The generation of images and drawings

The analysis of the structure and behavior of designs

Computer graphics workstations can be used to integrate all these design activities. In the past, each step in the design process was separate and mostly manual.

Today it is possible to use automated computer graphics from the initial concept of the design through all the stages of engineering, documentation, and manufacturing.

This chapter will focus on how CAD systems work and are used in design applications. The related engineering analysis tasks will be addressed in more depth in Chap. 12. Although there are many different types of CAD systems, the most widely used and most highly developed are for mechanical and electronic applications. This chapter will therefore deal primarily with them. Other special CAD systems which are not as closely aligned with manufacturing (e.g., architecture, facilities, piping, wiring) will not be covered here.

11-2 AUTOMATING THE DESIGN PROCESS

WHAT IS THE DESIGN PROCESS?

Designing a product for manufacturing involves several basic stages of activity:

1. **Specification.** Defining the requirements for the product which is to be designed.

2. **Design strategy.** Selecting the approach to be taken for the design solution.

3. **Design solution.** The actual detailed design work.

4. **Formalization.** Final checking and documentation of the design.

5. **Implementation.** Translating the design into manufacturing processes for fabrication.

The actual design and drafting work is usually the most visible part of the design process and has been the stage generally automated first. Some of the detailed activities involved during this stage include:

Geometric construction

Line work

Lettering

Symbol creation

Dimensioning

Cross-hatching

Drawing manipulation (i.e., changing views and scales)

Checking and correcting

Modifying and changing

An automated design system must be capable of handling all these activities.

WHY AUTOMATE THE DESIGN PROCESS?

Automating the design process can result in many benefits, including:

Increased productivity of the designer

Reduced time from design to manufacturing

Lower design and production costs

Greater accuracy and quality of designs

More standardization of designs and documentation

Improved control of design changes

Since the prime user of a CAD system is the designer, the ability of the system to make the design job easier and to improve productivity is very important. Automating the design process can give the designer the ability to:

☐ Analyze the feasibility of ideas quickly

☐ Try out more alternatives with less time and effort

☐ Solve complex design computational problems quickly and accurately

☐ Respond faster to design changes

When automated design systems are used for the first time, they impose constraints and discipline on the designer who could be more flexible in a manual process. Using them requires more attention to detail, which can result in improved quality, consistency, and standardization in designs. As products become more complex and subject to change, it is very difficult to maintain accurate documentation with a manual design system. An automated design system also requires the design to be checked and verified early in the process. This can replace the need for prototypes, which may be impractical or uneconomical for complex products. Automated design systems can also be used for other tasks beyond the initial design work itself. Once the geometric model has been created, it can be used by engineers in other disciplines to perform their jobs. This includes engineering analysis, simulation, process planning, and NC programming which will be covered in Chap. 12.

The implementation of integrated CAD systems has traditionally been inhibited by a number of factors, such as:

☐ Limited computer power available at the workstation

☐ Differences in data structures between systems

☐ Slow communications between workstations and host computers

☐ Continued reliance on hard copy for the documentation and exchange of design information

☐ User skill requirements for geometric modeling and engineering analysis tasks

The recent advances in computer graphics technology, however, have helped to overcome these inhibitors by providing:

☐ Increased local computer power in workstations

☐ Standards for graphics data exchange

☐ Efficient data communications networks

☐ Improved quality and performance of graphics displays

☐ Automated tools for engineering analysis using solid models

Automated design systems are one of the key elements in the competitive strategy of the manufacturing industry. The cost of CAD systems has come down relative to the cost of engineers. It is therefore even more important as well as economical to improve the productivity of engineers. Perhaps just as important as the economics involved is the fact that CAD systems make it possible for industry to overcome the problem of longer design efforts in the face of shorter product life cycles.

WHAT IS INVOLVED IN AUTOMATED DESIGN?

One of the principal differences between an automated design system and a drafting system is the modeling that is involved. Models are used to represent the structure and behavior of objects, not just to generate views and images. A number of ingredients are required to create a model in a design system:

☐ Data elements and their relationships

☐ Geometric descriptions of objects (e.g., layout and shape)

☐ Connectivity (i.e., relationships between objects or topology)

☐ Application data (e.g., electrical, mechanical, or physical characteristics)

☐ Processing algorithms (e.g., electrical or mechanical design analysis programs)

During the design stage, the user interacts with the system to perform the basic design tasks, including:

1. Constructing the design
2. Data handling (input, transfer, storage, and retrieval)
3. Design analysis
4. Checking and correction
5. Design modification

An automated design system provides a number of aids to the designer, which make the job easier and permit accurate and error-free results. Examples include:

Design constraints (e.g., on size, orientation, placement)

Grid on the display screen

Rounding of distances

Placement of numerical coordinates

Measurement of dimensions

Comparisons and checks

CAD systems can also provide data security and control features that would be difficult to implement in a manual system. The use of access codes and approval requirements can assure both the security and the integrity of a design. This feature can permit access to only those authorized to review or change a design. It can also prevent a design from being finalized until all required approvals are received.

The specific characteristics of how a design system works are usually tailored to the particular type of application involved.

CONSIDERATIONS IN AUTOMATING THE DESIGN PROCESS

Before implementing an automated design system, there are a number of important factors to consider:

1. Characteristics of the application:
   Types of products and parts involved
   Quantities of parts to be manufactured
   Nature of the design process
   Frequency and types of changes involved
   Characteristics of the designs (e.g., in terms of size, complexity, and features)
   Types and styles of drawings required
   Engineering analysis requirements
   Documentation requirements
   Use of standard parts or designs
   NC output requirements
2. Selection of the system:
   Software characteristics
   Hardware characteristics
   System functions and features
   Performance capabilities
   User-friendliness
   Flexibility
   Portability
   Availability and type of support
3. Personnel considerations:
   Existing CAD experience and skills
   Training requirements
   Selection criteria
   Working conditions and human factors/?
4. Economic justification:
   Productivity improvements
   Return on investment (ROI)
   Absolute cost and cash flow requirements

Even though the cost of computer power has been coming down, interactive graphics systems are still very expensive. Substantial improvements in productivity or other benefits are necessary to justify their cost. Produc-

tivity improvements of a least 2:1 are normally required, but this is quite feasible for a CAD system.

11-3 BASIC CAD SYSTEMS

TYPES OF CAD SYSTEMS

CAD systems can be characterized in several different ways:

1. Applications:
   Electrical (e.g., wiring, power systems)
   Electronic [e.g., integrated circuits (ICs), printed circuits]
   Mechanical (e.g., parts, assemblies)
   Architectural (e.g., buildings, structures)
   Facilities (e.g., layouts, utilities)
2. System configurations:
   Centralized (i.e., host computer with remote terminals)
   Satellites (i.e., host computer tied to smaller local computers with clusters of terminals)
   Distributed or stand-alone (i.e., workstation with internal processor)
3. Mode of operation:
   Batch (all data is processed in batches by central computer—not interactive)
   Interactive (local processor permits user to interact during design process)
   Integrated (design data shared with all phases of engineering activity from development through manufacturing)

Most CAD systems today are interactive and provide at least some output for other engineering functions. Figure 11-1 illustrates the architecture and operation of a stand-alone minicomputer-based design system which provides NC data for manufacturing as well as design drawings.

SYSTEM ARCHITECTURE

The architecture of a CAD system should provide the following capabilities:

☐ Functions that are useful in the design application

☐ Modular software that can be transported, upgraded, or replaced

☐ Efficiency of operation (e.g., response time and use of memory)

☐ Interactive features

The architecture of the system can be described in terms of the environment seen by the system as well as that seen by the user (Fig. 11-2). The system environment is determined by the application support programs and the operating system. Application support programs provide the services required by the application, such as communicating with terminals or saving data. The user

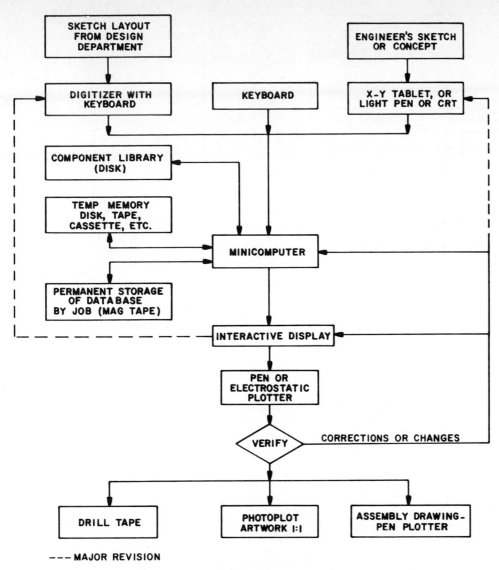

--- **MAJOR REVISION**

Fig. 11-1 CAD system: functional flowchart of the operation of an interactive stand-alone minicomputer design system. *(Julius Dorfman and Jack Staller, "Surveying Today's CAD Systems," Electronic Packaging and Production, April 1979)*

environment is determined by the configuration of the workstation and application program. The application program contains a command language which provides the user interface to perform the interactive tasks. It also contains the program with the design algorithms for the application. The command language should be easy to use so that experienced programmers are not required to perform design tasks. At the same time it should be powerful enough to perform a variety of functions, such as:

Displaying of menus for user selection of tasks and functions

Communication with I/O devices

Access of databases

Allocation of computer resources

Editing capabilities

Support for subroutines and commands for other systems

A device-independent graphics software architecture permits a number of different types of equipment to be used with the system and permits portability between systems. This can minimize software development needs and promote the sharing of software between design groups. It can also permit an application to migrate to new hardware without the need to rewrite the program. Figure 11-3 shows how such a system handles the operations involved in generating graphics output. The graphics primitives created by the application program go through a sequence of processing steps which transform the coordinate data into an image displayed on the screen. The initial clipping and mapping operations are performed independent of the output device. A device driver program then transforms the data in the display file into a form compatible with the unique device involved.

There are usually several major groups of subroutines in a graphics software package:

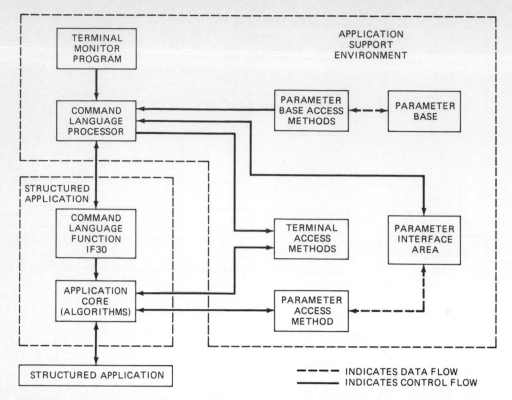

Fig. 11-2 Architecture of a CAD system. *(Richard L. Taylor, "A Software Architecture for a Mature Design Automation System," IBM Journal of Research and Development, Sept. 1984; © 1984 by International Business Machines Corp., reprinted with permission)*

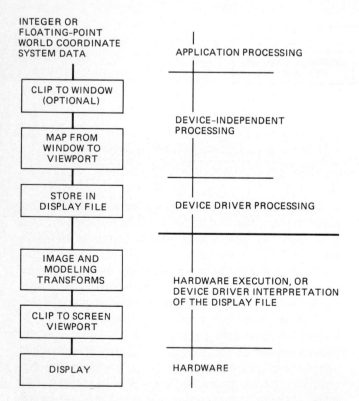

Fig. 11-3 Graphics output operations in a device-independent system architecture. *(R. B. Capeli and G. C. Sax, "A Device-Independent Graphics Package for CAD Applications," IBM Journal of Research and Development, Sept. 1984; © 1984 by International Business Machines Corp., reprinted with permission)*

Control for initializing and terminating the system as well as handling errors

Graphic input to provide logical input devices which interface with real input devices

Graphic output to generate primitives, attributes, and transformations

Display files to permit the selection and manipulation of objects

Local processing to allow interactive functions to be performed by the workstation processor

Like any other data system, the graphics system architecture must provide some data management functions. To allow interactive use, the system must provide that the data required during use be resident on the workstation storage file. Most large design systems, however, must also provide a means to store long-term data in a central file off-line until it is needed. Such a central database can be used by both the design and manufacturing organizations to store and retrieve geometric and alphanumeric data. This can help to assure the timely and accurate communication of design information.

CAD SYSTEM OPERATIONS

Three major types of activities are involved in the operation of a CAD system:

1. Building a model. Graphics primitives and symbols are used to create and modify a geometric model of an object.

2. Viewing a model. Viewing commands are used to define the type of projection and the point of view for displaying the image. Special output functions, such as hidden line removal or windowing, may also be involved.

3. Data handling. Support processing functions are used for data retrieval, computations, measurements, and so on.

Graphics applications are very data processing intensive. Most operations involve a great deal of data retrieval, transfer, and computation. The actual input and output activities that are the most visible to the user only require a small portion of the computer power necessary to support all the system operations.

CAD systems also provide some design control functions for the user. These may include:

Data management

File storage

Tracking of the design progress

Message prompts

Access security control

CAD TECHNIQUES

CAD systems use a variety of different techniques to make the design job easier for the user. These tend to vary by the type of application involved, but they often include the following:

1. Interactive input. Usually, several alternative techniques are provided with each system, such as:

☐ Menu selection (by program function keys, light pen, or digitizer tablet)

☐ Macro commands (single commands that initiate a standard sequence of operations)

2. Dynamic visual feedback. These are techniques that allow the user to see the effects of actions to create and modify images. They may include:

☐ Rubber banding

☐ Dragging

☐ Digital coordinate readouts

3. Object hierarchy. This approach creates models from geometric building blocks to improve the efficiency of use of computer resources and save design time (Fig. 11-4). The model is made up of a hierarchy of primitive graphics commands and tranformations. This is particularly useful for a complex design that has many details replicated a number of times (e.g., wheels and bolts on a car).

11-4 MECHANICAL DESIGN SYSTEMS

AUTOMATING THE MECHANICAL DESIGN PROCESS

Mechanical designs usually require detailed drawings with dimensions, notes, and often multiple views, so that parts can be fabricated from them (Fig. 11-5). The minimum that a mechanical design system must do, therefore, is to provide an automated drafting tool for the designer. Since most systems evolved from that base, many of the operations are similar to those in manual drafting. However, automated design functions are used, such as:

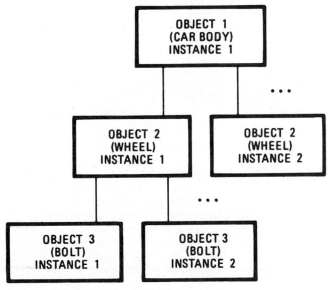

Fig. 11-4 Object hierarchy: nesting of parts which make up a complex assembly, such as a car. *(James H. Clark and Tom Davis, "Work Station Unites Real-Time with Unix, Ethernet," reprinted from Electronics, Oct. 20, 1983; Copyright © 1983, McGraw-Hill Inc., all rights reserved)*

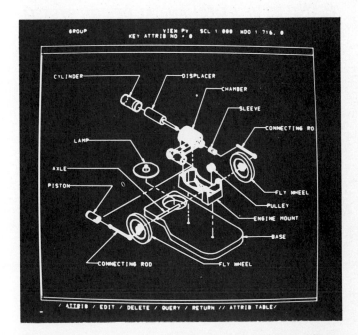

Fig. 11-5 Mechanical drawing from a CAD system. *(International Business Machines Corp.)*

☐ Generating primitive graphic elements (e.g., points, lines, arcs, surfaces)

☐ Performing geometric transformations (e.g., translation, rotation, scaling)

☐ Constructing alternative views (including assemblies)

☐ Editing (e.g., moving, changing, deleting)

☐ Cross-hatching

The entire mechanical design process may be automated with a graphics system, from the initial design concept through the programming of manufacturing tools, in the following sequence of steps:

1. A "sketch" of the basic design concept may be put on the screen using manual drawinglike techniques with a light pen or a digitizer tablet

2. A more formal, detailed design is constructed in much the same fashion as a drafter would work on a drawing board

3. Checks may be conducted for interference and fit with mating parts in an assembly

4. Engineering analyses may be performed to determine physical or structural properties

5. The design may be edited to correct errors or problems

6. Drawings are detailed and finalized with dimensions, notes, and symbols

7. The finished drawings are stored in a database and hard copies printed

8. Other engineers use the design data to prepare the parts for manufacturing (e.g., tool design, process planning, NC programming)

FEATURES OF A MECHANICAL DESIGN SYSTEM

A mechanical CAD system uses techniques which are familiar to designers and drafters and provides features that make the design job much easier and more accurate, such as:

☐ Selection of geometric construction functions from a menu

☐ Replication of details

☐ Transformation of views

☐ Selection of standard symbols from a library

☐ Assigning attributes to parts (e.g., physical properties or descriptions)

☐ Dimensioning

In some cases these features offer more than the automation of manual tasks. They may provide capabilities that are not practical or even feasible with manual methods, such as:

☐ Moving or creating views to visualize the mating or interference of parts

☐ Zooming in or out of any part of a large, complex design to view the level of detail desired

☐ Animating views to enhance design perspective or to conduct engineering analyses

DRAWING FUNCTIONS

A designer can construct the geometry of an object by using the DRAW mode of the graphics system. In this mode, the designer selects basic drawing functions from a number of different commands (Fig. 11-6). These typically include functions such as:

Create a line. Lines can created by specifying their length, starting point, and orientation.

Move an object. An object that has been created can be moved by specifying the amount and direction.

Rotate an object. An object can be rotated about a specified axis or point.

Scale elements. Geometric elements can be scaled by specifying a point about which, and the factor by which, it is to be scaled.

Reflect elements. Mirror images can be created about a line.

Break an element. An element such as a line or arc can be selected and broken into two at a specified point.

Edit elements. Elements can be manipulated in two or three dimensions.

Create primitive objects. Basic solid shapes can be selected and their critical dimensions specified.

Merge. Primitive objects can be joined to form new single objects with the shape resulting from their merger.

There are also several convenient sketching functions which can make the construction job simpler and faster. These use light-pen techniques to modify or move objects on the screen. The most common sketching techniques are rubber-banding (i.e., stretching) and dragging (i.e., moving), which are illustrated in Fig. 11-7.

In addition to these functions used to construct the geometry, there are functions that can manipulate the views of objects. These viewing functions translate the image by changing the coordinate system orientation relative to the viewpoint selected (Fig. 11-8).

GEOMETRIC DESIGN PROCESS

To create a geometric model of a solid mechanical object involves a series of construction and manipulation steps, such as:

☐ Selecting primitive shapes and merging them to create more complex objects (Fig. 11-9)

☐ Creating primitive volumes by sweeping polygon shapes (Fig. 11-10)

☐ Relating objects in the model to one another by creating a hierarchical structure or list, much like a bill of materials (Fig. 11-11)

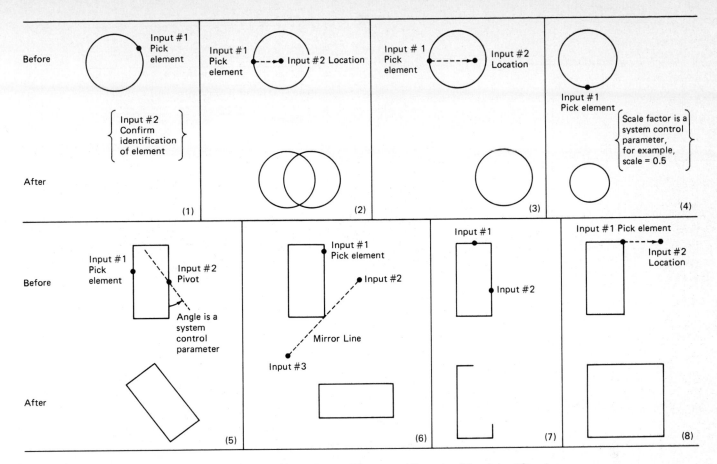

Fig. 11-6 CAD drawing functions: (1) delete, (2) duplicate, (3) move, (4) scale, (5) rotate, (6) mirror, (7) partial delete, (8) modify. *(Joan E. Scott, Introduction to Interactive Computer Graphics, Wiley, New York, 1982)*

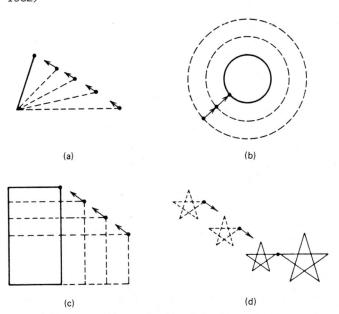

Fig. 11-7 CAD sketching techniques: *(a)* lines, *(b)* circles, *(c)* rectanges, *(d)* also "dragging." *(Joan E. Scott, Introduction to Interactive Computer Graphics, Wiley, New York, 1982)*

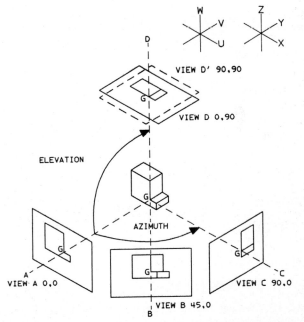

Fig. 11-8 CAD viewing functions: coordinate system orientation for display. *(International Business Machines Corp.)*

Once the design is created, a number of special functions are available to perform simple checks and analyses. These can help the designer detect problems, which can be fixed before finalizing the design. Examples of these functions include:

Alignment check. This can test the alignment of holes and cylinders to determine whether they match

Interference check. This tests for interference between

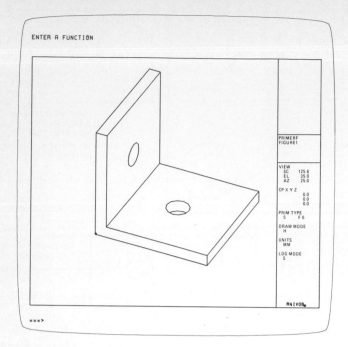

Fig. 11-9 Creating a mechanical design. *(International Business Machines Corp.)*

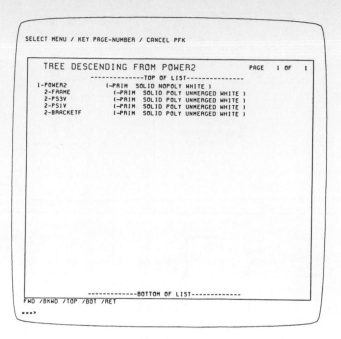

Fig. 11-11 List of assembly structure. *(International Business Machines Corp.)*

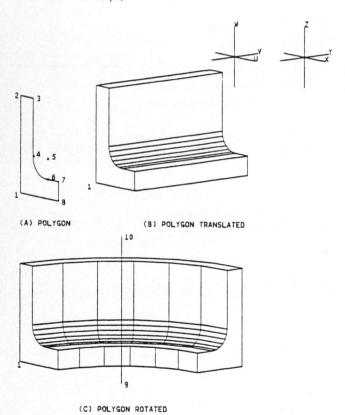

(A) POLYGON   (B) POLYGON TRANSLATED

(C) POLYGON ROTATED

Fig. 11-10 Creating primitive volumes by sweeping. *(International Business Machines Corp.)*

two objects by determining if any edges of one object touch or penetrate the face of another

Compare. This function can compare the lines and shapes of one drawing against another and display the differences

Physical properties. Functions can be used to compute properties of the object, such as its area, volume, weight, or center of gravity.

11-5 ELECTRONIC DESIGN SYSTEMS

AUTOMATING THE ELECTRONIC DESIGN PROCESS

Electronic design is performed at three major levels of the product technology:

1. System. This is the end product or piece of electronic equipment, such as a computer. Design at the system level usually starts by partitioning into functional subsystems, such as the processor or memory in a computer.

2. Printed circuit board. Most electronic equipment uses printed circuit boards for packaging and interconnecting the electronic components in the system. The board design must be partitioned into those components and all the inputs and outputs wired to be compatible with the functional subsystem design.

3. Components. The basic building blocks of an electronic system are the components, such as ICs. At this level, the design process involves the detailed logic circuits, the physical layout, and the generation of test and mask data.

The typical design process at one of these levels involves a series of steps, each of which can be extremely complex tasks (Fig. 11-12):

1. Specification. The basic requirements of the design are established in terms of functions and performance

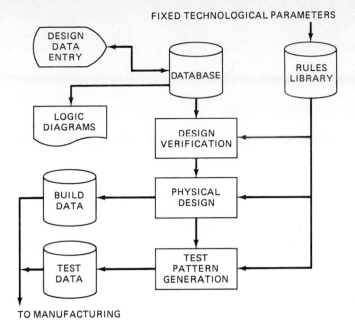

FIXED TECHNOLOGICAL PARAMETERS

TO MANUFACTURING

Fig. 11-12 Basic electronics design process. *(T. C. Raymond, "LSI/VLSI Design Automation," Computer, July 1981; © 1981 IEEE)*

characteristics (e.g., timing, output voltages, and logic functions). The design tools used at this stage are usually text editors and documentation systems to generate files of text and graphics data.

2. Functional design. This involves the generation of circuit schematics and interconnections which satisfy the functional specifications. Design programs are used, which provide libraries of standard circuits and functional simulation tools as well as interactive graphics capabilities.

3. Design verification. The circuit design is tested using logic simulation tools to verify that it meets the functional and performance specifications.

4. Test generation. The data required to perform electrical tests on the finished product is generated from the design using special test analysis programs.

5. Physical design. During this stage, the electrical design is translated into physical layouts of component elements and wiring interconnections. A number of design tools are used to generate standard graphics shapes and perform placement, interconnection, and checking routines.

6. NC data generation. As in mechanical design systems, the final output is NC data to program manufacturing tools which fabricate the parts involved. In the case of electronic designs, this data may drive the fabrication of photo masks for ICs, or it may be instructions for drilling holes in printed circuit boards.

Electronic design systems were first developed in the 1950s to eliminate human errors in manual documentation systems. The initial applications were for automating logic diagrams and back-panel wiring. As the electronics technology evolved into ICs and printed circuit boards,

the complexity of the design job increased substantially. This created the need for computer aids to perform most of the design tasks. Today it would not be practical (or perhaps even possible) to design high-density IC devices and packages using manual techniques. Not only are these designs extremely complex, but they must be perfect. Unlike earlier, discrete electronics, chips cannot be reworked or rewired. In addition, the packaging challenge at all levels of the electronic system is to minimize the number of I/O interconnections for the number of circuits involved. This requires a level of design optimization that could not be achieved manually.

PRINTED CIRCUIT DESIGN SYSTEMS

Today's large, high-density printed circuit boards require automated design systems with programs for:

Generating schematic diagrams

Automatically placing components

Automatically wiring interconnections

Checking physical layouts against design rules

Generating detailed drawings and artwork

Generating test data

Generating NC data (e.g., for drilling)

For the system to perform these design functions, the designer must provide input data on the electrical specifications (e.g., circuits and networks) and mechanical specifications (e.g., geometry of the board). The circuit schematic data is the basis for generating the physical layout of components and wiring. A set of design rules must be established so that the system can check the layout for manufacturability. The design verification process ensures that none of the design rules are violated and that all the necessary interconnections have been made for the circuits to function properly. Once the design is completed and checked, post-processing software generates the data required for documentation, artwork, testing, and NC programs.

The major computation task for a printed circuit design system is automated wiring. A printed circuit board can be thought of as a matrix of points on many levels. Some of these points are I/O pins from components, and others are plated through holes (or "vias") in the board which are used to interconnect the various layers. Between all these points are spaces which can be used for routing wires (actually printed lines) to interconnect the circuits. The challenge of the wiring program is to find a way to lay out all the lines in the board without requiring external wiring to complete the interconnections. Several different computational techniques can be used to perform this wiring task. One, for example, treats the board like a maze and runs wires in all possible directions until a satisfactory connection is found. Other techniques try to analyze the total wiring requirements first and anticipate possible problems in order to determine preferred wiring paths. To deal with overflow wires in high-density and complex designs, it is necessary to use interactive wiring techniques. It may also be desirable to use color graphics so that several layers of the board may be displayed simultaneously.

Another computation task is the electrical analysis performed as part of the design verification process. Signal delays which are critical to circuit performance must be computed on the basis of wiring lengths and electrical characteristics (e.g., capacitances and inductances caused by spacings between wires and layers). This is normally an iterative process, and before intelligent workstations it required batch processing.

SEMICONDUCTOR/INTEGRATED CIRCUIT DESIGN SYSTEMS

The design of IC devices involves two distinct stages of activity:

1. **Functional design.** This is basically the circuit design process. It starts with the definition of functional requirements and includes the steps which generate logic diagrams, circuit schematics, and design verification routines.

2. **Physical design.** This is the physical layout of the functional design. It includes the design and placement of circuit elements (e.g., transistors) and their interconnections.

Traditionally, these design activities have been performed by two separate systems of hardware, software, and users. Today, these systems can be integrated to automate the design process from circuits to physical layouts (Fig. 11-13). The logical and physical design processes are interdependent and interactive. The physical layout and the logic design must be compatible when checked against the functional specifications and the design rules. In addition, the graphics software which is used to create logic diagrams can also be used to generate physical designs. Some workstations even use two displays: one for logic design and the other for layout. This allows the designer to compare the logical and physical designs continuously.

Several considerations influence the approach and type of system used by a designer of ICs:

1. **Design methodology.** Factors such as the quantity to be produced, the performance required, and the time allowed for design can determine the approach selected to design an IC device. To optimize performance, cost, and size, for example, one might choose a custom design. However, if the chip is to be produced in relatively small quantities and is required in a short time, a semi-custom design approach might be used. This takes advantage of standard physical designs for typical circuit functions and wires them together to satisfy the requirements of each specific application.

2. **Process technology.** A number of different processes may be used to fabricate ICs such as for field-effect transistor (FET) and bipolar devices. The performance characteristics and design constraints for the device will be determined by which process is selected.

3. **Design system.** Many different systems have been developed to design ICs. Most have the same basic elements, but they may use different design techniques and

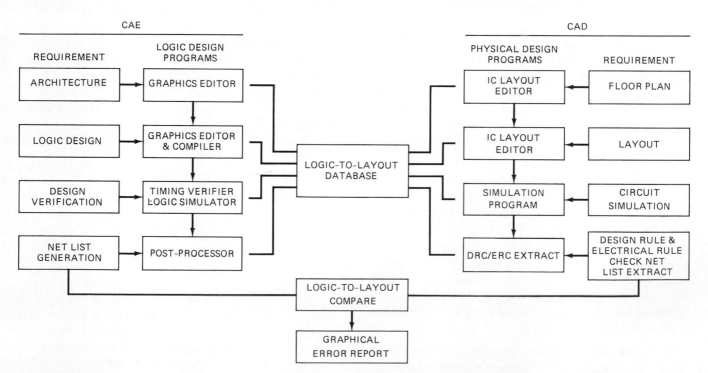

Fig. 11-13 An integrated-circuit design system. *(Peter H. Singer, "CAD Systems: Mapping Out Tomorrow's ICs," reprinted with permission from Semiconductor International Magazine, Aug. 1984; © 1984, Cahners Publishing Co., Des Plaines, IL)*

programs. One common approach is to use a structured hierarchical design. This partitions functional designs of circuits, components, or wiring into smaller blocks which can be designed more easily. These building blocks are then replicated or combined and interconnected to achieve the larger design function.

The typical design process for an IC device proceeds along a sequence of steps as follows:

1. **Design rules.** These describe the logical and physical building blocks available to the designer. Standard circuit designs as well as physical structures of circuit elements are stored in a library and retrieved when selected for use in the design. Also included are constraints in dimensions and tolerances to assure that the design is manufacturable and reliable.

2. **Design data.** The designer specifies the functional requirements of the application. This may be done at a very high level or a very detailed level, depending on the design methodology and system used. The input may be in terms of circuit elements (e.g., transistors), standard logic functions (e.g., AND, OR, INVERT), or macro circuit functions [e.g., arithmetic logic unit (ALU), microprocessing unit (MPU), programmable logic array (PLA)]. Some systems use structural inputs which define the logic elements and their interconnections to specify the design. Others use behavioral inputs which define the logic functions and their performance characteristics. Design data may be input as either lists or graphics. Graphics input has the advantage of being more understandable, and it provides documentation at the same time (Fig. 11-14).

3. **Database.** In order to proceed through the entire design process and permit iterations to the design, a design database must be established. For complex designs this database will be large and must be organized for random access. The data structure must be adaptable to the various stages of the design process which will require the use of the data. It should allow both batch and interactive processing and permit data to be available for the next level of design (e.g., packaging).

4. **Design verification.** This step usually involves a number of different tasks which are performed to determine how well the logical design satisfies the original design specifications. It typically includes a program which simulates the operation of the circuit design to verify its functional performance. It also includes software tools which analyze the timing and delay characteristics of the circuits. Depending on the complexity of the design, these may be large computational tasks. This step is usually repeated several times as the design is changed to correct errors or improve its electrical or physical characteristics.

5. **Partitioning.** This is the first step in converting the electrical design into the physical design. It separates the circuits into sections which will be handled as individual units during the physical design stage. Different approaches may be used depending on the program selected for the design system. Some organize these units by logical function (e.g., registers). Others try to separate the design into the fewest physical units for layout.

6. **Checking.** Preliminary checks are performed at this stage to ensure that basic physical constraints have not been exceeded (e.g., chip size, number of I/Os).

7. **Assignment.** Each circuit element in the design must have its input and output terminals identified to prepare the design for wiring.

8. **Placement.** The location for all the components is determined through an iterative layout process. A number of different techniques are used depending on the design methodology and process technology chosen. The objective is to place the components such that they will be easily wirable and will still meet the electrical and physical design constraints. Some programs use a standard grid approach to partition the design for regular rows and columns of wires. Others cluster circuit groups on the basis of their logical relationship to other groups to optimize performance.

9. **Wiring.** This is an iterative process that can be very time-consuming for complex designs. Unlike discrete electronic assemblies and printed circuit boards, an IC chip can have no overflow wires. All of the many thousands of interconnections that may be involved must be wired and their performance maintained. Many different techniques may be used, and typically several are employed for any one design. The wiring process usually starts at the "global" level of the total chip to establish a general pattern for wiring channels. Vertical and horizontal wiring runs are then made to pick up all the I/O connections possible. Follow-up runs are then made with special algorithms and interactive techniques to capture the unwired connections and overflow wires (Fig. 11-15).

10. **Analysis.** Final checks are performed to verify that the both electrical and physical designs meet the specifications and design rules. At this point accurate circuit performance calculations must be made and output functions simulated. Some design systems have graphics capabilities which display these outputs for the designer (Fig. 11-16).

Fig. 11-14 Graphics input for circuit design. *(Cadnetix Corp.)*

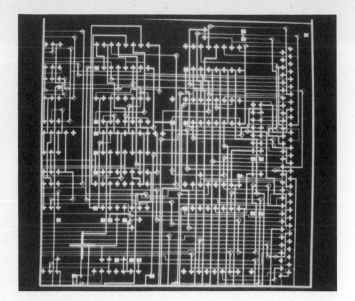

Fig. 11-15 Physical design of a printed circuit board: multi-layer wiring. *(Robert Keeler, "Benchmark Testing a Design Automation System," Electronic Packaging and Production, Nov. 1985)*

11. Test generation. When the design is complete, the designer must ensure that the device can be tested after it is manufactured. Programs using design data generate test patterns which can be used by production test equipment to detect faults if they occur. The challenge for a complex design is to generate the fewest number of test patterns which can detect the largest number of faults. For many complex ICs it is actually not possible to assure that 100 percent of the circuits and functions are tested and that all faults are detectable. The testing process for ICs can involve a great deal of data handling and analysis.

12. Mask data. The physical design is used to generate NC data for tools which fabricate photo masks for IC processing. This is analogous to the drilling data generated for printed circuit boards. It is also another example of how an automated design system must span the scope of activities from conceptual design through driving manufacturing tools.

OTHER ELECTRONIC DESIGN APPLICATIONS

Although printed circuit boards and ICs are the most common type of electronics applications for automated design systems, there are a number of others, such as:

☐ **Printed circuit–like technologies.** These include flexible circuits, hybrid ICs, and multilayer ceramic substrates.

☐ **Assemblies.** These include electronic subassemblies, such as power supplies, amplifiers, and controllers, which typically package a variety of electronic components onto special printed circuit boards.

☐ **Systems.** These are higher-level assemblies of electronics, such as computer equipment, which have complex wiring and interconnection of many subassemblies.

11-6 TRENDS IN COMPUTER-AUTOMATED DESIGN

DESIGN SYSTEMS

To increase the effectiveness and broaden the application of automated design systems, advances are required in the following areas:

☐ User-friendly interfaces and more automated design tools can minimize the skills and training required to use CAD systems.

☐ Solid models can serve as a common database for both development and manufacturing.

☐ Central libraries of standard instructions for common design features can serve as a source for automating programs for manufacturing tools.

☐ Integrated CAD systems can permit complete control over the design and manufacturing process (e.g., logical and physical design in electronics).

☐ Full automation of the physical design process in electronics can eliminate the need for designer interaction and can reduce the time required for design.

☐ High-level tools for logic design will permit the designer to describe the functions required in behavioral terms which the system can convert to primitive circuits.

☐ More efficient handling of batch operations (such as wiring and test generation) can significantly reduce design time.

☐ Common interfaces between different design tools (software) and devices (hardware) can promote the exchange of design data and the integration of computer-aided design and computer-aided manufacturing (CAD/CAM) systems.

☐ The integration of more support functions with CAD systems, such as word processing and electronic mail, can improve documentation and communication during the design process.

☐ The use of CAD systems will migrate from the expert part designers (e.g., IC chip designers) to the designers of the end products (e.g., equipment designers) as the design tools become more automated. CAD systems will require less detailed knowledge and less interaction at the part design level.

CAD HARDWARE

Design applications are driving advances in computer hardware in the following areas:

☐ Higher-performance and lower-cost workstations can provide more efficient design tools and integrated functions to more designers

☐ Intelligent terminals with 32-bit processors and large amounts of memory can increase the power and capabilities of the total CAD system when they are tied together in networks

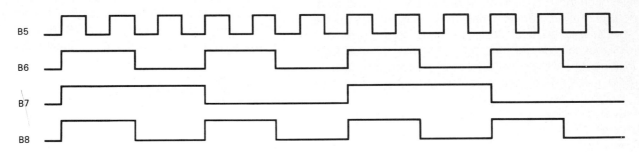

	B1	B2	B3	B4	B5	B6	B7	B8	B9	B10	B11	B12
1	0	0	0	0	1	1	1	1	0	0	0	0
2	0	0	1	0	0	1	1	1	0	0	0	1
3	0	1	1	0	1	1	1	1	1	0	0	1
4	0	1	1	0	0	0	1	0	1	0	0	1
5	0	1	1	0	1	0	1	0	1	0	0	1
10	0	1	1	0	0	0	0	0	1	0	0	1

WAVEFORM REPRESENTATION

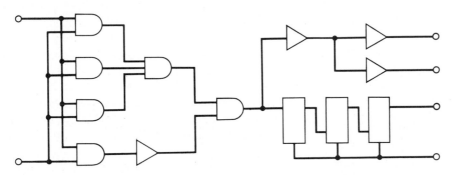

INTERACTIVE DISPLAY OF SIMULATED CIRCUIT

Fig. 11-16 Design analysis: simulation of circuit performance on a design system. *(Jeff Crowley, "Automation Smoothes CAD Process," reprinted from Electronics Week, Dec. 3, 1984; Copyright © 1984 McGraw-Hill Inc., all rights reserved)*

☐ Higher-resolution and lower-cost displays are essential for wider use of realistic graphics and detailed text applications

☐ High-density data storage (e.g., hard disks) in workstations can provide faster access to design data

☐ High-performance computers with parallel processing and high-speed memory can increase the efficiency of graphics computation tasks

☐ Special graphics processors can accelerate the speed at which data is converted to images written on the display screen

SILICON COMPILERS

A recent development in software tools to increase the efficiency of designing ICs is called "silicon compilation."

It got its name from the traditional software compilation process, which creates assembly language programs from instructions written in high-level languages. A silicon compiler is a software tool that automatically generates detailed logical and physical designs of an IC from high-level functional specifications. These specifications are usually expressed in terms of descriptions of standard macro functions, such as random access memories (RAMs), arithmetic logic units (ALUs), programmable logic arrays (PLAs). Rules for the circuit design and physical structure of these functions are stored in the compiler. This eliminates the need for the designer to perform each of the steps normally required in between the functional specification and the final layout (e.g., logic design, logic simulation, circuit design, circuit simulation, physical layout, wiring checks). This makes it possible for designs to be implemented in only a small fraction of the time previously required (Fig. 11-17). It also allows the designer to explore more alternatives without significant redesign effort.

ARTIFICIAL INTELLIGENCE

Artificial intelligence (AI) technology, and expert systems in particular, are finding some applications in CAD. In the area of electronics design there are a number of cases where rule-based systems using the knowledge of expert designers can perform physical layout tasks. By incorporating a large number of design options and layout strategies, complex physical designs can be generated automatically. This can eliminate the need for a separate specialist in physical design, and the logic or systems designer can complete the entire design process. AI techniques can also be combined with silicon compilers to help in high-level decisions for architecture and floor planning based on rules from an experience (or knowledge) database. AI languages and logic programming also lend themselves to many of the IC design tasks (e.g., placement, analysis, test generation). They are relatively easy to use, can deal with symbolic programming techniques, and allow frequent modification.

STANDARDS

The exchange of design data between different CAD systems requires a standard format. In mechanical design systems, the Initial Graphics Exchange Specification

(IGES) has served that purpose. It specifies a standard file structure and language format for communicating geometric data. Further extensions of this standard address more complex mechanical data sets, such as finite element, surface, and solid models. A similar standard, the Electronic Design Interchange Format (EDIF), has been developed for electronics design systems. EDIF specifies a standard format for communicating design data for ICs using symbolic representations similar to the LISP language. The objective of this standard is to permit design data for semicustom ICs to be exchanged between electronic equipment manufacturers and semiconductor manufacturers (Fig. 11-18). It also offers the advantage of allowing the designer to select different hardware and software tools without compatibility constraints.

11-7 SUMMARY

Computer graphics technology provides the base for computer-automated design (CAD) applications, which have made the automation of the design process possible. CAD systems increase the productivity of designers and speed up the process of product development. They are used throughout the design process, including concept development, detailed design, analysis, documentation, and manufacturing process programming. CAD permits changes to be made quickly and complex designs to be implemented without errors.

The principal difference between an automated design system and a drafting system is the modeling and analysis capabilities of the automated system. The major activities involved in the operation of a CAD system are building a model, viewing the model, and handling data. CAD is a data-intensive process. Much more computer power is

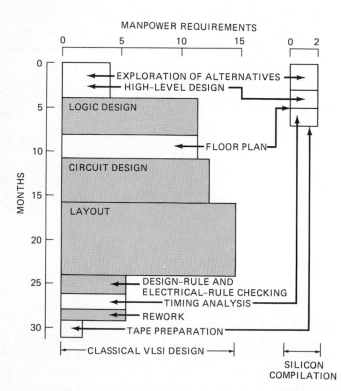

Fig. 11-17 Silicon compilation versus traditional IC design process. (Stephen C. Johnson, "VLSI Circuit Design Reaches the Level of Architectural Description," reprinted from Electronics Week, May 3, 1984; Copyright © 1984, McGraw-Hill Inc., all rights reserved)

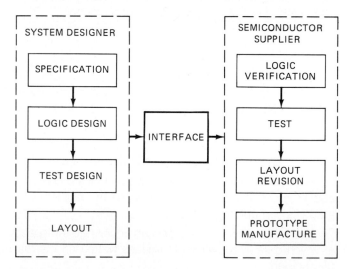

Fig. 11-18 Electronic design interchange format (EDIF): the goal is to enable semiconductor suppliers to fabricate semicustom chips designed at different types of workstations. (George Leopold, "Design Norm Takes Shape for Standard Cells," reprinted from Electronics Week, Sept. 24, 1984; Copyright © 1984, McGraw-Hill Inc., all rights reserved)

required for computation and data handling than for generating graphics output. CAD systems establish a database which can be shared by other organizations in development and manufacturing to perform engineering tasks. They produce outputs for manufacturing in addition to the design itself (e.g., tests, NC data).

A variety of tools and techniques are used in CAD systems to make the design job easier and faster. CAD systems provide tools for the analysis and checking of designs. They can also provide capabilities which are not practical or even feasible with manual methods (e.g.,

animation, zooming). It would not be practical, or in some cases even possible, to design some of today's high-technology products, such as integrated circuits (ICs), without a CAD system. Not only are the designs extremely complex, but they must be perfect.

Although mechanical and electronic design are the most common applications for automated design systems, a variety of other areas of manufacturing also employ CAD. As hardware and software advances continue, the scope of CAD applications will expand further, and truly automated CAD/CAM systems will result.

REVIEW QUESTIONS

The answer to each question can be found in the section(s) indicated at the end of the question.

1. Why are CAD systems used? [11-1]

2. What are the basic tasks performed by a CAD system? [11-1]

3. Describe the basic stages of the design process. [11-2]

4. Identify some of the benefits in automating the design process. [11-2]

5. Identify some of the factors to be considered before implementing a CAD system. [11-2]

6. Characterize the different types of design systems which exist. [11-3]

7. Identify the major components in the architecture of a CAD system and describe their functions. [11-3]

8. What are some of the techniques used to aid the designer during the operation of a CAD system? [11-3]

9. Describe the typical steps involved in using a mechanical design system. [11-4]

10. Identify some of the different techniques that can be used to draw on a CAD system. [11-4]

11. Describe the basic steps involved in creating a geometric model of a mechanical design. [11-4]

12. Describe some of the analysis and checking functions which are used in a mechanical design system. [11-4]

13. Describe the basic steps involved in the electronics design process. [11-5]

14. Describe how an integrated circuit is designed on a CAD system. [11-5]

15. Identify some examples of outputs from a CAD system other than the design itself. [11-4 and 11-5]

16. Identify some of the trends which will influence the capabilities and applications of CAD systems. [11-6]

COMPUTER TOOLS FOR ENGINEERING ANALYSIS

12-1 INTRODUCTION

Engineering tasks in a manufacturing environment do not stop with the design of the product. Computer-automated engineering (CAE) offers tools that can be applied to solve a wide variety of engineering problems. All the potential benefits of automated design systems can only be realized if they are integrated into the automation of the total engineering process. CAE can not only improve the productivity of the designer; it can also ease the transfer of information from development to manufacturing.

To bring a product from the design stage to production involves a number of engineering activities (Fig. 12-1). Engineers can use the geometric model created for the design, together with analytical tools to evaluate the funtionality and manufacturability of the design. This can include mechanical and kinematic analyses of stresses, rigidity, stability, and thermal performance. These tasks traditionally involve a lot of computation and require a great deal of data. By using computer tools with a design model, engineers can find problems in a relatively short time without ever having to build and test a prototype.

The design data for each part can also be used to determine which tools and process sequence should be used to fabricate them. Manufacturing analysis activities such as process planning and numerical control (NC) programming can be conducted on the same CAE system that was used to create the design. This can lead to improved product quality and shorter overall time needed to introduce new designs into manufacturing. CAE tools can also be used by manufacturing engineers to create models of the process and simulate the operation of the line in order to optimize its performance.

This chapter will cover all of these engineering activities and discuss the types of computer tools which can be applied to them. The emphasis will be on identifying the capabilities of CAE systems rather than on describing how to solve any specific engineering problems in detail.

12-2 DESIGN ANALYSIS

PHYSICAL PROPERTIES

Geometric models of objects can be used to calculate their physical properties. These may include:

Dimensions

Surface area

Volume

Weight

Center of gravity

Moments of inertia

It is often necessary to determine properties like these during the design process in order to properly specify the object or to conduct other engineering analyses. Depending on the shape of the object, the calculations involved in determining these properties may be very complicated. With the aid of a geometric model stored in a CAE system, an engineer can have the calculations done automatically when the design is complete. In some cases, this ap-

```
┌─────────────────────────────┐
│           DESIGN            │
└─────────────────────────────┘
       Geometric Model
       Engineering Drawings
              │
              ▼
┌─────────────────────────────┐
│      DESIGN ANALYSIS        │
└─────────────────────────────┘
       Physical Properties
       Finite Element Analysis
       Kinematic Analysis
              │
              ▼
┌─────────────────────────────┐
│   MANUFACTURING ANALYSIS    │
└─────────────────────────────┘
       Process Planning
       Group Technology
       NC Part Programming
       Tool Design
       Test Data Generation
              │
              ▼
┌─────────────────────────────┐
│  MANUFACTURING SIMULATION   │
└─────────────────────────────┘
       Tool Operation
       Line Layout
       Process Flow
       Human Factors
```

Fig. 12-1 Engineering activities.

proach makes it possible for the first time to accurately determine some properties of complex designs. For example, it is necessary to precisely determine the critical dimensions of the complex, contoured inner surfaces of the combustion chambers of engines in order to verify their ability to meet environmental emission requirements.

The mathematical methods used to calculate the physcial properties of objects depend on the type of geometric model involved. In some cases, the model may have to be converted to a different type of representation in order to determine a particular property. Wire-frame models, for example, cannot be used to automatically compute the properties of solid objects. In order to determine such properties, it is essential to distinguish where all the lines of the design are relative to the surface of the object. Since solid models are a true and unambiguous representation of solid objects, they permit such properties to be calculated directly by the design system. Even with solid models, different techniques may be required for different representations. For example, boundary representation models use surface-integration methods to determine the mass properties of objects. Constructive solid geometry (CSG) models, however, use other methods, such as ray casting.

Some CAE tools can perform sophisticated engineering tasks involving the physical properties of a design. Solid models give the designer the ability to cross-section an object, compute all of its dimensional characteristics, and determine all of its key mass properties automatically. Special programs have even been developed which can optimize a design by determining the minimum mass which is capable of satisfying the structural requirements of the application. This can be particularly useful in such weight-sensitive applications as aerospace. Such programs involve an iterative design process using other engineering tools, such as finite element analysis (FEA).

FINITE ELEMENT ANALYSIS

Finite element analysis is a mathematical technique that is used to calculate stresses in mechanical structures. The technique involves separating an object into many small uniform pieces or elements, then using stress and deflection equations to describe the behavior of each one when forces are exerted on the entire structure (Fig. 12-2). The behavior of the entire structure is then determined by solving all the equations simultaneously. This typically involves hundreds and even thousands of equations, and a large amount of computer power is required to solve them. FEA is not practical for complex structures using manual techniques. Until low-cost, high-performance computers were available, it was not even economical except for the most sophisticated applications (e.g., nuclear power and aerospace). During the early uses of FEA, the mesh patterns had to be produced and measured manually and the data input to the computer for batch processing. Today, the mesh can be generated

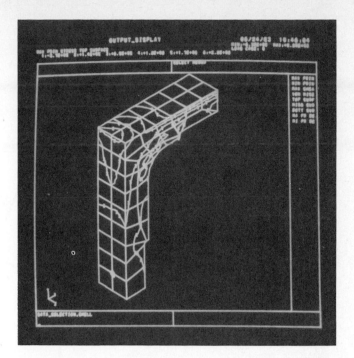

Fig. 12-2 Finite element analysis. *(International Business Machines Corp.)*

automatically from a geometric model and the equations set up and solved by an FEA program, all on a CAE workstation. This has made it practical to use FEA in a wide variety of design applications. In the automotive industry alone, these techniques are used routinely for such applications as:

Determining the stresses on pistons

Analyzing the effects of impacts on bumpers

Identifying deflection and stress in tires

Evaluating the behavior of vehicles in crashes

The basic method used for FEA involves the following steps:

1. Building the model. This is usually an iterative process. To get a quick approximation of the major deflections and points of highest stress, a simplified "stick-figure" representation of the object may be used first. The geometric model of the object is then divided into simple elements, such as cubes, rods, or shells. This creates the finite element model or mesh which is used for analysis. A coarse model with relatively few elements is usually created first and then refined with more elements in the critical areas until sufficient accuracy is achieved. This iterative process is a tradeoff between accuracy and computer time and cost. In Fig. 12-3, for example, as the number of nodes in the model increases, error is reduced but modeling time and computer costs rise sharply. Other techniques are also used to minimize the number of elements required to achieve a given accuracy. One uses a library of "isoparametric elements" which can be selected to fit complex shapes with fewer elements than would be required with conventional uniform shapes. In Fig. 12-4, for example, models constructed with isoparamet-

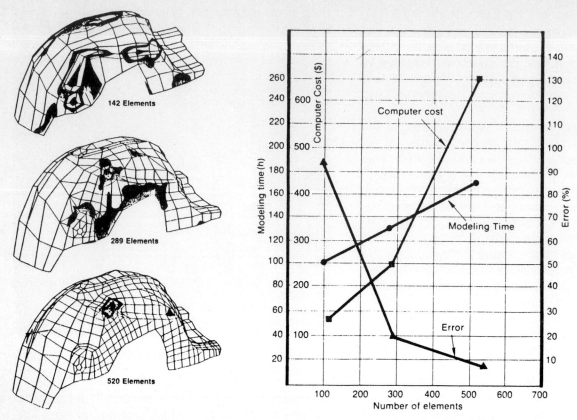

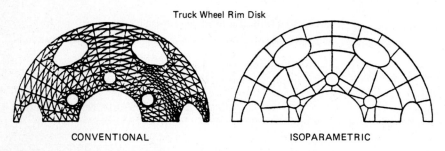

Fig. 12-3 Iterative process for finite element analysis modeling—analysis of a wheel well done with increasingly defined models. *(Structural Dynamics Research Corp.)*

Truck Wheel Rim Disk

CONVENTIONAL   ISOPARAMETRIC

Fig. 12-4 Isoparametric elements for finite element analysis. *(Structural Dynamics Research Group)*

ric elements may require only about 25 percent as many elements as those made with conventional elements. Another technique takes advantage of symmetries in part designs to model only portions of the structure. Some FEA programs can then combine these repeated patterns into a complete model.

2. Data entry and computation. The elements of the model are assigned physical properties on the basis of their materials (e.g., weight, strength, stiffness). Forces are identified which are to affect the entire structure in the particular application involved. The coordinates of each of the elements, together with their physical properties and the forces applied, are fed to the FEA program. Each element is treated like an individual beam, for which the stress and deflection characteristics are described in

terms of classical equations from physics and mechanics theory. The computer program then calculates the deflection and distribution of stress throughout the structure by solving all the equations simultaneously.

3. Output. The results of the calculations involve large amounts of data which must be condensed into a meaningful form to be useful for design analysis. Post-processing routines are typically used in FEA programs to display the results in graphic form. Contour plots can identify lines of constant strain; color codes can indicate the areas of highest stress (Fig. 12-5). Graphics output allows the designer to see deformations and patterns of stress without having to analyze large volumes of data. Potential weak spots that may need reinforcement can be easily detected. Changes can then be made to the design or physical properties and tried again using the FEA pro-

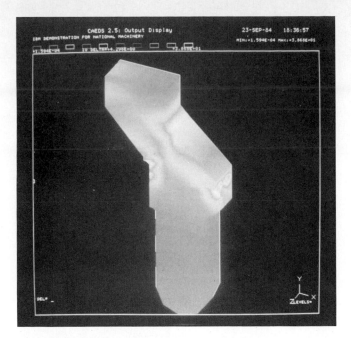

Fig. 12-5 Color graphics output for finite element analysis. *(International Business Machines Corp.)*

gram. This avoids the need to build and test prototype models.

KINEMATIC ANALYSIS

Kinematic analysis, which is sometimes also referred to as dynamic analysis, is the simulation of the behavior of objects when they are in motion. Kinematic programs use geometric models and FEA models to create animated images of the movements of mechanisms and complex structures. They can also determine the velocities, accelerations, and loads involved. This can give a designer the ability to see how a design will behave when it is actually used. Slow-motion animation can reveal potential problems, such as unwanted vibrations and dangerous deflections. By rotating 3D models, parts can be tested for fit and clearances in an assembly. The paths of mechanisms that undergo cyclical, repetitive motion (e.g., cams, gears, linkages) can be traced. Computer simulation can give a designer a view of events that occur too rapidly to see in real life, such as a car crash (Fig. 12-6). It can also provide more data than could be obtained by actually

testing a prototype. Kinematic analysis also makes it possible to test objects that would not be practical or even feasible to test in real life, such as artificial limbs and artificial organs, deep-sea oil-drilling platforms, spacecraft, and satellites.

Dynamic analysis programs can be used to simulate the motion of complex mechanical systems, such as entire automobiles or machine tools. They can handle both small and large displacements to analyze deflections and vibrations in mechanisms which are subjected to a wide range of loads and stresses, such as suspension systems (Fig. 12-7). This kind of analysis permits the designer to determine the greatest loads that components such as bearings and bushings can be expected to experience in an application. FEA techniques are then used to determine the peak stresses that would be associated with those loads.

Modal testing is a form of kinematic analysis that is used to predict the behavior of large, complex structures, such as automobiles and aircraft, that are subjected to heavy vibration stresses. It combines the analytical data from a finite element model with experimental data collected from testing physical models. Test data is collected by the computer system from test points on the structure, using sensors which can detect stress and strain. With slow-motion animation, the system then displays the deformations that would result from applying specific forces. The so-called mode shapes that are simulated illustrate how the structure would deform and vibrate, and the designer can identify those areas which exhibit the largest deflections. Animated graphics are often exaggerted on purpose to reveal the most severe deformations and evaluate the effects of design changes. This can help the designer find the source of the problem and modify the design accordingly. Modal testing is a powerful tool that can significantly reduce the time and effort involved in testing and changing designs.

Another example of dynamic design analysis combines structural and acoustic computer models to predict what parts of a structure create noise during motion. As illustrated in Fig. 12-8, the use of such an engineering tool allows the designs of automobiles to be tested and modified to reduce noise caused by vibrations without building and testing prototypes. This can save the time and expense of the traditional trial-and-error "road-test" approach.

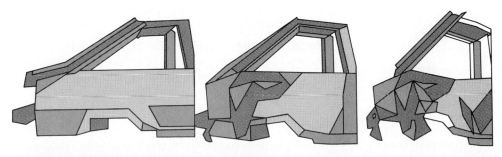

Fig. 12-6 Kinematic analysis simulation (car crash).

151

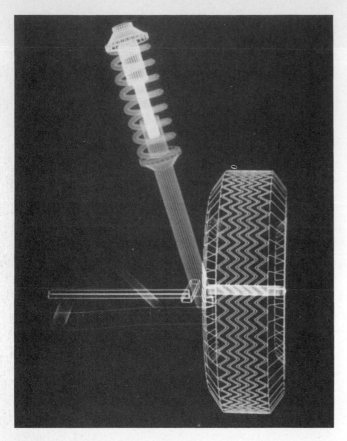

Fig. 12-7 Dynamic analysis of automobile suspension system. *(Evans and Sutherland Computer Corp.)*

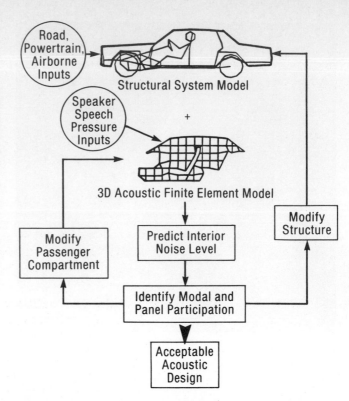

Fig. 12-8 Computer models to predict noise in car designs. *(General Motors Corp.)*

12-3 MANUFACTURING ANALYSIS

PROCESS PLANNING

After a product is designed, the manufacturing process to be used to produce it must be planned. In the case of machined parts, for example, this involves determining:

☐ The sequence of operations required

☐ The machine tools or workstations to be used

☐ The machining parameters to be set (e.g., cutting speeds and feed rates)

In the past, many of these decisions were left to the skilled machinist on the factory floor. For numerically controlled tools, however, these decisions must be made in advance. After a process plan has been developed, NC programs must be generated to drive the production tools. In most cases, the process planning and NC programming is done by people who do not have the production experience of skilled machinists. As a result, they do not always understand all the implications of their decisions. Errors in process planning can lead to inefficiencies or even parts damage if the speeds and feeds specified are not optimum. The solution to this dilemma between relying on the experience of individual machinists or the inexperience of process planners is the use of the computer. In computer-aided process planning (CAPP), the skills and experience of machinists are built into a computer program. There are two major types of CAPP systems:

1. Variant or retrieval. This uses a standard process which has been developed for existing parts, which can be reused or retrieved for similar parts.

2. Generative. This operates like an expert system to develop process plans. It uses rules and experience to make decisions based on an analysis of the geometry and materials involved.

An automated process planning system may involve a number of operations and subsystems (Fig. 12-9):

Macroplanning. A specific sequence of operations on selected machine tools is developed.

Microplanning. The details of each process operation for each tool are specified, including the selection of cutting tools, the cutting sequence, the cutting speed, the feed rate, and the depth of cut.

Model building. Data is collected on machine performance to build a mathematical model of machining parameters which can be used to optimize operations (e.g., depth of cut and cutting speed versus tool wear or breakage).

Decision making. Process plans and optimized operations are determined using the manufacturing database and the model.

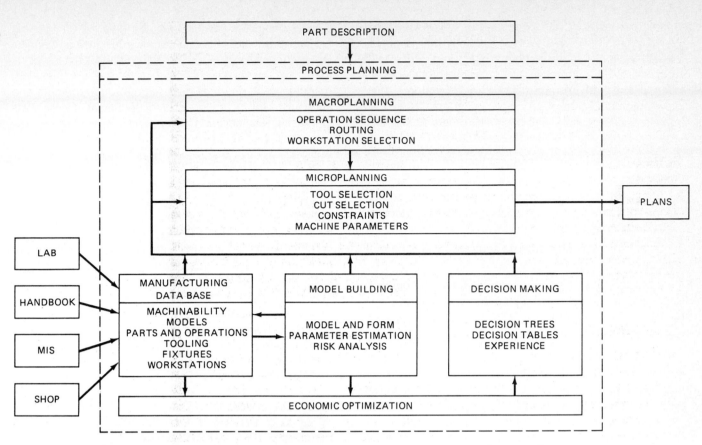

Fig. 12-9 Automatic process planning system. *(William J. Zdeblick, "Process Design by Computer," IEEE Spectrum, May 1983; © 1983 IEEE)*

To use a CAPP system, the process planner inputs data in response to questions about the types of operations, the dimensions, and the tolerances involved. The program then develops a process plan based on this data, using its decision-making capability, from the manufacturing database and operational model. Some programs can even instruct the operator when to change a tool or how to modify the machining conditions as the tool wears. CAPP can provide benefits to manufacturing, such as:

Increased productivity of process planners

Improved consistency of process plans

Optimized machining operations

Reduced process cycle time

Better-quality documentation

Greater use of existing programs

GROUP TECHNOLOGY

Group technology (GT) is an approach to classifying standard parts on the basis of their physical characteristics so that they may be grouped together with other similar parts that could be manufactured using the same process. Most CAPP systems use a data retrieval method based on GT where the parts are organized into families with similarities in shape and machining operations. GT provides a link between design systems that specify the geometry and materials of a part and the manufacturing systems that determine the production process.

GT can be used to avoid the use of duplicate parts as well as to provide a grouping of parts for common processing. Since many large manufacturers use hundreds of thousands and even millions of different parts, it may require a substantial effort to build a GT database. However, automated design systems can provide the capability to automatically code parts into GT classifications using the information in the design database. The actual coding and classification process involved in creating a GT database will be discussed in Chap. 18.

The GT concept extends beyond the classification of parts. It also includes the arrangement of tools in manufacturing to optimize the work flow and throughput of the production operation. By grouping machines into "cells" that perform all the necessary machining operations on similar parts, one may achieve higher efficiency than is obtained by routing the parts through a variety of machines throughout the factory floor which are working on many different types of parts. The ability to arrange a manufacturing floor in this manner depends a great deal on the quantities of parts, the number of groups, and the stability of designs that are involved. In any event, GT provides the ability to identify opportunities for improved efficiencies in both design and manufacturing. The applications for GT will be covered further in Part 5.

NC PART PROGRAMMING

Numerically controlled machine tools were originally programmed manually from design drawings. Computer-automated design (CAD) systems now make it possible for NC programs to be developed directly from the geometric model of the part to be manufactured. Since solid models provide a complete and unambiguous representation of the part geometry and all of its features, they allow automatic generation of NC programs as well as improved visualization. The basic NC programming process has the designer follow these steps:

1. Establish a geometric model of the part in the design database.

2. Establish a tool-parameter database with all the unique characteristics and capabilities of each machine available to production.

3. Select the tools and process sequence from the process plan.

4. Input the machining requirements for each operation (e.g., type of cutting tool and area to be machined).

5. The NC programming system generates a tool-path program based on the input requirements and geometric model.

6. Verify the tool-path program by simulating it on the graphics display (Fig. 12-10). By displaying the cutter path and part outline during the advance of the tool, potential problems can be identified, such as interference or incorrect tool movements. This avoids the need to verify the program on the actual machine. NC verification involves not only graphics but also computations to check for clearances, cutter constraints, and collisions.

7. Translate the high-level NC program into the language used by the machine tool.

8. Download the NC program to the machine tool on the factory floor for operation.

TOOL DESIGN

By using the capability of a graphics design system together with an NC programming system, it is possible to eliminate the need for some of the master tools that were traditionally needed to fabricate the final production tools. This can avoid an extra step and source of error in the

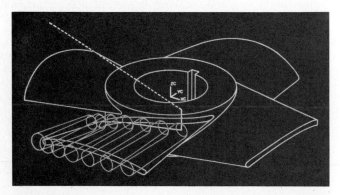

Fig. 12-10 NC tool-path verification. *(McDonnell-Douglas Manufacturing Industry Systems Co.)*

tooling process. It can also result in lower overall tooling costs and a faster response to new designs or changes. By working directly off NC programs and process plans that have been automatically generated and verified, manufacturing does not have to go through the time and effort involved in trying out and debugging the process first, using real tools and parts.

TEST DATA GENERATION

As we saw in the design process for integrated circuits (ICs) in Chap. 11, data for testing the final product can be generated by design systems. This is also possible for mechanical parts. The design data stored in the CAD system can be used to check the accuracy of parts before they are assembled. This same data can be used to generate test routines for the inspection of parts in manufacturing. Dimensional data can be fed to computer-controlled coordinate-measurement machines which can automatically inspect large, complex parts with extreme precision. Programs also exist which can take the results of these measurements and calculate adjustments that may be necessary in the machine tools to correct errors.

12-4 MANUFACTURING MODELING AND SIMULATION

WHY SIMULATE MANUFACTURING OPERATIONS?

Simulating a manufacturing operation on a computer before it is actually tried out on the factory floor can avoid the time and effort of trial-and-error experimentation. In the case of a new operation or process that has not yet been implemented, simulation can test its feasibility and modify its design before money is spent on installing tools and facilities. Modeling and simulation can also help to improve the efficiency of an existing operation by trying alternatives without interrupting production. Some of the advantages of using computer simulation tools for solving manufacturing problems include:

☐ The effects of changes on manufacturing operations (e.g., quantities and mix) can be evaluated rapidly

☐ Specific causes of problems can be identified in a complex process

☐ The process can be speeded up to show manufacturing performance over a long period of time in just a few minutes

☐ Decisions on manufacturing controls and product flow can be tested before implementation

☐ The activities involved in laying out the line and planning the process can be tied together

☐ Performance data as well as a visualization of performance can be provided

☐ A number of alternatives can be tried and variables tested to reach an optimum solution

Simulation tools take advantage of the conceptual thinking of a human by allowing an interactive decision-making process. Previous simulation methods produced large volumes of data that were difficult for management to interpret and use. Although modern simulation tools have limitations, their speed of computation, interactive capabilities, and graphics displays make them practical for planning and decision-making tasks in manufacturing.

BASICS OF MODELING AND SIMULATION

Simulation is the use of a model of an operation to evaluate its behavior under varied conditions. A number of different mathematical techniques can be used to create models, but most of them do not lend themselves well to complex manufacturing processes. Since manufacturing consists of many discrete operations that behave in a random or nonlinear fashion, it cannot easily be represented by equations that are directly solvable. Manufacturing models use symbols that represent objects, such as parts, and resources, such as machines. A computer can simulate the operation of a process represented by such a symbolic model by keeping track of all the objects and resources during their operation over a period of time. It monitors their behavior by identifying changes in their status as they interact with each other.

This "discrete-event simulation" represents intervals of time in discrete steps and updates the status of the system at specified time intervals or when scheduled events occur. Using decision rules, the simulation system tracks the flow of objects and the operation of resources from one state to another as time progresses The decision rules define whether and how certain events follow each other. Manufacturing simulation is a convenient way to relate output to input. It can provide detailed information on the behavior of complex operations with accuracy, if the model is representative of reality.

The basic process involved in simulation involves:

☐ Defining the problem to be solved

☐ Creating a symbolic model of the manufacturing operation

☐ Gathering data to build the model

☐ Developing or selecting a computer program to run the simulation

☐ Debugging the program

☐ Validating the model against actual operations

☐ Designing experiments for evaluation

☐ Exercising the model against the experimental conditions

☐ Analyzing the results of the simulations

☐ Using the results to develop recommendations or decisions

☐ Modifying the model to accommodate changes or improvements

TYPES OF MANUFACTURING MODELS

Models can be used for simulating a variety of different types of manufacturing operations and problems:

1. Machine or tool operation. Perhaps the simplest, or at least the most limited in scope, is a model of an individual machine operation. This can be used to debug a new or proposed process step or tool. It may involve a complete work cell with materials handling equipment as well as the manufacturing tool itself. One example of a simulation in which the process, materials, and tool interact is plastic injection molding (Fig. 12-11). A simulation of this operation can help to optimize throughput and material usage while maintaining the product specifications. The data for the mold design can be obtained from the geometric model of the part design. The user must then input the material characteristics and tool parameters.

2. Line layout. Models can be used to simulate the physical layout of manufacturing lines. This permits the user to try different locations for the placement of tools and materials handling systems in an attempt to optimize the use of space as well as the efficiency of product flow. Alternative layouts and configurations of tools or work cells can be evaluated before they are actually installed on the factory floor, avoiding the need to change them later to fix problems. An example of this is the placement of robots in relation to a moving conveyor line for the assembly of automobiles (Fig. 12-12). This simulation can determine the speeds and clearances required to perform the operation efficiently and avoid problems such as interferences and collisions.

3. Process flow. One of the most common and valuable applications for manufacturing models is to simulate the operation of an entire production line. A model of a production process must describe it in terms of a system

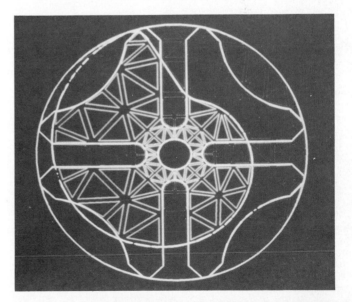

Fig. 12-11 Simulating plastic injection molding. *(McDonnell-Douglas Manufacturing Industry Systems Co.)*

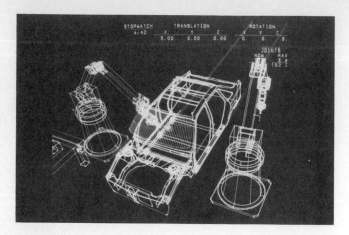

Fig. 12-12 Simulating robot assembly of automobiles. *(Mc-Donnell-Douglas Manufacturing Industry Systems Co.)*

(Fig. 12-13). The production system would have inputs and outputs, such as information, material, and resources. It would also have functions, such as the control and execution activities in manufacturing (i.e., rest, move, make, and verify). The model of a process flow permits the scheduling, routing, and utilization of equipment to be evaluated before implementing or changing the system. The performance of the line can be determined under different conditions in order to identify potential bottlenecks or capacity constraints. Such simulations can lead to changes in scheduling techniques or materials handling schemes. They may also identify the need for additional tools or improvements in the performance of specific tools. Graphics output can visualize the flow of parts through the production operation, making it

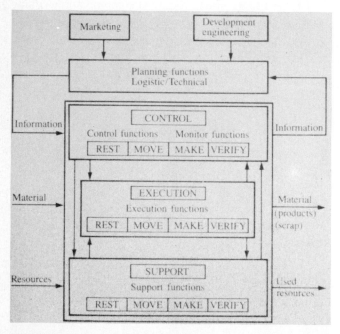

Fig. 12-13 Model of a manufacturing line operation. *(H. Engelke et al., "Integrated Manufacturing Modeling System," IBM Journal of Research and Development, July 1985; © 1985 by International Business Machines Corp., reprinted with permission)*

easy to detect problem areas. The system can also be used to determine the key control factors for the performance of the line [e.g., work in process (WIP), job priorities, cycle time, batch sizes].

4. Human factors. Computer models can also be used to simulate the motions of operators in the production environment. This can help to identify potential problems with clearances, ease of movement, and strain. The design of a workstation can be done on a graphics system with the movements of the human operator simulated by a "software manikin" (Fig. 12-14). Workstations which have been evaluated for human factors will usually permit more efficient operation and reduce fatigue and injuries.

USING SIMULATION TOOLS

Special programming languages are used to create simulation models which are then compiled into a commmon computer language. Over the years, many different simulation languages have been developed, each of which has its own merits and intended types of applications. Some of the most popular are:

General Purpose Simulation System (GPSS)

General Activity Simulation Program (GASP)

Simulation Language for Alternative Modeling (SLAM)

Research Queueing Package (RESQ2)

Most of these do not require programming experience to use. Some do not even require that the user be familiar with simulation techniques. They may only need data to be supplied to the system and not a description of the process. Modeling program packages have been developed for certain types of applications that avoid the need to build and program a model. However, for simulators to be most effective, they should be tailored to the specific application and reflect the real nature of the operation involved.

Once a model has been built or selected, data must be collected to run the simulation. The validity of the results depends not only on the reality of the model but also on the accuracy and completeness of the data used. Detailed data is normally required to describe the sequence and operation of each step in the process. This includes the timing, delays (e.g., queues), utilization of tools, and performance factors (e.g., throughput rates, capacities). The types of information that are typically needed are:

1. Characteristics of the parts involved. This typically includes an identification and description of all the tools and operations needed to fabricate each part as well as the sequences and time values involved.

2. Characteristics of the tools involved. This includes the number of tools available of each type as well as their performance capabilities.

3. Characteristics of the materials handling system. The type of system, together with its layout and performance, must be described.

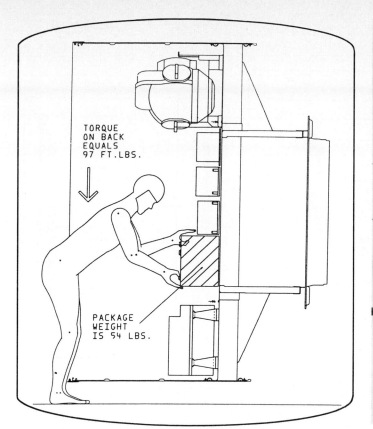

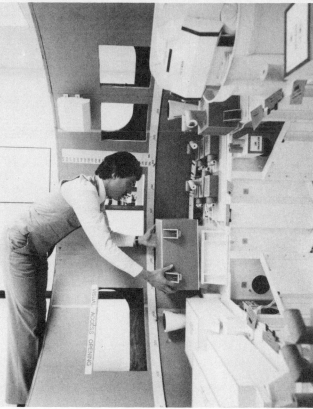

Fig. 12-14 Simulating human factors in a workstation. *(Cadam, Inc., a subsidiary of Lockheed Corp.)*

4. Characteristics of manufacturing operation. This identifies the way in which the line is staffed and scheduled (e.g., number of shifts, hours, number of workers).

Once these details are input to the model, the production requirements or problem conditions can be entered and the simulation executed. The effort required to obtain all the data necessary to perform a manufacturing simulation can be substantial. However, all of this information will eventually be needed to implement the actual operation. The simulation exercise can therefore not only prevent problems before they are experienced; it can also prepare manufacturing better for the implementation process. Aside from the effort involved in building and quantifying the model, there is a major challenge in convincing the owner of the problem to make changes on the basis of the simulation results. This requires confidence in the model and the desire to make improvements in the operation.

The output of a simulation run can be expressed in quantitative as well as graphic form (Fig. 12-15). Traditional methods of simulation yielded large amounts of data that were not easy to interpret and did not permit dynamic interaction with the user. Simulators that are available today display the results so that they can be visualized as well as quantified in summary form. They also permit the user to speed up the process to accelerate the time base or stop it to make changes. In addition, the use of animated graphics makes it easier to identify potential problem areas.

12-5 INTEGRATED CAE SYSTEMS

SYSTEM ARCHITECTURE

CAE systems are not just design tools. They can help to integrate the entire engineering process from the concept of a product design through the programs that drive the machines that fabricate the product on the factory floor.

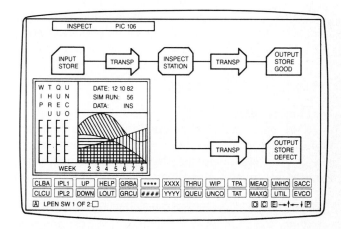

Fig. 12-15 Graphics output of a manufacturing simulation run. *(H. Engelke, et al., "Integrated Manufacturing Modeling System," IBM Journal of Research and Development, July 1985; © 1985 by International Business Machines Corp., reprinted with permission)*

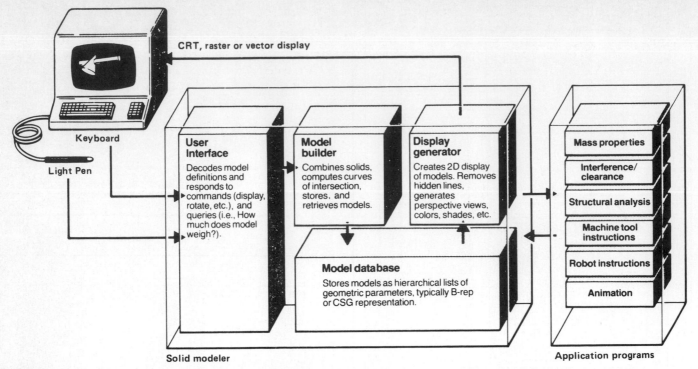

CRT, raster or vector display

Keyboard

Light Pen

User Interface
Decodes model definitions and responds to commands (display, rotate, etc.), and queries (i.e., How much does model weigh?).

Model builder
Combines solids, computes curves of intersection, stores, and retrieves models.

Display generator
Creates 2D display of models. Removes hidden lines, generates perspective views, colors, shades, etc.

Model database
Stores models as hierarchical lists of geometric parameters, typically B-rep or CSG representation.

Mass properties

Interference/ clearance

Structural analysis

Machine tool instructions

Robot instructions

Animation

Solid modeler

Application programs

Fig. 12-16 Architecture of an integrated CAE system. *(Paul Kinnucan, "Solid Modulers Make the Scene," reprinted with permission, High Technology Magazine, July/Aug. 1982; copyright © 1982 by High Technology Publishing Corp., 38 Commercial Wharf, Boston, MA 02110)*

The key to achieving an integrated engineering system is the creation of a geometric model. Solid models, in particular, permit the user to perform a variety of other engineering tasks after the basic design is created. The architecture of an integrated CAE system, therefore, is built around the solid modeler (Fig. 12-16). A graphics display system obviously provides the user interface. To that is added the application programs for each of the types of engineering analysis tasks to be performed as part of both the development and manufacturing processes. Such a CAE architecture can permit a single user to perform all or many of the engineering tasks. For complex products, it permits all the engineers involved to use a common system and share the same database.

APPLICATION EXAMPLES

Sheet-metal parts. The aerospace industry was one of the early users of computer-aided design and manufacturing systems. Because of their experience and the nature of their operations, they have become a major user today of integrated CAE systems. The manufacture of an aircraft, for example, involves a wide variety of precision fabrication operations on relatively small batches of parts. Computer tools are used to minimize the cost and time required while maintaining the precision and quality necessary.

One good example of an integrated CAE approach in aerospace manufacturing is the nesting and machining of sheet-metal parts (Fig. 12-17). In aircraft manufacturing

there is often a need to machine many odd-shaped parts from sheet metal. Done manually using templates, this can be time-consuming and can waste a lot of material. This application was ideally suited for the capabilities of a CAE system. The part designer can use the system to perform the following operations:

☐ Assemble the part designs into a nesting pattern that will optimize the use of material as well as the machining time

☐ Establish a sequence of machining operations to cut the parts to shape

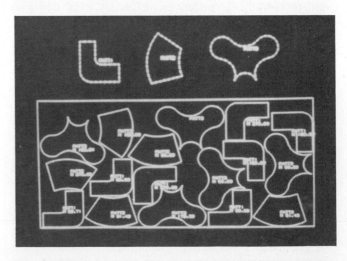

Fig. 12-17 Nesting of sheet-metal parts for machining. *(McDonnell-Douglas Manufacturing Industry Systems Co.)*

158

☐ Create a riveting pattern for holding a stack of sheets together during the machining operations

☐ Convert the design and nesting data into NC programs to drive the machine tools

☐ Determine the utilization of material on the basis of the nesting pattern and machining program created

Machined parts. Machining applications have become a major user of CAE tools. This is particularly true of high-volume manufacturing operations like the automotive and heavy-equipment industries. These obviously are not

(a)

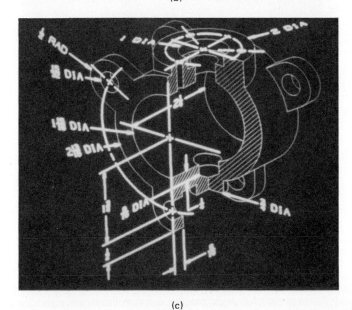

(b)

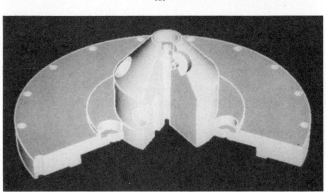

(c)

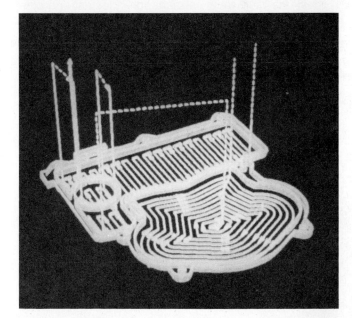

(d)

(e)

Fig. 12-18 Integrated approach to computer-automated design and manufacture of a machined part. *(McDonnell-Douglas Manufacturing Industry Systems Co.)*

characterized by the precision and low-volume nature of the aerospace industry. The driving forces for CAE in these applications are cost and time to implementation. CAE has become an important tool for the design and manufacture of such mechanical products at all levels, including chassis, engines, transmissions, and suspension systems.

An excellent example of an integrated approach to a CAE application is the design and manufacture of critical machined parts (Fig. 12-18). In this case, the system is used to perform all the engineering tasks from design through manufacturing of the part, including:

1. Using interactive computer graphics to describe the geometry of the part (Fig. 12-18*a*)

2. Creating a solid model of the part which can be cross-sectioned and analyzed for physical properties (Fig. 12-18*b*)

3. Generating design drawings and alternative views for analysis and documentation (Fig. 12-18*c*)

4. Performing structural analyses using FEA (Fig. 12-18*d*)

5. Generating tool paths and NC programs for machining operations (Fig. 12-18*e*)

Plastic Molding. The plastic injection molding industry fabricates a wide variety of parts for industrial and consumer use. Many of these involve complex shapes and specially formulated materials. Computer tools have taken the guesswork out of the molding process by simulating the flow of plastic through a mold under specified conditions. The engineering analysis starts with the generation of an FEA model of the mold (Fig. 12-19). Data on the material, temperature, and other process information is then entered into the model. The program will then calculate the resulting volume, flow rates, temperature, and pressure. Using computer graphics, the pressure

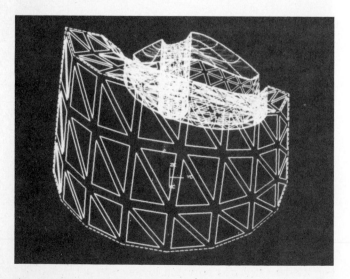

Fig. 12-19 Finite element model of a plastic injection mold. *(McDonnell-Douglas Manufacturing Industry Systems Co.)*

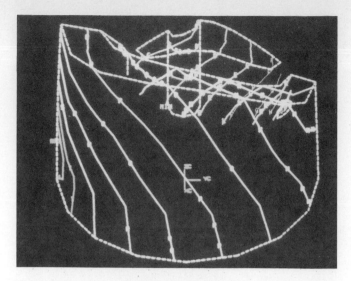

Fig. 12-20 Graphics display of pressure contours of a plastic mold. *(McDonnell-Douglas Manufacturing Industry Systems Co.)*

contours of the mold can be displayed for analysis (Fig. 12-20).

12-6 TRENDS IN CAE

The most significant trend for CAE is the integration of the hardware and software technologies that now exist into a complete system for design and engineering analysis. Some of the implications of this are:

☐ The development and manufacturing organizations will be linked by a shared engineering tool which will eventually cause their tasks to blend together. The analysis, decisions, and control data for manufacturing will be driven by the design of the product.

☐ Engineers designing parts will generate process plans and NC programs for manufacturing tools directly from the design model.

☐ Parts will be designed and manufactured without the need for any physical drawings.

☐ Artificial intelligence (AI) techniques will be added to CAE systems to make them easier to use as well as to provide expert skills for decision-making and problem-solving tasks.

☐ CAE systems will be expanded to provide automated manufacturing management systems which can collect and analyze data from the factory floor to adjust or correct the fabrication operations.

☐ More efficient hardware and software will be developed to improve the efficiency of user inputs and interactive capabilities in building and displaying geometric models.

☐ The application of CAE tools will be extended beyond the operations which have traditionally taken advantage of computer capabilities, such as metal casting and plastic molding.

☐ Powerful CAE workstations will eliminate the need for

remote or batch processing for most design and analysis tasks.

☐ GT techniques may be extended to materials selection. A database of materials properties can be used to evaluate designs and changes to avoid process problems and reduce costs.

12-7 SUMMARY

Computer-automated engineering (CAE) is a tool for a wide variety of engineering tasks in both development and manufacturing. It can be used for more than just graphics design. The entire engineering process, from design concept through manufacturing implementation, can be automated. Creating a geometric model is the key to making this possible.

In the design process, CAE tools can be used for analyzing physical properties, stresses, and kinematics. The availability of advanced CAE tools makes it possible to perform finite element analysis (FEA) and kinematic analysis that would not have been practical or economical in the past. These techniques permit the designer to detect potential problems and optimize the design without having to build and test prototypes.

In manufacturing, engineering tasks such as process planning, group technology (GT), numerical control (NC) part programming, tool design, and test data generation can be performed on a CAE system. Traditionally, these tasks required a number of different people with special skills. In addition, the effort involved was tedious and subject to errors. With modern CAE tools, many of these tasks can be performed efficiently by the same engineer without special expertise in any one area.

Manufacturing operations can be simulated on a CAE system to debug a new process or optimize the operation of an existing one. Simulations can be conducted on individual machines, the layout of production lines, or the flow of an entire process. Computer graphics technology permits the user to visualize a manufacturing operation as well as get quantitative results for decision making. Integrated CAE systems make it possible to establish a common database for development and manufacturing and to accelerate the introduction of new designs into production.

REVIEW QUESTIONS

The answer to each question can be found in the section indicated at the end of the question.

1. Identify some of the design analysis tasks that can be done on a CAE system. [12-2]

2. Why are solid models required to calculate the physical properties of objects? [12-2]

3. What is finite element analysis? [12-2]

4. Describe the basic steps involved in finite element analysis. [12-2]

5. What is kinematic analysis? [12-2]

6. Identify some examples of how kinematic analysis can be used. [12-2]

7. What is process planning? [12-3]

8. Describe the basic steps involved in developing a process plan. [12-3]

9. What is group technology? [12-3]

10. Describe the basic process involved in NC programming. [12-3]

11. Identify some of the reasons for simulating a manufacturing operation. [12-4]

12. What is "discrete event simulation"? [12-4]

13. Describe the basic process involved in simulating a manufacturing operation. [12-4]

14. What types of information are required to perform a simulation of a manufacturing operation. [12-4]

15. Identify some of the trends in the application and capabilities of CAE systems. [12-6]

ROBOTICS

CHAPTER 13
BASIC ROBOTICS TECHNOLOGY

CHAPTER 14
INTELLIGENT ROBOTICS SYSTEMS

CHAPTER 15
ROBOT APPLICATIONS

CHAPTER 16
IMPLEMENTING ROBOTICS IN MANUFACTURING

Robotics is no longer confined to the world of science fiction. It is a rapidly growing manufacturing technology that is being widely used to automate production operations. Robotics technology evolved from numerical control (NC) technology. It is a form of programmable automation that is made possible by computer control. Although robots have been developed to exhibit some humanlike capabilities, they cannot equal the combined physical and mental abilities of humans. Robots are machines that can be used to improve manufacturing productivity. Their performance capabilities and especially their flexibility give them advantages over the traditional forms of automation in manufacturing.

The objective of Part 4 is to establish a familiarity with robotics technology and its capabilities. It is not intended to create experts in the design or programming of robots. However, the reader should understand how robotics technology fits into the world of computer-automated manufacturing (CAM). If the opportunity presents itself to get involved more deeply in the subject, the reader should also be able to identify the specific areas where additional research and study are necessary.

Part 4 is divided into four chapters. Chapter 13, "Basic Robotics Technology," will cover what a robot is, how it works, and what it is capable of doing. Chapter 14, "Intelligent Robotics Systems," deals with some of the more advanced technologies involved in modern robotics. These include sensing, vision, artificial intelligence (AI), and other techniques that have been developed to enhance the capabilities of robots. Chapter 15, "Robot Applications," addresses how robots can be applied in manufacturing operations. Chapter 16, "Implementing Robotics in Manufacturing," deals with how to develop and implement a robot application.

BASIC ROBOTICS TECHNOLOGY

13-1 INTRODUCTION

To understand the role of robotics in computer-automated manufacturing (CAM) and appreciate what robots can do in production operations, one must start with the basics of the technology. The fundamentals need to be addressed before we pursue some of the more exciting aspects of robotics that are based on advanced technologies and techniques. The objective of this chapter is to provide a basic understanding of the hardware and software that make up a typical robotics system. This chapter will show the reader what robots are, how they work, and what they can do.

The chapter starts by defining what a robot is, as well as what it is not. It then covers the various types of robots that exist and how they differ. A section is devoted to the typical performance capabilities and specifications of robots. Then the programming and software aspects of robotics technology are addressed. Finally, the ways in which the hardware and software work together in the actual operation of a robot is described.

13-2 WHAT IS A ROBOT?

DEFINITIONS

Many formal definitions have been developed to distinguish robots from other forms of automation. Most of the definitions have several key characteristics in common. Robots are:

Programmable

Automatic

Manipulators

Humanlike

A robot is not just a machine tool, nor is it a mechanical person. It is a special type of computer-controlled machine that can perform a wide variety of tasks by using some humanlike capabilities. Two popular definitions which have been adopted by industry attempt to be very specific about what a robot is from different viewpoints. The first was developed by Computer Aided Manufactur-

ing–International (CAM-I), a manufacturer's user group; it states that a robot is "a device that performs functions ordinarily ascribed to human beings, or operates with what appears to be almost human intelligence."

The second definition was developed by Robotics Institute of America (RIA), a robot manufacturer's group, which obviously focused more on the actual capabilities of the robot than on its similarity to humans. RIA defines a robot as "a programmable, multi-function manipulator designed to move material, parts, tools, or special devices through variable programmed motions for the performance of a variety of tasks."

The word "robot" itself was derived from a Czechoslovakian word, "robota," meaning compulsory service or work. It was first used in *R.U.R.* (for "Rossum's Universal Robots"), a science fiction play written in 1921 by Karel Capek. Later, it was popularized in *I, Robot*, a novel written by Isaac Asimov in 1950. But the robot of today is not a science fiction character and it does not look at all like a human. It is an automated tool with sophisticated capabilities that can be used in manufacturing applications.

HISTORY

Robotics is an extension of numerical control (NC) technology. The basic robot is a mechanical manipulator with programmable control. Many of the programming and control techniques were derived from those used for NC machine tools. The first patent for a robot was granted in England in 1957 to Cyril Walter Kenward. It was followed by a U.S. patent which was issued in 1961 to George C. Devol. This design, developed in conjunction with Joseph Engelberger, became the first industrial robot, the Unimate. It was first installed in a production operation by General Motors in 1959. Today, thousands of robots are working in factories around the world. They differ in size, shape, and capabilities, but they were all derived from the same basic technologies developed in the 1950s.

BASIC ELEMENTS OF A ROBOTICS SYSTEM

A robot is not a single entity. It is a system made up of several major elements of hardware and software (Fig. 13-1).

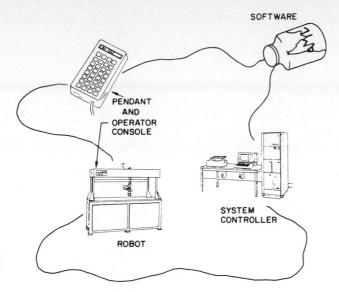

Fig. 13-1 Basic elements of a robotics system. *(Russell H. Taylor and David D. Grossman, "An Integrated Robot System Architecture," Proceedings of the IEEE, vol. 71, no. 7, July 1983; © 1983 IEEE)*

1. Mechanical components. These make up the physical robot that moves and performs the tasks. Components include:

A manipulator (the base and arm assembly)

End-of-arm tooling, such as a gripper or hand

Actuators (motors or drives that move the robot)

A power source (either a motor or pump or compressor, depending on the type of drive system involved)

2. Control system. This is the means used to control the movements of the robot. It includes:

Control techniques

Mechanical, electrical, or electronic controls

Sensors

Equipment interfaces

3. Computer system. This provides the data processing necessary to program and control the robot. It includes:

Processor

User interfaces (e.g., keyboard, display)

System software

Application programs

Robots can be configured in many different ways by using variations and combinations of these basic elements. Since robots are designed to perform tasks that might otherwise require humans, many of their basic features are like those of a human. They have controlled appendages (arm, hand, and wrist) to reach, grasp, and move objects. They are driven by a power source which provides the energy necessary for their movements. They have an intelligent control system, like the brain, which can store and process data so that actions can be programmed and controlled.

WHY ARE ROBOTS USED?

Robots are being used increasingly in manufacturing applications for many reasons. In certain circumstances, they have a number of advantages over alternative approaches using humans or special purpose hard or fixed automation:

1. Robots versus humans

☐ Robots can perform simple and repetitive tasks with better quality and consistency than unskilled laborers.

☐ Robots can perform tasks that are difficult and hazardous for humans because of factors such as size, weight, reach, precision, or environment (e.g., heat, chemicals, radiation, air quality).

☐ Robots do not have the same limitations as human workers, such as fatigue, need for rest, diverted attention, or being absent.

☐ Robots can be made with performance capabilities superior to those of humans in terms of strength, size, speed, accuracy, and repeatability.

☐ Robots can be used to perform tasks which humans do not want to do—jobs that are considered unskilled, undesirable, or low-paying or that involve poor working conditions.

☐ Robots can lower operating costs, such as materials usage, through their efficiency and consistency.

☐ Robots become more economical as labor costs increase.

2. Robots versus hard automation

☐ Robots are more flexible than hard automation since they can be reconfigured and reprogrammed. This permits them to be reused in new or modified applications, and they are less subject to becoming obsolete.

☐ Robots can make it easier and faster to implement automation since they are more flexible.

☐ Robots can make it easier to debug an application.

☐ Robots generally can be expected to have less downtime than complex hard automation systems.

☐ Automating with robotics can be more economical than hard automation in batch manufacturing operations.

Although robots may have all these advantages, they are not always the best choice for all manufacturing applications. Even though they may have a wide range of capabilities, they still cannot match the unique combination of mental and physical skills of a human.

13-3 TYPES OF ROBOTS

Industrial robots come in a variety of shapes and sizes, but they can be classified by several basic characteristics:

Mechanical configuration. The physical geometry of the manipulator arm and base unit.

Freedom of motion. The complexity of the path in which the manipulator can move.

Drive system. The type of power source used to move the manipulator.

Control system. The techniques and system used to control the motion of the manipulator.

MECHANICAL CONFIGURATIONS

Four basic robot arm geometries are used for industrial applications (Fig. 13-2):

1. Rectangular coordinate system. This is also known as the Cartesian coordinate system. A robot with this geometry has three linear axes using sliding joints which are typically arranged in a cantilever configuration whose motion traces a boxlike rectangular shape. This type of configuration is best suited for straight-line and side-to-side movements.

2. Cylindrical coordinate system. This is also referred to as a rectilinear coordinate system. A robot with this geometry has three axes of motion that trace the shape of a cylinder. It has a base unit which rotates, a vertical extension, and a horizontal arm that moves linearly. This type is best suited for movement around a base.

3. Spherical coordinate system. This is also known as the polar coordinate system. A robot with this geometry has three axes of motion that trace the shape of a sphere. It has a base unit that rotates, a main body that tilts, and a horizontal arm that slides in and out.

4. Revolute coordinate system. This configuration is also known by several other names: anthropomorphic (i.e., like a human arm), articulated arm, or jointed arm. Such a robot has three axes of motion involving a base that rotates, a "shoulder" that rotates, and an "elbow" that also rotates.

Several variations of these basic geometries have also been developed to provide improvements in performance for certain applications. Two of the most popular are:

1. "Gantry" or "box-frame." This is a rectangular coordinate configuration with all three linear axes of motion suspended above the work space (Fig. 13-3). It can be made very rigid, which allows high precision and high acceleration. It also lends itself to modularity in design for a variety of configurations, including multiple arms.

2. Selective compliant assembly robot arm (SCARA). This configuration, shown in Fig. 13-4, is horizontally revolute. A robot of this type moves by sweeping over the work space at a fixed horizontal distance before moving a vertical arm down. This permits a compact and relatively low-cost design for small assembly tasks.

FREEDOM OF MOTION

The primary function of the manipulator is to position the arm of the robot such that it can perform its intended tasks. This requires precise and sometimes complex control of the motions of each of the mechanical elements (i.e., base and joints). To actually perform useful tasks, however, the robot must also grip and apply force to objects. This requires the addition of end-of-arm tooling, such as a robot hand or gripper, whose motions must also be controlled.

The motion ability of a robot is usually expressed in terms of the number of "degrees of freedom" it has in which to move—the number of axes it uses or independent types of movement it can make. An axis of motion can be either linear or rotational. Six geometric parameters or axis coordinates are required in order to completely specify the location and orientation of an object. Three coordinates can locate the center of gravity of an object (e.g., x, y, and z coordinates in a rectangular coordinate system). Three more can determine its orientation (e.g., angles of rotation). Therefore, a robot requires six degrees of freedom in order to be completely versatile in its motions.

Since all of the basic mechanical configurations of manipulators provide only three degrees of freedom, additional axes must be added if more flexibility in motion is

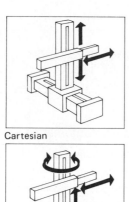

Cartesian

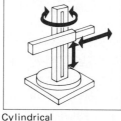

Cylindrical

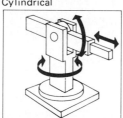

Polar

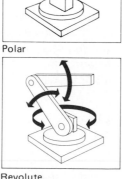

Revolute

Fig. 13-2 Mechanical configurations of robots. *(photo courtesy of Parker Hannifin Corp.)*

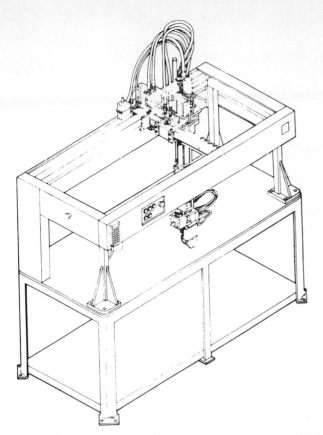

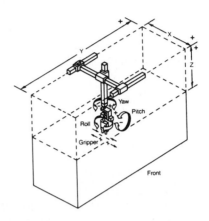

Fig. 13-3 Gantry or box-frame robot configuration. *(International Business Machines Corp.)*

Fig. 13-4 SCARA robot configuration. *(International Business Machines Corp.)*

required. This is normally achieved by adding wrist and hand movements with the end-of-arm tooling (Fig. 13-5). There are three basic types of wrist motions:

1. Roll. Rotational or swivel movement in a plane perpendicular to the end of the arm.

2. Pitch. Rotational or bending movement in a plane vertical to the arm.

3. Yaw. Rotational or twisting movement in a plane horizontal to the arm.

The motion ability of a robot is also described in terms of its "working envelope." This is the shape of the area that can be reached by the maximum movements of the end of the robot arm. Each manipulator configuration has a different shape for its working envelope, which will also vary in size with the size of the robot (Fig. 13-6).

DRIVE SYSTEMS

The motions of the mechanical linkages and joints of a manipulator are driven by actuators, which can be various types of motors or valves. The energy for these actuators is provided by some power source, which can be electric, hydraulic, or pneumatic. Actuators may be directly coupled to the linkages or joints that they drive, or they may be connected through gears, chains, or screw mechanisms. There are three major types of drive systems for industrial robots:

1. Pneumatic. These systems are driven by compressed air. Pneumatic-drive robots are usually small and have limited flexibility, but they are relatively inexpensive to build and use. The weight of the payload they can carry and the speed of their motion are limited by the compressibility and low operating pressure of air.

2. Hydraulic. These systems are driven by a fluid that is pumped through pistons, cylinders, or other hydraulic actuator mechanisms. Hydraulic-drive robots can be rela-

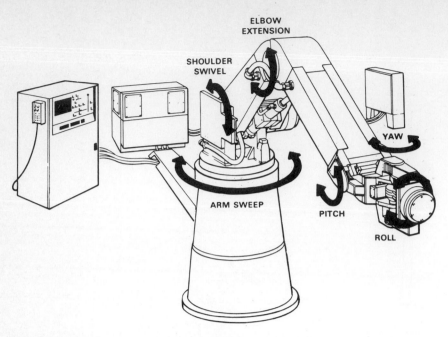

Fig. 13-5 Robot with six degrees of freedom, including wrist motions. *(Cincinnati Milacron)*

tively compact and yet provide high levels of force, power, and speed with accurate control. They can also be made very large for heavy payloads and large working envelopes. Because the power supply (hydraulic pump) can be isolated from the robot and no electric power is re-

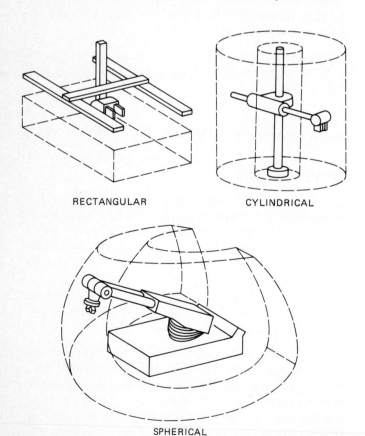

RECTANGULAR   CYLINDRICAL

SPHERICAL

Fig. 13-6 Working envelopes of three different robot configurations. *(Roger N. Nagel, "Robots: Not Yet Smart Enough," IEEE Spectrum, May 1983; © 1983 IEEE)*

quired in the manipulator, hydraulic robots are safe and reliable in wet, dusty, and potentially explosive environments.

3. Electric. These systems are driven by rotational electric motors. Electric-drive robots are best suited for applications involving light payloads which require high accuracy and cleanliness. They avoid some of the maintenance and reliability problems associated with pneumatic or hydraulic systems, which can leak. However, they require more sophisticated electronic controls and can fail in high-temperature, wet, or dusty environments. There are both ac and dc electric-drive systems. DC motors can provide high speed and accuracy with simpler controls, but they are more limited in their payload capacity.

CONTROL SYSTEMS

All the motions of a robot are controlled by a system of hardware and software that is programmed by the user. There are two basic types of robot control systems:

1. Non-servo-controlled. This is the simplest and least expensive type of control system, but it is also very limited in its flexibility and performance. It can be a purely mechanical system of stops and limit switches which are preprogrammed or positioned for specific repetitive movements. This can provide accurate control for simple motions at low cost. Such a system can also use some type of electromechanical logic, such as pneumatic valves or electrical relays, to control fixed sequences of movements. The motions of non-servo-controlled robots are controlled only at their endpoints, not throughout their paths. Due to the nature of the types of controls used, the number of points which can be programmed into a sequence of movements is also limited.

2. Servo-controlled. Such a system is capable of controlling the velocity, acceleration, and path of motion, as well as the endpoints of the path. It can store complex control programs which can be easily reprogrammed. Servo-controlled systems use electronic controllers or computers and sensors to control the motions of robots. They are more flexible than non-servo systems, and they can control more complex and smoother motions. Their capabilities and cost will vary depending on the type and sophistication of the controller and sensors involved.

Sensors are used in servo-control systems to track the position of each of the axes of motion of the manipulator. These sensors may be located internally, in the robot joints, or externally, in the work space. Many different types of sensors can be used (e.g., electronic, optical, or magnetic), depending on the nature of the task and performance requirements involved. Sensors will be covered in more depth in Chap. 14.

The major functions of a control system involve:

Generating the path of motion for the manipulator. A robot's arm accelerates until it reaches a maximum velocity. It then continues at that velocity until it begins to approach the position it is programmed to reach. It then decelerates until it stops at that position (Fig. 13-7). The control system must provide the direction for this motion.

Monitoring sensors. Servo control requires feedback from sensors which track the movement and actions of the manipulator. The signals from these sensors are read by the controller and used to make any adjustments required in the motion of the robot to achieve its task (Fig. 13-8).

Coordinate translation. The position of a robot can be expressed in terms of the angles of its joints and the lengths of its arms. This is the "joint coordinate system" with which the robot must be controlled. However, robot motions are usually programmed in terms of the rectan-

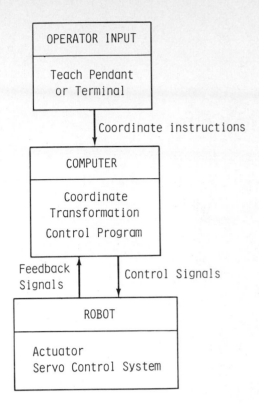

Fig. 13-8 Feedback control system.

gular coordinates of the real world (i.e., the "world coordinate system"). The control system must therefore convert the programmed path from one coordinate system to the other (Fig. 13-9). This can involve a lot of computation and is not a perfect process. The translated path may not be precisely the same as the one described in the original coordinate system program.

Safety controls. To assure that failures or errors are detected before any damage or harm can be done, the actual motion of the robot must be constantly monitored and compared to allowable limits. This can involve a variety of monitoring or measuring techniques, which will be addressed further in Chap. 15.

Interfaces. The control system must communicate with the user or programmer. This can be done through terminals, keyboards, control boxes, or switches (Fig. 13-10). The system controller must also have interfaces to the sensors and actuators that control the motion of the robot. In some applications, the system must also communicate with other equipment, such as production machines, materials handling devices, or even other robots. The communication channel provides all these interfaces to the control system.

13-4 PERFORMANCE CAPABILITIES

The performance of a robot can be described in terms of many different parameters or characteristics. Their magnitude and relative importance vary, depending on the inherent capabilities of the robot design and its intended

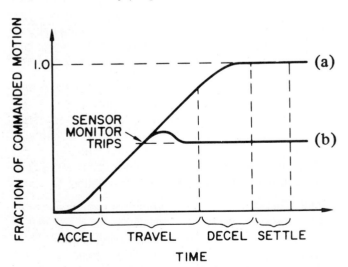

Fig. 13-7 Motion control: (*a*) normal and (*b*) interrupted trajectories. *(Russell H. Taylor and David D. Grossman, "An Integrated Robot System Architecture," Proceedings of the IEEE, vol. 71, no. 7, July 1983; © 1983 IEEE)*

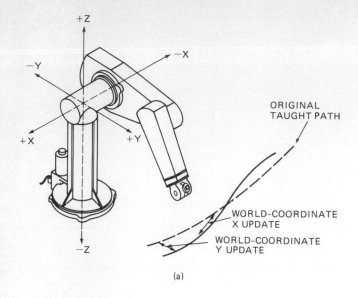

(a)

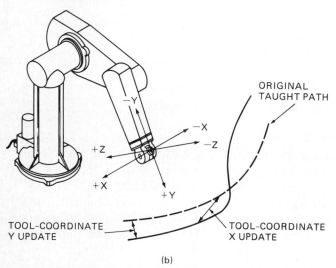

(b)

Fig. 13-9 World versus tool coordinates: robot paths can be modified by world coordinates—fixed x, y, z directions predefined for each robot—and tool coordinates, which indicate the direction in which a robot tool is pointing at any given time. *(Thomas M. Larson and Anthony Coppola, "Flexible Language and Control System Eases Robot Programming," reprinted from Electronics Week, June 14, 1984, copyright © 1984, McGraw-Hill Inc., all rights reserved)*

applications. In general, a robot user would look for desirable characteristics such as high reliability, high speed, programmability, and low cost. However, there may be tradeoffs between even these basic characteristics. For example, increasing the speed of the robot or lowering its cost may lead to lower reliability. There is no one robot which is best for all applications. To select one that is well suited for a particular application, the key features and specifications required must be understood and prioritized.

SPECIFICATIONS

Axes of motion. The number of degrees of freedom of the robot's joints determines the complexity of motion of

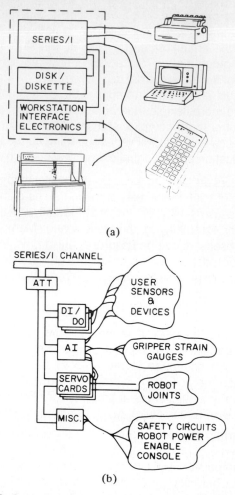

(a)

(b)

Fig. 13-10 Control system interfaces: *(a)* overview, *(b)* workstation interface electronics. *(Russell H. Taylor and David D. Grossman, "An Integrated Robot System Architecture," Proceedings of the IEEE, vol. 72, no. 7, July 1983; © 1983 IEEE)*

which it is capable. Three degrees of freedom are provided by all basic manipulator configurations; this is adequate to locate the position of an object. Additional degrees of freedom (up to seven with some configurations, including the gripper) can permit more complex motion, which can adapt to the orientation and shape of objects.

Work envelope. The maximum reach or range of arm movement varies in shape and size, depending on the configuration and size of the manipulator. The work envelope can be described in terms of degrees of rotation, vertical motion, and radial arm extension. The work envelope of the robot must obviously be compatible with all the paths and positions intended in the application.

Speed. The speed of a robot—which is measured at the end of the arm, where the task is to be performed—determines how fast the end-of-arm tooling or gripper can get from one position to another. This speed will vary, depending on payload, position in the work envelope, and axis of movement.

Acceleration. This is the rate at which the manipulator can increase its speed. It can be important for complex paths, in which the time to reach maximum velocity can be a significant portion of the total travel time. The acceleration capability of a robot is determined by the power that can be developed by the drive system. Hydraulic drives can provide very high accelerations by storing energy for bursts of power.

Payload capacity. The maximum weight the manipulator can carry is normally specified at low or normal speeds. It depends on the size and configuration of the robot. High payload capacity is a tradeoff with some other performance specifications, such as speed and accuracy. Some very large robots can carry payloads of several hundred kilograms (Fig. 13-11).

Accuracy. How closely the end of the robot arm can be moved to a specific position is a function of several factors:

☐ The basic geometric configuration of the manipulator

☐ The mathematical model used by the control system to convert position specifications from world coordinates to joint coordinates

☐ The effects of the payload on deflections and distortions in the joints and arms

☐ The effects of temperature on the joints and arms

☐ The accuracy of the specification of the end position or objects to be reached

☐ Degradation or drift in the performance of the actuators and sensors

Small robots which have been designed for precise work can be accurate to within 0.1 mm in their motions.

Resolution. This may also be referred to as the precision of the robot. It is not the same as accuracy. Resolution is the smallest increment of motion for which the robot can be controlled; it is normally a function of the sensors used.

Repeatability. This should also not be confused with accuracy. Repeatability is the ability of the manipulator to repeatedly return to the exact same position. It depends

on the stability of the control system and is affected by temperature and load.

Reliability. This is measured in terms of the average time between failures. Reliability depends on the speed and load during operation as well as on the quality of the robot design. Industrial robots typically perform for hundreds of operating hours between failures.

KEY FEATURES

In addition to the typical specifications which define a robot's performance capabilities, a number of other features can also influence its performance. These features are more subjective and therefore difficult to specify. They are:

Quality. Generally, one can expect higher accuracy and reliability from robots that are built with better and more rugged electrical and mechanical components.

Serviceability. Many features can be added to the design of the robot to minimize both the frequency of failure and the time it takes for repair. Such features include conservative designs for electrical and mechanical parts, interchangeable parts, quick-release mechanisms, and long-life seals.

Safety. A wide variety of features can be used to prevent damage or injury during robot operation, including sensors, restraints, and barriers. This subject will be covered in depth in Chap. 14.

Modularity. Using simple modular design for the manipulator may permit it to be reconfigured for different applications.

Dexterity. The ability of the manipulator to perform delicate, precise, or complex tasks is often limited by the gripper. Advanced designs and control techniques that can provide capabilities such as a dextrous or humanlike hand will be covered in Chap. 14.

13-5 PROGRAMMING ROBOTS

The major advantage that robots have over fixed automation is their ability to be programmed. They can perform complex tasks under the control of stored programs which can be modified. In addition, they can move in response to real-time inputs from sensors. Programming is an important feature of robots, and it is also a significant component in their design and application. The process of robot programming involves "teaching" it the task to be performed, storing the program, executing the program, and debugging it. A number of different techniques and software approaches are used to program robots. No one of them is best or even compatible with all robots and applications.

The principal task of the robot program is to control the motions and actions of the manipulator. However, due to the inaccuracies and uncertainties involved in the posi-

Fig. 13-11 Large gantry-type robot for heavy payloads. *(Cincinnati Milacron)*

tions and movements of robots, much of the programming involved deals with the detection and correction of errors. Programs must also be written to deal with both people and machines. The users may include professional programmers, as well as operators, maintenance personnel, and application engineers.

PROGRAMMING METHODS

Three basic methods are used to program robots:

1. Guiding. This is also known as the "walk-through" or "playback" method. It involves counterbalancing the manipulator arm so that it can be moved manually through the intended motions while its path is being recorded by the control system (Fig. 13-12). This is a simple technique that does not require the operator to write any program code, but it is limited to relatively short and simple motions. With this method, all of the motion is under the control of the operator who is physically guiding it through its movements. The stored "program" of robot motion can then be played back, at a different speed if so desired, during the actual performance of the task.

2. Teach pendant. This is also known as the "lead-through" method. It uses a control panel, called a "teach pendant," which has buttons or switches that control the motion of the robot through a cable connected to the control system (Fig. 13-13). The operator or programmer can lead the manipulator through the task one step at a time, recording each incremental move along the way. Only the major points in the path of the robot's movements are recorded, so the intermediate points must be interpolated or calculated by the control system. The control system then generates a program for the complete path of the robot. Like the walk-through method, this technique does not require code writing, but it can be used for more complex and precise tasks. It is often used for developing applications or training users.

3. Off-line programming. This is the traditional form of computer programming. A high-level language is used to write a control program which describes all the move-

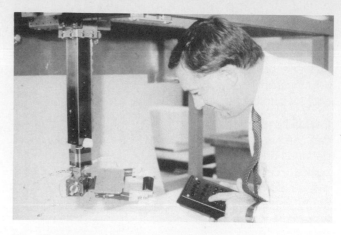

Fig. 13-13 Teach pendant/lead-through progamming. *(International Business Machines Corp.)*

ments and actions of each of the actuators in the manipulator. The program can involve many steps, requiring a large number of lines of program code. In addition, a great deal of computation is normally associated with the translation and generation of the path of motion. This method of control is the most flexible and can also enable the robot to respond to signals from external sensors to modify its movements. Although many types of users can program relatively simple movements using high-level languages, a complex task normally requires the skills of a professional programmer.

A variation of these programming methods is on-line programming using the operator console (Fig. 13-14). This requires the availability of a robot, but it also gives the programmer the ability to see the robot actually executing the program as it is being developed. In most cases, this approach is used either as a substitute for a teach pendant to write relatively simple programs, or to debug programs that have been written off-line.

ROBOT PROGRAMMING FUNCTIONS

A robot programming system involves several major functions to make the execution of the actual task possible:

1. World modeling. A significant part of specifying the robot's task is defining the positions of the objects and features involved in the application. When the task environment is not known precisely, the position of the robot must be specified relative to the objects. Position data may be obtained from several sources, such as robot sensors, geometric models, or external sensing systems.

2. Path generation. The path of motion is normally specified by interpolating intermediate points from a sequence of motions and positions identified in the program. As mentioned earlier, the position specifications must be translated from world coordinates to joint coordinates which the robot control system can understand. The type of path generated may not be coordinated between the joints. That is, each joint may move indepen-

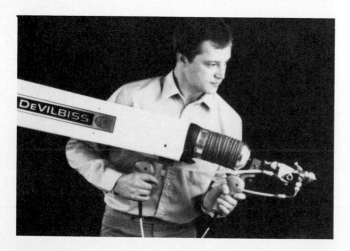

Fig. 13-12 Guiding/walk-through programming. *(DeVilbiss Co.)*

Fig. 13-14 On-line programming. *(International Business Machines Corp.)*

dent of the others. This will normally result in a nonuniform, curved path. If straight-line motion is desired, the motions of the joints must be coordinated.

3. Sensing. The use of sensors permits the robot to deal with uncertainties that cannot be preprogrammed. These uncertainties may include:

☐ The detection of errors or failures

☐ The identification of objects or positions

☐ The initiation or termination of actions

To use sensors, the robot control system must be able to collect and interpret sensory data from internal or external sources and generate control signals that will modify the actions of the manipulator. The data acquisition, computation, and output signal generation activities must be built into the control program. A sensor-based robot must have a program that allows it to choose between alternative actions on the basis of data from the sensors and its model of the environment.

4. Programming support. Some of the other functions that are necessary in a programming system for robots include:

☐ Editing tools for modifying programs

☐ Debugging tools for finding problems during program development

☐ Diagnostic tools for finding problems during operation

☐ Simulators for trying out robot programs without actually operating the robot

☐ Interfaces to controllers, peripheral equipment, and other computers

ROBOT PROGRAMMING LANGUAGES

Most robot programs are written in what are referred to as "robot-level" programs. That is, they specify the task the

robot is to perform in terms of the individual motions required of each of its actuators. Robot programming is usually done with high-level languages that use a limited number of basic commands with motion and position specifications. They are normally executed by an interpreter rather than a compiler for ease of use. High-level interactive languages provide functions for data processing, computation, and sensing as well as manipulation.

The commands used in a programming language are subroutines that are provided by the system which define frequently used functions. They may be used to control the actions of the robot, to perform computations, or to process data. Some of the basic types of commands are:

☐ Motion and sensing functions (e.g., MOVE, SENSIO, MONITOR)

☐ Computation functions (e.g., ADD, SORT)

☐ Program-flow control functions (e.g., RETURN, BRANCH)

The most common command in most robot programs is to move the manipulator. An example of a simple MOVE command is:

MOVE (<1,3,5> <5,10,20>)

or, in general terms:

MOVE (<JOINTS> <GOALS>)

The first command means that joints 1, 3, and 5 are to be moved by 5 inches, 10 inches, and 20 degrees, respectively.

There are also a number of frequently used functions that are not involved in the path itself (e.g., WAIT, CONTINUE, TOOL, BRANCH, OUTPUT).

Many robot languages have been developed. Each is unique in its design and capabilities. Some are more powerful and versatile than others. Some were developed by robot users, others by robot manufacturers. Some of the most widely used are:

VAL (Unimation)

AML (IBM)

HELP (General Electric)

RAIL (Automatix)

AL (Stanford University)

MCL (McDonnell-Douglas)

RPS (SRI)

JARS (Jet Propulsion Laboratory)

A subroutine from a real robot program written in AML is listed in Fig. 13-15.

13-6 ROBOT OPERATION

OPERATING METHODS

Four basic modes of operation can be used to control the movements of a robot:

1. Pick-and-place. As the name implies, this mode involves a very limited sequence of moves. Usually, the robot simply moves to a fixed position where it grasps a

```
pick_up_slug: SUBR(fdr,tries);
 cc: NEW STRING(8);
 t: NEW 0;

 step_1:                           -- Move to grasping position.
 CMOVE(<feeder_loc(fdr),feeder_orient,.5>);
 IF DCMOVE(<<0,0,-.75>,ANY_FORCE(2*OZS),<.5>) THEN
   BEGIN                           -- Hit something on way in
   DCMOVE(<<0,0,2>>);              -- Back out
   RETURN('jammed');              -- Return error
   END;

 step_2:                           -- Attempt to grasp slug.
 cc = GRASP(0.1,<-.04,.04>,PINCH_FORCE(1*LBS));
 IF cc NE 'ok' THEN
   BEGIN                           -- Hit something on way in
   MOVE(GRIPPER,.5);              -- Readjust gripper
   DCMOVE(<<0,0,1>>);             -- Back out
   RETURN(IF cc EQ 'toosmall' THEN 'empty'
       ELSE if cc EQ 'toobig' THEN 'jammed'
       ELSE cc);
   END;

 step_3:                           -- Update feed location.
 fy = HAND_GOAL()(1,2);           -- "Y" position at grasp
 IF ABS(fy-feeder_loc(fdr),2) GT 0.04 THEN
   feeder_loc(f,2) = fy;

 step_4:                           -- Pull out slug and reverify hold.
 DCMOVE(<0,0,1>);
 IF ABS(SENSIO(<LPINCH,RPINCH>)) LT 15*OZ THEN
   BEGIN                           -- Dropped part.
   IF tries GE t = t+1 THEN        -- If not too many,
     BRANCH(step_1);              -- then try again.
   ELSE                            -- Otherwise.
     RETURN('dropped');           -- give up.
   END;

 RETURN('ok');
 END;
```

Fig. 13-15 Robot program: AML subroutine to pick up a type slug. *(Russell H. Taylor and David D. Grossman, "An Integrated Robot System Architecture," Proceedings of the IEEE, vol. 71, no. 7, July 1983; © 1983 IEEE)*

part ("pick"), then moves to another position where it "places" the part. Some applications may involve several move positions. The controls involved are generally the simplest. A non-servo-control system, with either mechanical stops or pneumatic logic, is adequate. Because of the simplicity of motion and the fixed positions involved, the pick-and-place mode can be accurate and capable of high speed.

2. Point-to-point. This is used for more complex movements where the arm is controlled in a series of steps that have been stored in memory. The programming is usually performed by the use of a teach pendant. Although the movement is normally under servo control, there is no coordinated motion between the axes. Each axis operates at its maximum rate until it reaches the desired endpoint position. The intermediate path, velocity, and relative motion between axes are not controlled. This is adequate for many applications where only the activity at the endpoint positions is important.

3. Continuous path. This mode is required when the control of the manipulator's path is critical, such as in a spray-painting application. The robot's path is not determined by a series of preprogrammed points. The path and movement of each axis is stored during a walkthrough programming session. Although this creates a continuous path, it is not precise. All the movements of the operator, intended or not, are recorded. A large amount of memory and a high-speed sampling system are needed to record all the path data.

4. Controlled path. Where the total control of the robot's motion is desired, a detailed control program and sophisticated servo-control system must be used. This provides coordinated control of all the axes in terms of their position, velocity, and acceleration. The program can optimize the movements of the manipulator to reduce cycle time, minimize forces, eliminate jerky motions, and improve precision.

END-OF-ARM OPERATION

The robot control program must also control any operations involving the end-of-arm tooling or gripper. For simple one-step operations, a single output from an on/off switch may be all that is necessary to activate the gripper. For movements involving a sequence of several steps, multiple outputs may be required, including those from support fixtures and tools. In complex applications, the control program must integrate all the signals which can influence the actions of the robot. These may include inputs from other machines to determine if they are ready to operate. They may also include signals which notify the system that a part is present or that an operation was performed successfully.

CONTROL SYSTEM OPERATION

During the operation of a robot, the control system is very busy, even for relatively simple movements. A real-time servo-control system executes a series of tasks at a fixed frequency, normally a high-speed sampling rate that has been preprogrammed into the control system (e.g., every 20 ms). A typical sequence of events goes as follows:

1. Read input status (e.g., safety interlocks, power on)

2. Execute program commands (e.g., MOVE)

3. Check safety conditions

4. Monitor sensor values

5. Interpret motion commands (e.g., interpolate points in trajectory path)

6. Output updated joint position goals

7. Log and terminate processing

13-7 SUMMARY

Robots are programmable machines with some human-like capabilities. They are automation systems made up of mechanical components, a control system, and a computer. These elements can be arranged in different ways and varied in size and complexity to perform different tasks. Robots are available in a wide variety of types which vary in their mechanical configuration, freedom of motion, and drive and control systems.

The motion ability of a robot is determined by the degrees of freedom and working envelope provided by its geometric configuration. There are a number of different types of drive systems, each has its own advantages in cost and performance capabilities. No one drive system is best for all applications. Robots can be controlled by different hardware and software systems. However, the more complex tasks usually require servo-control systems which use sensors and computers. The control system performs monitoring, computation, and interfacing functions to govern the robot's motion. Robots have a wide range of performance capabilities. Robot specifications must be matched to the needs of a particular application in order to select a type which will perform the required tasks.

Robotics systems have a number of economic and performance advantages over human labor or hard automation in many applications. This is particularly true for those applications involving batch manufacturing operations. The major advantages are due to their programmability. Robots can be programmed by several different techniques, of which some do not require the experience of professional programmers. Robot programming is usually done with high-level interactive languages that provide functions for data processing, computation, sensing, and manipulation. Many different robot languages have been developed by both manufacturers and users. Although they are similar, there is no common standard. Robot programs can be very simple or extremely complex, depending on the nature of the tasks and type of motion control involved. However, a robot control system must perform a lot of tasks at high speed, even for relatively simple movements. Robotics involves a great deal more than the visible mechanical manipulator mechanism which is normally thought of as the robot.

REVIEW QUESTIONS

The answer to each question can be found in the section indicated at the end of the question.

1. Define what a robot is. [13-2]

2. Identify the basic elements of a robotics system. [13-2]

3. What advantages do robots have over manual or other automated approaches? [13-2]

4. Identify the basic characteristics that distinguish different types of robots. [13-3]

5. Describe the basic types of mechanical configurations of industrial robots. [13-3]

6. How is the motion ability of robots described? [13-3]

7. Identify the differences between the major types of drive systems used for industrial robots. [13-3]

8. Describe the basic types of robot control systems. [13-3]

9. What are the major functions of a robot control system? [13-3]

10. Identify and define the typical specifications which describe a robot's capabilities. [13-4]

11. What is the difference between the accuracy, resolution, and repeatability of a robot? [13-4]

12. Describe the basic methods that can be used to program robots. [13-5]

13. Explain the major functions of a robot programming system. [13-5]

14. Describe the basic modes of operation that can be used to control the movements of a robot. [13-6]

INTELLIGENT ROBOTICS SYSTEMS

14-1 INTRODUCTION

Now that we have learned about the basics of robotics technology, we can address some of the advanced robot functions and capabilities that exist and are being developed. The early robots used by industry were extensions of numerically controlled machine tools. Although they were programmable, they had limited flexibility and had none of the truly intelligent capabilities of humans. Such "nonintelligent" or "dumb" robots merely tried to mimic the physical movements of humans. However, humans' sensing and decision-making capabilities permit them to adapt to uncertain conditions and modify their actions.

Most manufacturing applications for robots still only require limited capabilities, such as pick-and-place movements. Complex tasks, such as assembly applications, however, require some of the additional capabilities of humans. Forms of sensing and adaptive control, in particular, separate intelligent robotics systems from their more basic relatives. This chapter will cover the advanced features and capabilities for industrial robots in the following topics:

Robot intelligence

End-of-arm tooling

Sensors

Vision

Control systems and software for intelligent robots

Artificial intelligence (AI) in robotics

Trends in robotics

14-2 WHAT IS AN INTELLIGENT ROBOT?

Robots can be classified into several categories by relative intelligence:

1. **"Slave manipulators."** These are controlled directly by a human operator through some type of physical or communication connection.

2. **Limited-sequence robots.** These are the typical pick-and-place devices that use mechanical stops with limited ability to be adjusted.

3. **Teach/replay robots.** These include robots that are programmed by walk-through or lead-through methods to perform specific tasks.

4. **Computer-controlled robots.** These can be preprogrammed off-line to perform complex tasks, but they are restricted to the specific set of instructions in the control program.

5. **Intelligent robots.** These computer-controlled robots are not restricted in their actions to only the preprogrammed instructions.

The classes of nonintelligent robots are all capable of performing useful work, but limitations in their capabilities restrict their application in a manufacturing environment. These restrictions may include:

☐ The inability to detect and correct errors

☐ Insufficient materials handling flexibility to accommodate variations in physical dimensions

☐ Lack of mobility

☐ Inability to respond to changes or variations in the workplace

The term "intelligence" in relation to robots implies the use of information to modify actions. That information is typically obtained from some form of sensing technique. An intelligent robotics system is able to interact with its environment and adapt its behavior to uncertain or changing conditions. It may be flexible enough to perform a variety of tasks or to handle objects which have variations in size and shape (e.g., dimensional tolerances). Intelligent robots may use humanlike senses, such as touch, vision, and hearing, to provide them with information about their environment. They may then use humanlike thought processes to reason and make decisions which will control their actions.

Intelligent robots are actually automated manufacturing systems which integrate manipulation, sensing, computation, and control functions. The physical system includes (Fig. 14-1):

☐ "End effectors" (or end-of-arm tooling), which act as the arms and hands of the robot

☐ Sensors, which provide data about the environment by contact or noncontact techniques

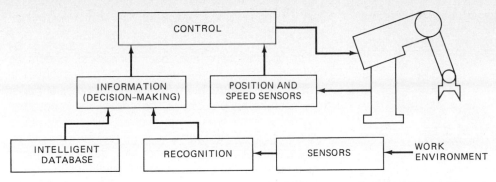

Fig. 14-1 Functional block diagram of an intelligent robotics system. *(Masaki Togai, "Japan's Next Generation of Robots," Computer, March 1984; © 1984 IEEE)*

☐ Computers, which typically include a hierarchy of controllers, processors, and communication channels

☐ Auxiliary equipment, which performs special tasks as part of the application, such as tools, fixtures, conveyors, feeders, and pallets

As such systems have been developed, a number of advanced features have been incorporated which distinguish them from the more traditional nonintelligent industrial robots. These include:

Contact and noncontact sensing

User-friendly operator interfaces

Powerful, high-level programming languages

AI techniques

Communication links to the factory floor control system

Multiple arms

Dextrous hands

Intelligent robotics systems offer potential benefits to a manufacturing operation beyond the performance of specific tasks. For example, they may:

☐ Reduce the requirements for precise parts or special fixturing by adapting to variations and misalignments

☐ Permit the automation of a variety of custom designed products

☐ Perform a complex combination of tasks that would otherwise require multiple tools and operations

Intelligent robots are perhaps the best example of how many of the computer-based technologies can be integrated into a truly automated manufacturing system.

14-3 END-OF-ARM TOOLING

CONSIDERATIONS

A wide variety of devices have been developed as end-of-arm tools or end effectors for robots. In most cases they have been designed to perform specific types of tasks. However, recent efforts have been directed toward developing general purpose devices that can be used for a variety of different tasks. Some of the factors to consider when designing or selecting end-of-arm tooling for robots are:

☐ The type of task to be performed (e.g., grasping, assembling, or using a tool)

☐ The size, shape, and weight of the objects involved

☐ The geometry and layout of the work space

☐ The accuracies and tolerances involved

☐ The work environment (e.g., temperature, surface conditions)

☐ Other tools and fixtures involved

☐ The speed and complexity of motions required

☐ The performance capabilities of the robot (e.g., payload, reach, accuracy)

No one end effector is suitable for all tasks. None can be as versatile as the human hand. Often tradeoffs are necessary because of application considerations (e.g., payload versus accuracy and speed). However, advanced techniques can make the end-of-arm tooling more flexible so that it can deal with variations in the application environment that traditional fixed automation could not handle, such as:

☐ Tolerance variations on objects and surfaces that are greater than the accuracy required for the robot task

☐ Changes in the dimensions and orientation of objects

☐ Lack of precise alignment of objects to be handled

☐ The need to handle delicate objects or surfaces (Fig. 14-2)

TYPES OF END-OF-ARM TOOLING

The types of end effectors used normally fit into one of the following function categories:

1. Grippers or hands. Their major function is to grasp objects. They may take a variety of forms. Most are simple vicelike jaws (Fig. 14-3). Others are more complex and may even have humanlike fingers.

2. Tools. These include a wide variety of devices that one might find human workers using in the factory to

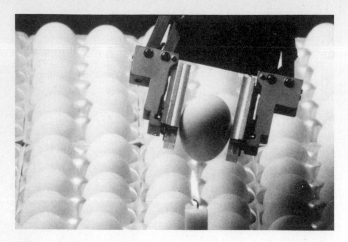

Fig. 14-2 Robot grippers with sensors can hold delicate objects, even something as fragile as an egg. *(International Business Machines Corp.)*

perform manufacturing tasks. Some typical examples are welding guns and torches, drills, screwdrivers, and spray guns (Fig. 14-4). The interesting concept involved here is that robots can do more than just hold things—they can use tools to do work. They become tools using tools!

3. Hand and tool holders. These are mechanisms which are mounted on the end of a robot arm to permit it to change the hands or tools that it uses. This provides flexibility in an application that can result in higher throughput, fewer setups, and even fewer robots. Since there are many different styles of end effectors and robot designs, there is no universal tool holder. However, some have been developed with standard interfaces (i.e., electrical, mechanical, and pneumatic connections) and self-aligning features to provide for quick changes with a variety of different tools and grippers (Fig. 14-5). Robots equipped with such mechanisms can quickly change from one tool to another to perform multiple tasks at a single workstation (Fig. 14-6).

4. Micromanipulator. This is a special device used to perform precise alignments or accurate probing tasks (Fig. 14-7). It normally requires better spatial resolution

than can be provided by the arm or wrist motion alone and uses a sensing scheme for locating positions.

End-of-arm tooling may use mechanical, pneumatic (including vacuum), magnetic, or electrical techniques to perform their unique functions. Several sources of power may need to be brought to the end of the robot arm. Depending on the type of tool involved, this can result in a relatively heavy weight, which can detract from the performance capabilities of the robot. An ideal end effector would be small, light, fast, accurate, multifunctional, and inexpensive.

COMPLIANCE DEVICES

When robots were first tried in assembly tasks, it was found that the mating of parts was extremely difficult to repeat reliably. Either a great deal of precision had to be built into the parts and fixturing involved, or sophisticated sensory control systems had to be used. A simpler approach developed later was "passive" in nature. By taking advantage of the compliance (i.e., the ability to conform or adapt) of springs and tapers, these devices enabled robot grippers to accommodate misalignments and variations in the positions and dimensions of parts. These "compliance devices" are spring-mounted fixtures on the end of the robot arm that allow parts with chamfered edges to be grasped quickly and easily even when they are misaligned by several millimeters (Fig. 14-8).

DEXTROUS HANDS

The ultimate form of a robot gripper is the human hand. Nothing in nature can match its dexterity, agility, and sensory capabilties. A truly general purpose gripper would have to emulate some of the same capabilities. The two key areas required to achieve such capabilities are:

1. Sensing. By using sensing schemes, a robot gripper can obtain information about the parts it is handling or the environment in which it is working. It can then modify its movements, as in the alignment of parts for mating.

(a)

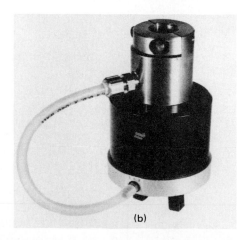
(b)

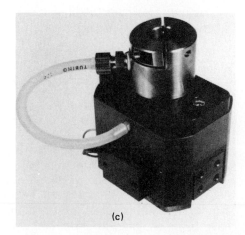
(c)

Fig. 14-3 Different types of robot grippers: (*a*) pneumatic parallel gripper, (*b*) pneumatic scissors motion gripper, (*c*) pneumatic parallel motion part presence gripper. *(International Business Machines Corp.)*

Fig. 14-4a Robot with power screwdriver. *(International Business Machines Corp.)*

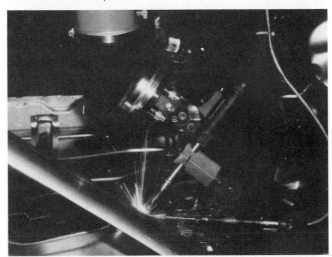

Fig. 14-4b Robot with arc-welding torch. *(American Cimplex Corp.)*

Fig. 14-4c Robot with routing tool. *(American Cimplex Corp.)*

Fig. 14-4d Robot with water-jet tool. *(American Cimplex Corp.)*

Fig. 14-4E Robot with paint-spraying nozzle. *(DeVilbiss Co.)*

Fig. 14-4 Robot end-effector tools.

2. Articulated hand. The addition of multiple fingers with humanlike joints can permit a robot to grasp objects of varying shapes, sizes, and orientation. This can avoid the need to change hands or tools in order to perform multiple tasks. A dextrous hand provides additional degrees of freedom for performing complex motions and tasks with great precision that would not otherwise be possible with an automated system.

Much work has been done in order to develop dextrous hands for robots using a variety of designs (Fig. 14-9). Using the human hand as the analogy, robot hands incorporate such features as:

☐ Multiple, articulated "fingers" attached to a "palm"

☐ "Bonelike" structure for dexterity and rigidity

☐ "Skinlike" touch sensors

☐ "Fleshlike" surfaces for object-grasping compliance

☐ Proximity sensors for collision avoidance (the only human equivalents may be temperature sensing and sight)

To build such a hand requires advanced approaches in several areas:

☐ Lightweight, high-strength materials

☐ Small, high-speed, precision actuators and motors

☐ Multiple sensory systems for position, force, proximity, and vision

☐ Small, high-strength, tendonlike cables to connect fingers to motors

☐ Substantial computer power

14-4 SENSORS

TYPES OF SENSING

The five basic senses of humans are often taken for granted (i.e., touching, hearing, seeing, smelling, and tasting). However, they make it possible for us to cope with the unpredictable world around us and to perform tasks. Robots without some kind of sensory capabilities can only perform preprogrammed actions. They cannot deal with uncertainties, variations, or changes. Robot sensing is used to obtain information about the work environment that will permit uncertainties to be dealt with, such as:

Variations in the dimensions and shapes of parts

Variations in the positions and orientations of parts

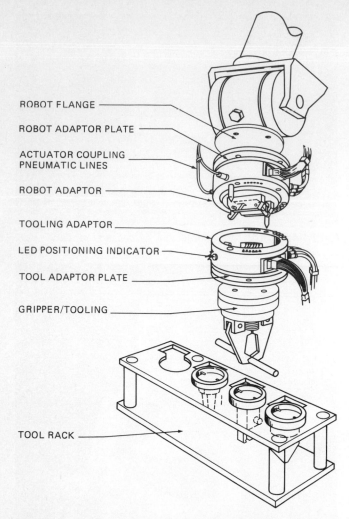

ROBOT FLANGE

ROBOT ADAPTOR PLATE

ACTUATOR COUPLING PNEUMATIC LINES

ROBOT ADAPTOR

TOOLING ADAPTOR

LED POSITIONING INDICATOR

TOOL ADAPTOR PLATE

GRIPPER/TOOLING

TOOL RACK

Fig. 14-5 Robot tool changer. *(Applied Robotics)*

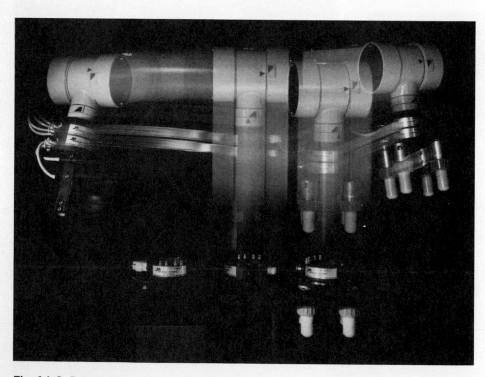

Fig. 14-6 Robot changing tools. *(Applied Robotics)*

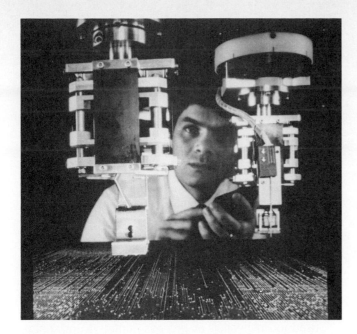

Fig. 14-7 Micromanipulator. *(International Business Machines Corp.)*

Unknown surfaces or obstacles

In manufacturing applications, sensing may be used for a variety of reasons, such as:

Locating parts

Detecting errors

Verifying quality

Several basic types of sensing are most often used for robot applications:

Force or torque sensing

Position sensing

Motion or displacement sensing

Pattern recognition

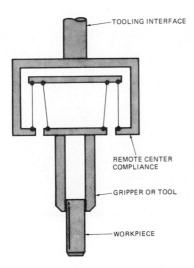

Fig. 14-8 Remote center compliance device. *(Roger Allan, "Busy Robots Spur Productivity," IEEE Spectrum, Sept. 1979; © 1979 IEEE)*

Fig. 14-9 Dextrous robot hand. *(J. K. Salisbury and D. Lampe, Massachusetts Institute of Technology Artificial Intelligence Laboratory)*

If a robot lacks some of these sensing capabilities, it cannot deal with such common situations as a part being misoriented, misplaced, or damaged. The alternative for many automated manufacturing operations is to build relatively expensive and inflexible support fixtures and tools which will eliminate some of these potential variables. Intelligent robots can avoid such costs through the use of sensing capabilities.

TYPES OF SENSORS

A variety of techniques that can be used to simulate the human senses and provide the sensory capabilties that are often required by intelligent robots that perform manufacturing tasks. Most sensing techniques involve a mechanism which converts the parameter or physical property of interest into an electrical signal which can be transmitted to a computer control system and interpreted into useful information. This process requires some type of "transducer" to provide that sensing mechanism. The signals generated by these transducers are a function of the physical properties which are being sensed, such as pressure, strain, displacement, velocity, acceleration, temperature, or light intensity. Many different types of transducers and sensing techniques can be used, depending on the property to be sensed, the performance required (e.g., accuracy), and the constraints of the application (e.g., environment, layout). The most common types of sensing schemes fall into one of the following general categories:

1. **Contact sensors.** This category includes all those types of sensors that require physical contact with an object in order to determine the particular parameter to be sensed or measured. It includes:

   Tactile or touch sensors. These normally take the form of single or multiple contact probes at the end of the robot gripper to determine the location, orientation, shape, or presence of an object. This may involve a simple microswitch or an array of pressure-sensitive elements, depending on the amount of information desired. For example, the gripper in Fig. 14-10 is equipped with strain gauges which are used as tactile sensors. The gripper operates like a human thumb and forefinger to detect the forces of pinching and slippage of the gripped object. A light-emitting diode and a phototransistor are used to detect the gripped article.

Force or torque sensors. These measure forces or torques applied to the gripper or manipulator arm by using transducers that convert deflections into electrical signals, such as piezoelectric devices and strain gauges. Force sensing makes it possible to control the physical contact and relative motion of the manipulator and other objects. It may provide information on the forces between the gripper fingers during grasping; along the axes of the object being grasped; or in the joints, wrist, or fingers of the manipulator.

Position or displacement sensors. These are usually located in the joints of the manipulator to measure their movements so that the position of the robot can be controlled. Many different sensing techniques can be used. Among the contact types are strain gauges, potentiometers, transformers, and "resolvers" (i.e., inductively coupled devices that measure the angular position of a rotating shaft).

Compliance devices. These are used to control the forces and relative motions between the gripper and an object when the direct measurement of the forces is not practical. This may be accomplished by deriving force information from displacement data or by using compliant mechanisms to assist in the grasping or mating of objects (e.g., using springs or vibration techniques).

2. Noncontact sensors. This category includes all the types of sensors that generate signals using transducers that are not in physical contact with the objects involved (Fig. 14-11). The most common among these are:

Optical sensors. These typically use pairs of photocells mounted on the end of the gripper or application tooling to detect the presence or absence of an object by the interruption of a light beam. By using an optical encoder, the relative positions of objects and the manipulator can be measured.

Range and proximity sensors. These determine the relative distance between the manipulator or gripper and objects. A variety of noncontact sensing mechanisms can be used, including acoustical, ultrasonic, optical, and magnetic devices.

Imaging systems. These are complex sensing systems which simulate human sight capabilities. A variety of different techniques can be used, with varying degrees of performance capability. Among them are optical systems (Sec. 14-5) and x-ray, thermal, ultrasonic, and infra-red imaging. Such systems can be used for inspection or process control or for controlling a robot's operations in a remote or hazardous environment.

Another way to classify sensors is by their relative intelligence capabilities. Some sensing systems are only passive mechanisms. They limit their interpretation to the information as obtained from the sensors. Some active sensing systems have the ability to alter the object or the sensor to obtain additional information for a more complete interpretation (Fig. 14-12). The human analogy is the coordination of grasping and eye actions to obtain more visual and tactile information about an object.

USE OF SENSORS

Some of the principal uses for sensors in robot applications are:

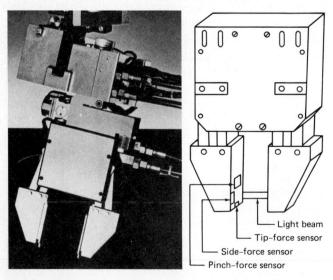

Fig. 14-10 Tactile sensors for a robot gripper. *(Paulo Dario and Danilo De Rossi, "Tactile Sensors and the Gripping Challenge," IEEE Spectrum, Aug. 1985; © 1985 IEEE)*

- Light beam
- Tip-force sensor
- Side-force sensor
- Pinch-force sensor

Fig. 14-11 Noncontact sensors for a robot gripper. This device uses five linear proximity sensors and one optical sensor to check the accuracy and repeatability of the movement of the robot arm. *(Ulrich Griebel and Paul Cotnoir of the Worcester Polytechnic Institute Manufacturing Engineering Applications Center)*

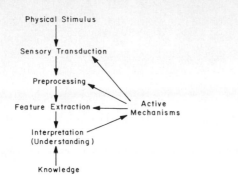

Fig. 14-12 Active perception with sensors: coordinating visual and tactile sensing to perform a task. *(Arthur C. Sanderson and George Perry, "Sensor-Based Robotic Assembly Systems: Research and Applications in Electronic Manufacturing," Proceedings of the IEEE, vol. 71, no. 7, July 1983; © 1983 IEEE)*

☐ Controlling the actions of the manipulator (e.g., grasping, aligning, inserting, turning)

☐ Locating objects or positions

☐ Inspecting objects for recognition, verification, or defects

☐ Avoiding obstacles and collisions

☐ Performing assembly or materials handling operations

☐ Adjusting the path of the manipulator to comply with physical constraints or conditions

Assembly applications are particularly demanding of a robot's sensing capabilities. The forces and movements involved in tasks such as grasping, inserting, mating, aligning, and turning must be measured and controlled. This often requires a number of different types of contact and noncontact sensors. Some manufacturing applications involve many operations and tasks that may require or benefit from sensory controls. The sensors may be mounted on the robot, on the workstation, on materials handling equipment, or on other machines (Fig. 14-13).

Some sensing tasks are relatively simple binary controls. For example, many movements can be controlled by a contact sensor which permits the manipulator to move until it touches an object. This technique is frequently used to reach for an object, avoid collisions, or

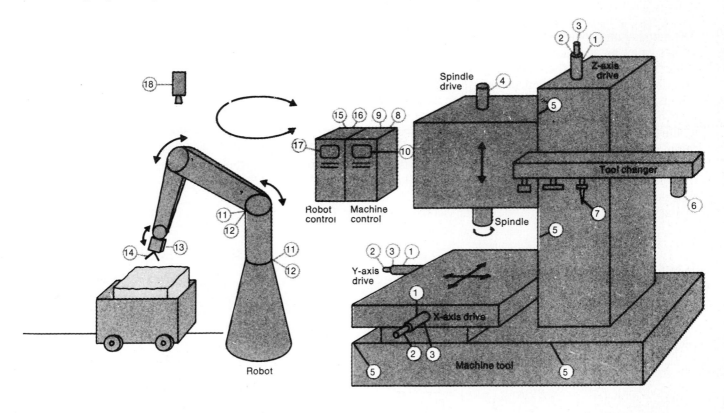

An automatic milling machine with a loading–unloading robot relies on diverse sensors, actuators, and displays. **On the machine tool,** dc motors (1) provide movement on the *x, y,* and *z* axes; tachometers (2) sense the speeds of the axis motors; resolvers (3) sense axis–motor shaft position; an ac motor (4) drives the tool spindle; and limit switches (5) sense when the milling table is approaching its maximum allowable bounds and thus prevent overtravel. A stepping motor (6) positions the tool changer so that the spindle can accept a new tool at the appropriate moment, and a tactile probe (7) measures the dimensions of the workpiece at each machining step. In the machine–control unit, servo amplifiers (8) regulate the machine drives, a computer (9) exercises overall control, and a display (10) keeps a human supervisor informed of the machine status. **On the robot,** hydraulic servo valves (11) actuate the arm, optical encoders (12) sense the position of the arm, a pneumatic control valve (13) actuates the robot's gripper, and a tactile sensor (14) measures the gripper force. The robot control contains servo amplifiers (15), a computer (16), and a display (17). **Overhead,** a TV camera (18) identifies parts and guides the robot.

Fig. 14-13 Sensor applications for robotics. *(John G. Bollinger and Neil A. Duffie, "Sensors and Actuators," IEEE Spectrum, May 1983; © 1983 IEEE)*

grasp objects. A term often associated with this type of sensing capability is "guarded move." The robot may use touch, force, or proximity sensors to detect the relative position of its gripper and an object. The guarded-move technique can be used to avoid obstacles, locate undefined surfaces, or calibrate the path of the manipulator. Other sensing tasks, such as recognizing or inspecting objects, require much more data and interpretation.

Another important use of sensing in control systems is in "line-tracking" applications. This is where a robot must perform a task in which the object is moving—for example, parts on a conveyor. The control system must continuously track the movement of the conveyor or objects and adjust the manipulator's path and speed to perform its preprogrammed task independent of the object's motion. Such systems typically use resolvers, tachometers, or encoders to track the conveyor motion, often together with vision systems to detect objects (Fig. 14-14).

SELECTING SENSORS

Although sensors are used by robots to deal with the uncertainties involved in performing manufacturing tasks, robots cannot match the capabilities of human workers. The human hand, for example, has approximately 20,000 nerves which permit it to sense a wide range of parameters such as pressure, temperature, texture, direction, and vibration. However, only a limited amount of sensory data is actually required for simple grasping and moving actions.

Contact and noncontact sensing should complement rather than compete with each other in robot applications. Each has its own unique capabilities. Touch sensing is usually of greater value in assembly applications, to provide accurate data about the forces involved with the interaction of the manipulator and objects. Contact sens-

ing can be more precise for delicate operations. It is also necessary for close and complicated tasks which may obscure a vision system. However, contact sensors are subject to wear and damage that noncontact sensors usually avoid.

The performance capabilities of sensors vary considerably, depending on the type of sensor and technology involved. Many of the most modern sensors employ solid-state technologies, which provide improved signals and accuracy in small, rugged, low-cost devices. Sensors are usually specified in terms of:

Sensitivity. The smallest increment of the measured parameter that can be sensed.

Accuracy. The absolute tolerance at which the measurement is considered correct.

Dynamic range. The measurable variation in the magnitude of the parameter.

Response time. The speed of the measurement.

Advances in sensor technology have greatly expanded robots' capabilities, which begin to approach those of humans for specific types of applications. Tactile sensor arrays, for example, use several contact techniques, such as pressure-sensitive switches or conductive materials, to simulate the area-sensing capabilities of skin. High-density versions of these can be used in imaging systems for object recognition, just as a blind person learns to "see" through touch.

There is no one best sensor for robot applications. Like a human worker, a robot usually needs several senses to perform a task. The selection of the best sensor system for a given application depends on the task involved and the performance requirements.

14-5 ROBOT VISION

WHY DO ROBOTS NEED VISION?

As we saw in Sec. 14-4, different types of contact and noncontact sensors can give robots the capability to perform a wide variety of tasks. No matter how good those sensors are, however, the best they can do is make a robot perform as well as a blind worker. Vision makes a great deal more information available, which can be used to perform tasks more efficiently or do things that would not otherwise be possible. Robot vision systems are used to:

Identify or verify objects

Locate objects

Inspect objects

Guide manipulator motion or robot navigation

Vision is more efficient than tactile sensing for tasks involving the identification and acquisition of parts. Vision also makes it possible for robots to perform tasks on parts that vary in size and shape. A robot with vision can also adjust its movements to accommodate variations or

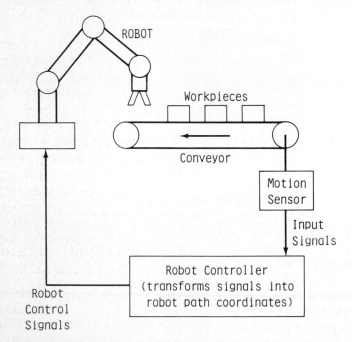

Fig. 14-14 Line-tracking robotics system.

changes in the workpiece or the work environment. Most manufacturing applications for robots today do not require vision capability. Those that do usually require other types of sensors as well, to work in conjunction with the vision system just as the eyes must work together with the hands.

WHAT IS A ROBOT VISION SYSTEM?

A robot vision system is basically a large-area optical sensing scheme. It uses visible light and some type of image sensor, such as a television camera, to acquire data which describes an object or scene. This data is then processed and interpreted by a computer which drives the robot control system. A typical vision system consists of (Fig. 14-15):

☐ A source of light directed at the object or scene

☐ An image sensor or camera to record the image

☐ An analog-to-digital converter to digitize the image

☐ A high-speed image processor to extract features from the image and compact the data

☐ A computer system that controls the robot

☐ Interfaces to the user, the robot, and other equipment

A robot vision system acquires data about the object or scene it is intended to work with from the image recorded by the camera. Key features, such as edges, area, or perimeter shape, are extracted from the data and identified by image-processing techniques using some form of geometric model. Some vision systems are more sophisticated than others in terms of the amount of information they can use to recognize objects. The simpler systems

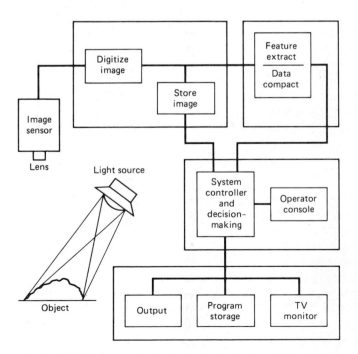

Fig. 14-15 Basic elements of a robot vision system. *("Vision Systems Make Robots More Versatile," Mechanical Engineering, Jan. 1985)*

either measure and match features against a standard image or sense a threshold of light or dark. Higher-level vision systems use knowledge about the objects together with expectations about the image and the objective of the task to aid in the interpretation of the image data.

TYPES OF ROBOT VISION SYSTEMS

Robot vision systems can be categorized in several ways:

1. **Dimensions**
   2D systems. These are the most common and least complex. They are designed to recognize 2D features of objects, such as edges or outlines. In most applications this is sufficient to identify objects. There are a number of different kinds of 2D systems that provide a range of capabilities.
   3D systems. These are much more complex and more expensive than 2D systems. They are capable of recognizing 3D shapes and contours as well as depth and distances, much like the human eye. Several different techniques have been developed which provide some form of 3D capability.

2. **Contrast**
   Binary. This is the simplest type of vision system. Objects are recognized in terms of 2D black or white images only. No shades or 3D features can be distinguished. This approach is adequate for many applications which involve simple or uniform shapes.
   Gray scale. These more complex systems are capable of distinguishing shades of light and dark. This enables them to detect edges and features within the image of an object. More complex geometric algorithms and a great deal more computation are needed in order to process and interpret gray-scale images.

3. **Image analysis technique (2D systems)**
   Edge finding. This is a binary approach which detects the edges of objects as transitions from white to black or black to white (Fig. 14-16). It requires a high-contrast lighting setup, but not much computation. The image is identified by recording the transition line in the vertical direction and a clock count in the horizontal direction.

   Silhouette matching. This is also a binary approach requiring high contrast (Fig. 14-17). The light reflected off an object is used to create an outline or silhouette of its 2D shape. That outline image is then compared to a reference image or "template" which is stored in memory. The system attempts to align the reference image with the observed image to find matches in shape and size. This is a computational process based on rules and algorithms. The geometric center of the image is then calulated to direct the robot motion to its position. Although silhouette matching is a very effective and common approach used for many applications, it has significant limitations:

Objects must contrast sharply with their background
Objects cannot touch or overlap

Partially obscured objects may not be identified
Objects with similar silhouettes, but different features cannot be distinguished (e.g., internal edges

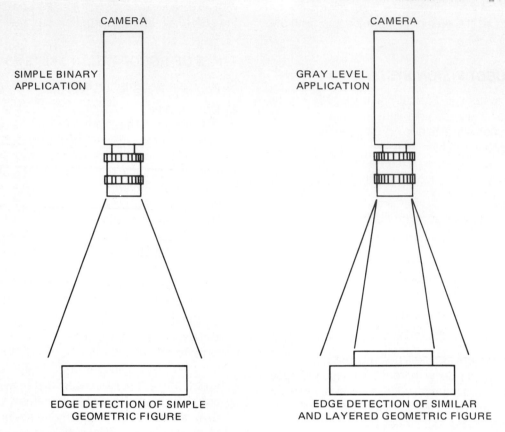

CAMERA

SIMPLE BINARY
APPLICATION

EDGE DETECTION OF SIMPLE
GEOMETRIC FIGURE

CAMERA

GRAY LEVEL
APPLICATION

EDGE DETECTION OF SIMILAR
AND LAYERED GEOMETRIC FIGURE

Fig. 14-16 Edge-finding technique. *(Larry Werth, "Selecting Machine Vision Products: The Performance Perspective," Production Engineering, Aug. 1985)*

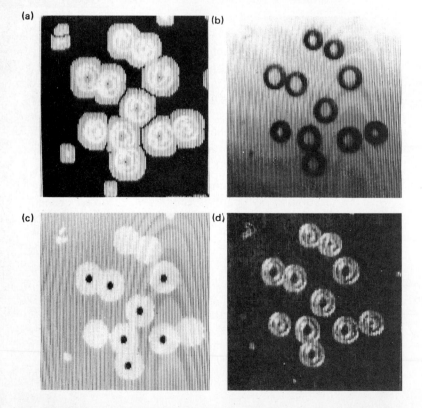

(a)

(b)

(c)

(d)

(a) First camera frame: The vision system records a frame of 13 washers, four of which are undersized. The washers are nonuniformly illuminated and rest on a dirty conveyor belt. Some are touching.

(b) 12 milliseconds: The vision system has identified the outline of each washer and has suppressed the background to a uniform black.

(c) 84 milliseconds: The vision system has visually "filled in" the hole in each washer. A single pixel dot remains in the center of each correctly sized washer. (The hole-size fault tolerance is programmable).

(d) 129 milliseconds from the first camera frame: The identification is complete. Black spots locate the centers of correctly sized washers. A microcomputer will now transfer the coordinates to the materials handling equipment.

Fig. 14-17 Silhouette-matching technique. *(Applied Intelligent Systems Inc.)*

and shapes, surface markings, textures)

Feature matching (Fig. 14-18). This is a more complex image analysis technique that requires gray-scale interpretation. Design-rule criteria and shape characteristics are used to match and identify objects on the basis of their recorded images. Several different schemes can be used for recognition. Some classify features on the basis of the nearest match found to stored reference shapes. Others use a decision-tree approach for narrowing down the alternatives. It is also possible to look for distinguishing characteristics of the overall shape first, instead of comparing each individual pixel in the image. Although this technique is very powerful and can be useful in many manufacturing applications, it is slower and more expensive than silhouette matching.

3D VISION SYSTEMS

For a robot to perform some tasks in manufacturing, it must be able to see and move in three dimensions. For example, picking parts out of bins may be simple for humans, but without vision, this task would be all but impossible for either a robot or a human worker. Three-dimensional vision systems are needed to increase the ability of robots to locate and grasp objects as well as avoid obstacles in their path. To be effective, such systems require 3D imaging and AI techniques for interpretation and recognition.

Three basic techniques are used for 3D vision systems:

1. Structured light. Also known as "light striping," this technique involves projecting controlled beams of light onto an object (Fig. 14-19). The pattern formed by the intersection of the light stripes with the edges of the object is used to determine the distance and shape of the object. This technique can distinguish objects without high contrast and even can identify parts that touch or overlap.

2. Range finding. This technique is similar to radar. It uses light reflected from an object's surface as a depth cue to determine its distance and shape. One approach is to measure the time a precisely focused beam of light (e.g., laser) takes to reflect from an object back to its source. Alternatively, distances can be calculated by triangulation techniques using the angular differences between the light source and the detected reflection (Fig. 14-20).

3. Stereo vision. This is also known as binocular vision, just as in human sight (Fig. 14-21). We use the separation between our eyes to create a slight disparity in the images we see as a depth cue. Stereo vision systems use two cameras to obtain images from two different positons. These binocular images permit the system to compute the distances to all the edges and features on the object's surface. The map of the resulting 3D image is then matched against geometric models of reference objects.

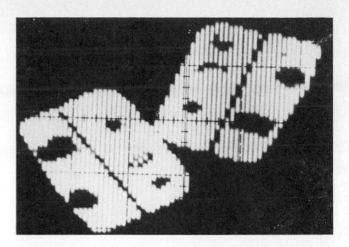

Fig. 14-18 Feature-matching technique: workpieces are identified by edges and contours, making selection possible even when pieces touch or overlap. *(General Motors Corp.)*

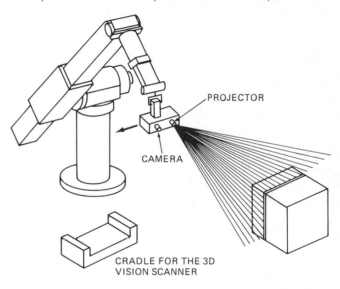

Fig. 14-19 Light striping for 3D vision system: Robot vision system scanning structured light image. *(A. C. Kak, et al., "A Knowledge-Based Robotic Assembly Cell," IEEE Expert, Spring 1986; © 1986 IEEE)*

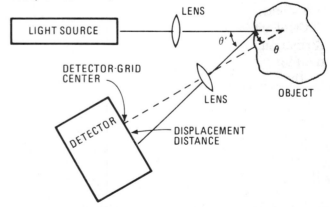

Fig. 14-20 Range finding for 3D vision systems: triangulation is used to determine how far an object is from the light source by measuring the "displacement distance" of a deflected beam of light from the center of the detector grid. *(Erik L. Zeller, "Robotics: Clever Robots Set to Enter Industry en Masse," reprinted from Electronics, Nov. 17, 1983; Copyright © 1983, McGraw-Hill Inc., all rights reserved)*

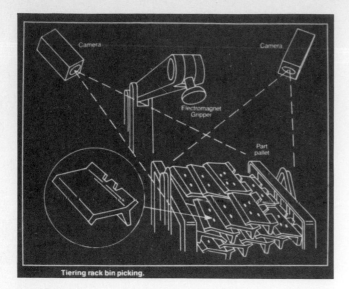

Fig. 14-21 Stereo (binocular) 3D vision system. *(Automatix Inc.)*

HOW DO VISION SYSTEMS WORK?

Although each of the techniques addressed uses different approaches to detecting and recognizing objects, the processes involved are similar. A typical vision system will operate in the following sequence of steps:

1. Image sensing, using a light source and camera in one of the methods already discussed

2. Preprocessing of the image, to enhance its quality and clarity

3. Segmentation, to separate the image from its background

4. Interpretation and recognition, by matching or classifying features

Image processing can be very complicated, depending on the technique and algorithms used. It typically involves many steps and computations, such as:

Detecting edges

Determining whether the edges connect

Fitting edges together into shapes

Classifying shapes

Matching shapes with reference models

Verifying the matches

The recognition process may be based on a correlation or matching scheme, as in template matching. It may also use rules and a decision process based on knowledge and experience, as in expert systems. The reference images used may be generated by computer design systems, particularly in the case of 3D systems, which need 3D geometric models for comparison. The computation task can be substantial in vision systems. General purpose computers are not well suited for image processing since they operate sequentially. The interpretation of visual data requires parallel processing if it is to be efficient.

USE OF VISION SYSTEMS IN ROBOT APPLICATIONS

Most manufacturing applications for robots do not require vision systems. When the task is simple and the objects involved are not complex or disorganized, other sensor systems are usually adequate. However, there are many cases where a vision system has a distinct advantage and may even be essential for a robot to perform a task. Some typical examples are (Fig. 14-22):

☐ Arc welding seams that vary in length, shape, and position

☐ Spray-painting a series of parts of different sizes and shapes

☐ Selecting parts from a mixed group or pile

☐ Picking parts from a moving conveyor line

☐ Inspecting parts for defects

The selection of a vision system should be based on the needs of the application. A general purpose vision system with a wide range of image recognition capability may be unnecessary and too expensive for most applications. If the shape of the object involved is simple and the object's orientation or movement is controlled, a relatively simple vision system should be adequate.

14-6 CONTROL SYSTEMS AND SOFTWARE FOR INTELLIGENT ROBOTS

SENSOR-BASED CONTROL SYSTEMS

Data from sensors is used by adaptive robot control systems to:

☐ Monitor the execution of robot actions to verify that it is performing as expected

Fig. 14-22 Vision system application: electronics assembly application uses vision system for monitoring and inspection at a robot workstation. *(American Cimflex Corp.)*

☐ Provide servo control by measuring the errors in actual movements and correcting for them

Robots which use multiple sensors require sophisticated control systems and software. Tactile and vision sensors can generate a lot of data for the controllers to handle. Therefore, sensor-based robot applications may require a hierarchical system in order to provide closed-loop control efficiently (Fig. 14-23). Local controllers can manage the data collection from each sensor system while a central controller supervises the use of that data by the control system computer. The feedback from sensors is used to modify the task instructions and coordinate position models which are generated by the control system.

The prevention and detection of errors is an important function of a control system. Typical automatic error recovery techniques will restore the motion of the robot to its last error-free operation. However, it is much more difficult to handle errors that are irreversible, such as when incorrect or damaged parts are involved. To deal with these types of situations the robot requires additional data about the work environment, which may be obtained through sensors and communication with other operations. This data must then be used, together with knowledge about the application, to determine the cause of error and to initiate corrective action.

ROBOT DECISION MAKING

Sensor inputs are used by an adaptive control system to change the preprogrammed path of a robot so that it can adjust for changes or errors in the work environment. They can also be used to take dynamic conditions in the application into account in controlling the sequence of actions of the robot. One of the programming techniques used to deal with this type of robot decision making is called "branching." From a software point of view, "branching" means a command directs the computer to change the sequence of execution to a different step in the program. In terms of the robot's actions, branching means that the manipulator stops executing one sequence of operations and initiates another, perhaps at a different position (Fig. 14-24). Several types of branching may be used for robot control systems:

1. Standard branching. The robot changes its motion or position in response to inputs at specific decision points, such as the presence or absence of parts at a workstation.

2. Offset branching. The robot repeats the execution of an entire motion sequence at another position. This technique can save programming by merely specifying the position move without repeating the program for the operation itself.

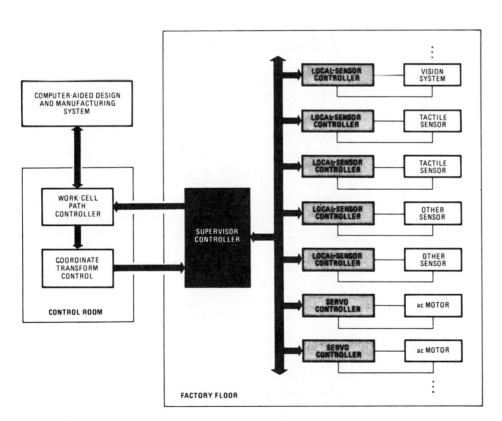

Fig. 14-23 Hierarchical sensor control system: sensor data from tactile and vision systems is integrated by supervisor controller for use by robots. *(Erik L. Zeller, "Robotics: Clever Robots Set to Enter Industry en Masse," reprinted from Electronics, Nov. 17, 1983; Copyright © 1983, McGraw-Hill Inc., all rights reserved)*

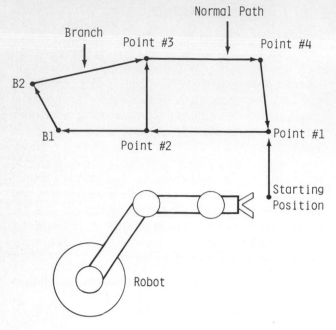

Fig. 14-24 Robot program branching routine.

3. Conditional branching. The branching occurs only when a specific combination of inputs is sensed, such as when one operation is complete and the next workstation is ready to accept a part. This is obviously a more complicated decision-making process for the robot control system. It requires multiple inputs, communication links between operations, and more complex logic processing.

4. Adaptive branching. This can be even more complex, since it involves changing a robot's motion solely on the basis of sensor inputs in order to accommodate changes or variations in the work environment. For example the robot might have to grasp parts which are randomly placed on a moving conveyor or stacked in a pile. Although the grasping movements may be preprogrammed, the searching movement to find the location of the parts is dependent on a sensor-based control system.

SENSOR-BASED ROBOT PROGRAMMING

In a sensor-based robot control system, the flow of program control is dependent on sensor inputs. The sequence of operations and the motion specifications can be modified by decisions the control system makes on the basis of inputs about changes, conditions, or errors. Sensor inputs are used in robot programming to:

1. Initiate or terminate motions
2. Choose between alternative actions
3. Locate and identify objects
4. Adapt to physical constraints

Sensory control can be used for application programming as well as in the execution of tasks. Instead of using a lead-through or walk-through technique to "teach" the robot a program, one can use a sensor-based control

system to guide the manipulator through a sequence of operations.

TASK-LEVEL PROGRAMMING

We have seen that robots can be programmed using several different techniques. There is a high-level approach which goes beyond the interactive teaching methods and the use of robot programming languages. It is called "task-level" programming, since it specifies the goals of the tasks involved rather than the specific motions of the robot required to achieve those goals. A task-level program can be written without having to specify any of the positions or paths that depend on the geometry or motion of the robot. This data must, however, be provided by a geometric model of the robot and its work environment. A task-level program is then automatically translated into a robot-level program using the task specification and the geometric model. This can greatly simplify the programming process and provide better interaction between the robot and the programmer for debugging applications. Task-level programming is new, and it requires the development of very high-level languages and sophisticated geometric models. Sensor-based robot control systems are obviously essential in the use of such a programming technique to deal with some of the more difficult tasks, such as grasping, mating parts, or avoiding obstacles.

14-7 ARTIFICIAL INTELLIGENCE IN ROBOTICS

WHY DO ROBOTS NEED ARTIFICIAL INTELLIGENCE?

Robots without sensor-based control systems are only capable of performing simple, repetitive tasks. They can only operate in a predictable work environment where all the positions are fixed and all the objects involved are known. The use of sensors for adaptive control permits robots to modify their actions on the basis of changes or variations in the work environment. If we take robot intelligence a step further, we need to introduce concepts such as perception, decision making, and planning. A robot can recognize, understand, and cope with a wide variety of uncertainties only if it uses AI techniques.

As the use of robots becomes more widespread in manufacturing, their application becomes limited by:

☐ The variations in the parts, the workplace, and the robotics system

☐ The predictability of the actions and motions required in the task

☐ The interrelationships between multiple tasks and equipment operations

☐ The effort involved in programming the robot to perform the tasks sucessfully

The robot user must make a choice. One alternative is

to create an artificial world for the robot to operate in, where the uncertainties are eliminated or minimized. Another is to provide the robot with AI which will permit it to deal with the real world. A truly intelligent robot does not just sense things in its environment. It percieves them! This capability allows it to understand the implications of what it senses so that it can reason and respond. Although this sounds a little like science fiction, it is merely a natural extension of computer science and control system technology. Robotics challenges AI to deal with physical tasks in the workplace in competition with humans.

AI FUNCTIONS FOR ROBOTS

AI encompasses a number of different functions which can be applied to robotics:

1. Expert systems that use experience or prior knowledge to make decisions and solve problems

2. Knowledge acquisition and representation to establish a knowledge base for expert systems or other AI functions

3. Heuristic search to try alternative solutions to solving a problem (or selecting a path)

4. Deductive reasoning to reach a conclusion based on interpretations of information in relation to prior knowledge

5. Planning or developing an approach to achieving a goal, based on experience and an understanding of the task

6. Image understanding to interpret and recognize sensory information about objects or scenes

7. Natural language understanding for the ability to translate human vocabulary into machine language

One key element of an intelligent robotics system is a geometric model. The model must describe the environment in which the robot is to operate. The robot must be able to modify the model in response to its perception of changes or variations in the real world. The robot's reasoning process must be based on the information in the model and its actions must be based on interpreting this information. The reasoning processes used by AI are symbolic in nature. Representations and relationships of the properties and characteristics of objects and the environment are manipulated in a logical process to reach a conclusion. In robotic systems, these AI techniques are used to provide geometric and spatial reasoning based on an interpretation of observations and on the knowledge in the geometric model. Figure 14-25 illustrates the conceptual relationship among the basic functional elements in an assembly system. The sensing representation and manipulation description provides a basis for allocation of tasks to specific devices in the system and programming of coordinated actions among devices.

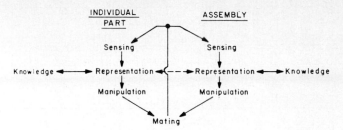

Fig. 14-25 Artificial intelligence in robotics. *(Arthur C. Sanderson and George Perry, "Sensor-Based Robotic Assembly Systems: Research and Applications in Electronic Manufacturing," Proceedings of the IEEE, vol. 71, no. 7, July 1983; © 1983 IEEE)*

AI APPLICATIONS IN ROBOTICS

AI techniques are beginning to be used for practical tasks in robotics applications. Following are some examples of how the various types of AI functions are used:

1. Intelligent vision systems. A robot can make use of knowledge it acquires about objects, such as their shapes and spatial relationships, to recognize images observed in a vision system. This may include the interpretation of complex 3D images which require a perception of depth and possibly even motion and color. AI techniques could also provide an understanding of the interaction between the image and other information about the scene, such as written descriptions. An intelligent vision system could provide the robot with an understanding of what objects look like rather than just a template for it to match. For a robot to begin to approach the vision capability of humans, it must have extensive use of AI functions.

2. Automatic process planning. As an extension of the task-level programming concept, AI techniques could be used to generate plans and programs for robot operations. A robot process planner would use an expert system with a knowledge base consisting of a geometric model together with rules for applying sensor data and generating robot actions. This process plan would be used to drive the control system and would be modified by inputs from the robot sensors.

3. Spatial planning. A more specific form of planning robot actions is to generate a path for a manipulator which is in the presence of obstacles or unknown variations (Fig. 14-26). This requires a representation of the problem (i.e., a model) as well as an algorithm to search for a solution path. Both sensory and compliance techniques could be used to provide control over position and force variables. One of the most dramatic examples of a sucessful spatial planning application was the development of an automatic robotic system for shearing sheep.

4. Hand-eye coordination. One of the simplest tasks for humans is perhaps one of the most complex for a robot. Coordinating the actions of the eyes and hands requires many of the capabilities of intelligent robots. A typical manufacturing operation requiring this would be to pick disordered parts from a bin or pile. This involves tactile and vision sensing, force control, and dextrous hands.

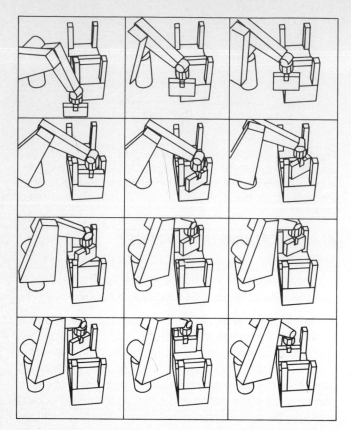

Fig. 14-26 Spatial planning of robot movements: path-finding for a robot through obstacles. *(reprinted from The International Journal of Robotics Research, vol. 2, no. 4, Winter 1983, Rodney A. Brooks, "Planning Collision-Free Motions for Pick and Place Operations," by permission of the MIT Press, Cambridge, MA)*

The robot must use its perceptions of what it sees and feels to adjust its actions continuously as it reaches and grasps the parts.

14-8 TRENDS IN ROBOTICS

MANIPULATOR HARDWARE

Some of the developments which will enhance the mechanical capabilities of robots include:

☐ Multifingered, dextrous hands for general purpose use

☐ Lightweight and powerful motors integrated within the arms and joints to eliminate the problems associated with gears

☐ Lightweight and powerful arms made of advanced materials, such as composites and high-strength metals

☐ Miniature high-resolution tactile sensor arrays to provide "smart fingers"

☐ Standard grippers and interchangeable fingers

☐ Multiple, coordinated arms

MOBILE ROBOTS

Most industrial robots today are stationary. This restricts their activity to a fixed work envelope. Some robots with extended work spaces are mounted on rails so they can move between workstations on the factory floor (Fig. 14-27). A great deal of development effort has been directed toward removing restrictions and permitting robots to move relatively freely in an unstructured environment. The mechanisms used to accomplish this depend on the surface on which the robot must move. Wheeled robots can maneuver on a flat surface with few obstacles. To deal with uneven surfaces and unpredictable obstacles, as a human worker must, a robot requires other approaches and a great deal of intelligence. There has even been some success with legged robots. These, however, require complex balancing and motion control schemes that will be difficult to perfect economically.

ROBOT PROGRAMMING

Most robot programming languages today are either very limited in their capabilities or complex to use. Some of the software developments that should relieve both restrictions are:

☐ Standard high-level languages

☐ Task-level programming

☐ Coordination of simultaneous tasks

☐ Incorporation of capabilities of general purpose data processing languages into robot programming languages

☐ Computer-aided robot programming using interactive graphics

☐ Automated process planning and robot programming

☐ Automatic error recovery

☐ Geometric programming to describe relationships between objects

Fig. 14-27 Mobile robot: robot moves between workstation positions on a track. *(ESAB North America Inc.)*

VISION SYSTEMS

Since vision can provide a substantial amount of data for robot control, substantial effort is being made to improve the capabilites of vision systems for robots, including:

High-resolution cameras

Adaptive focusing

Fast range finding

Multispectral sensing

High-speed parallel image processors

Inference capabilities for image recognition

3D vision systems

ROBOTICS SYSTEMS

Robots will no longer be stand-alone tools on the factory floor. Complex tasks require a sophisticated control system of hardware and software to support the manipulator. In an automated manufacturing environment, the robot must also be tied to the operation of other robots and production machines (Fig. 14-28). In the factory of the future, robotics systems will be integrated into the computer-automated manufacturing (CAM) system that runs the production operations. Individual robot control systems will be part of a hierarchical control system which ties all the tools together in a network. A truly "peopleless" factory will rely on the monitoring of a large number of sensors and self-diagnostic capabilties for fail-safe robot operation. The robotics systems of the future will use all the advances of computer science to improve their capabilities to the point that they can perform tasks with humanlike skills.

14-9 SUMMARY

Robots require advanced, humanlike capabilities in order to perform complex tasks. The ability to use information to modify actions is an attribute of intelligence. Intelligent robotics systems use sensing and data processing techniques to adapt their behavior to uncertain or changing manufacturing environments. The use of intelligent robots provides the flexibility which makes practical some applications that otherwise could not be automated economically.

Fig. 14-28 Automated manufacturing system: multiple robots working together in an automated manufacturing operation. *(American Cimflex Corp.)*

An end effector is the working end of the robot arm. End effectors can be customized for the unique needs of an application or designed as general purpose "hands." They may be grippers or tools. Some use compliance techniques to accommodate variations, while others use hands that have sensors to obtain information about those variations in the work environment. A number of different contact and noncontact sensing techniques give robots the flexibility to deal with uncertainty. Intelligent robots often use several different types of sensors on the application equipment as well as the robot itself.

Robot vision makes it possible for robots to perform tasks where contact sensing alone is inadequate. A number of different types of 2D and 3D vision systems can be used for relatively simple image matching or complex image interpretation tasks.

Sensor-based robot control systems are data intensive and require a lot of computer power. Such robotics systems are dependent on sensor inputs for the adaptive control of the manipulator's actions. Artificial intelligence (AI) expands the capabilities of robots to include perception, decision making, and planning. Although many manufacturing applications could take advantage of AI techniques, most robot tasks today are relatively simple. Advances in computer science and robotics technology will continue to expand the capabilities of intelligent robotics systems.

REVIEW QUESTIONS

The answer to each question can be found in the section indicated at the end of the question.

1. What does "intelligence" mean in relation to industrial robots? [14-2]

2. Identify some of the advanced features that are used in intelligent robotics systems. [14-2]

3. What are some of the advantages of using intelligent robots? [14-2]

4. What factors influence the design of an end effector? [14-3]

5. Identify the basic types of end effectors. [14-3]

6. Define the term "compliance device." [14-3]

7. Describe some of the features of a dextrous robot hand. [14-3]

8. Why are sensing capabilities important for robots and how are such capabilities used in manufacturing applications? [14-4]

9. Identify the basic types of sensors used for industrial robots. [14-4]

10. Describe the major elements of a robot vision system. [14-5]

11. Explain some of the ways in which vision systems can differ. [14-5]

12. Describe the different approaches to 3D vision systems and how they are used. [14-5]

13. How do vision systems work? [14-5]

14. Give some examples of robot decision making and how it can be used. [14-6]

15. What is task-level programming? [14-6]

16. How is artificial intelligence used in robotics? [14-7]

17. What are some of the trends that will expand the capabilities of intelligent robots? [14-8]

15

ROBOT APPLICATIONS

15-1 INTRODUCTION

Robots have been used in manufacturing for over 20 years, but only recently have they expanded into a large number and wide variety of applications. As robotics technology has advanced, manufacturing has been able to take advantage of its increased capabilities. At the same time, the cost of robots, relative to other alternatives, has become more attractive. Robots play a major role in the factory of the future. One key to sucessful automation is identifying and implementing applications for robots which can benefit production operations.

This chapter will cover how robots can be used in manufacturing applications. A robot application can be described in terms of the task or activity performed by the robot or by the type of manufacturing process in which it is involved. First we will focus on the basic types of tasks which robots can perform in a manufacturing environment. These tasks are common enough to apply to many different manufacturing operations. Robots are therefore usually designed with capabilities to perform specific types of tasks so that they can be applied in a variety of different industries. Only recently have robots been put to work in applications that have traditionally used either manual- or fixed-automation approaches to manufacturing. In the balance of the chapter we will review the major types of manufacturing operations that use robots.

15-2 MATERIALS HANDLING OPERATIONS

Perhaps the most common task for robots is some type of materials handling operation. The general purpose pick-and-place robot is the best example of a robot designed to perform materials handling operations. The basic tasks involved in materials handling operations are picking up, transferring, and moving parts or objects. Robots can be used to handle small parts as well as large assemblies (Fig. 15-1). The use of a robot for materials handling can be beneficial for a number of reasons:

1. Robots can improve the productivity of operations where the handling time is a significant portion of the total time required for the operation

Fig. 15-1 Materials handling robot for a large payload. *(GCA Corp.)*

2. Robots can perform materials handling tasks that would be too heavy or difficult for a human worker

3. Robots can be used when the materials handling task would present a hazard or safety exposure for a human worker

4. The use of robots may be able to better control the quality of the operation due to repeatability

5. A robot can relieve the tedium that human workers find in many repetitive materials handling tasks

To select the proper materials handling robot, the requirements of the application must be understood in terms of:

☐ Weight of objects involved

☐ Reach and working envelope of operation

☐ Speed of operation

☐ Accuracy and repeatability of motion and positions

☐ Design of the gripper to handle the objects

☐ Interfaces with other machines or equipment (e.g., conveyors)

Some of the most common materials handling applications are:

☐ Loading and unloading machines

☐ Presentation of parts for assembly operations

□ Palletizing or depalletizing parts (e.g., kitting or stacking operations)

Materials handling applications can get fairly sophisticated even though the basic tasks are relatively simple. Robots used for load and unload operations may be set up to service several tools (Fig. 15-2). Pick-and-place applications involving conveyors require line-tracking capabilities. Orienting and sorting parts may require the use of vision systems and tactile sensors. Automated manufacturing lines normally require the flexiblity of programmable materials handling robots to feed the operations and move the parts between workstations. Materials handling robots are a key part of integrated manufacturing operations.

15-3 TOOL HANDLING OPERATIONS

Robots can not only be used to move things; they can also use tools to perform manufacturing tasks. Specialized robotics systems have been developed to perform particular operations involving the use of manufacturing tools. Following are some of the most common types of tool handling operations for robots.

MACHINING OPERATIONS

Machining involves the removal of material and the shaping of objects. Many different machining tools and operations are used in the manufacture of parts from all kinds of materials. Some typical machining operations in the manufacture of metal and plastic parts are:

Drilling
Grinding
Routing
Deflashing
Sanding
Deburring
Profiling

Most machining operations in manufacturing involve relatively small quantities of parts. Machining therefore lends itself to a customized batch-type operation rather than a highly automated manufacturing line. The programmability and flexibility of a robotics system, however, makes it possible to automate such operations without the need for a lot of specialized tooling. Using robots for machining can eliminate the need for some manual operations and provide higher throughput with improved accuracy and repeatability. If the entire machining process is automated, the need for manual inspection and finishing operations can also be eliminated. Such a process might include the use of robots for loading and unloading parts, performing machining operations, and deburring.

For a robot to perform a machining task, it must either grasp or have a tool attached to its arm (Fig. 15-3). Such tools may be either electrically or pneumatically driven and are often compliantly mounted to the robot arm, using force sensors for feedback control. In some cases, vision systems may also be used to reduce the amount of programming and fixturing needed. Some of the requirements for robots in machining applications include:

Continuous path control
High accuracy and repeatablility
Wide range of speeds
Pneumatic or electric drive systems
Compensation for variable loads
Force-sensor feedback control systems
Compliance in end effector

The removal of burrs or flash is a common and tedious manufacturing task. Burrs can be created by machining or casting operations. Flash is caused by molding and casting operations. Grinding or sanding tools are typically used to remove burrs and flash so that the finished part

Fig. 15-2 Robot loading multiple machine tools. *(Cincinnati Milacron)*

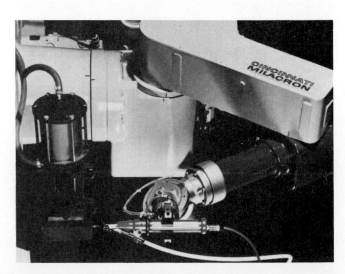

Fig. 15-3 Robot performing drilling task. *(Cincinnati Milacron)*

has uniform surfaces. Robots can be used to automate these tasks with relatively simple programming and fixturing (Fig. 15-4).

FINISHING OPERATIONS

One of the final operations in many manufacturing processes involves some form of finishing. This often is a spray-coating operation in which a paint, stain, or plastic material is applied to the surface of the part or object (Fig. 15-5). There are a number of advantages to using robots in such applications:

Fig. 15-4 Robot performing deburring task: finishing cooling fins on heat sinks. *(copyright Unimation Inc., a Westinghouse Co., Shelter Rock Lane, Danbury, CT 06810, used with permission)*

Fig. 15-5 Spray-painting application. *(General Motors Corp.)*

□ Robots replace manual labor

□ Robots improve quality and consistency

□ Robots save on materials consumption

□ Robots save on energy consumption

□ Using robots can prevent exposure of humans to potential hazards from materials, fire, or explosion

Robots designed for finishing applications must be versatile and precisely controlled. They are typically servo-controlled, continuous path units with at least six degrees of freedom. For safety in potentially explosive environments, they use hydraulic drive systems. The spraying tool is attached to the end of the arm and is usually programmed by walk-through teaching. Several different types of spray application systems can be used:

□ Air spray, which uses air to atomize the particles of material to be applied

□ Airless spray, which uses hydraulic pressure to create an atomized spray

□ Electrostatic spray, in which a charge induced in the material being sprayed attracts it to the part being finished

□ Heated spray, in which material is heated to reduce the pressure and amount of solvent required for applying it

Finishing applications may place some special requirements on the robotics system, such as:

□ Coordination of motion with movement of parts (e.g., on conveyors or overhead lines)

□ Sensing for the presence of parts

□ Controlling the timing for changing colors of spray

□ Precise alignment and placement of finish (e.g., applying pinstriping to car bodies, as illustrated in Fig. 15-6)

An automated finishing system can be expensive to install, but it can still be economical by reducing operating costs and improving product quality.

WELDING OPERATIONS

Welding has traditionally been a labor-intensive operation which involved relatively poor working conditions and inconsistent quality. The use of robots in welding applications has proved to yield significant benefits:

Lower operating costs

Increased throughput

Improved consistency and quality

Reduced hazards to human workers

A welding robot can repeat even complex welding tasks continually and consistently without the delays, interruptions, or fatigue normally involved in a manual operation. The path of the weld and the parameters controlling the process are programmed by walk-through or lead-through teaching techniques. Sensors may be required for process control, and in some complex applications,

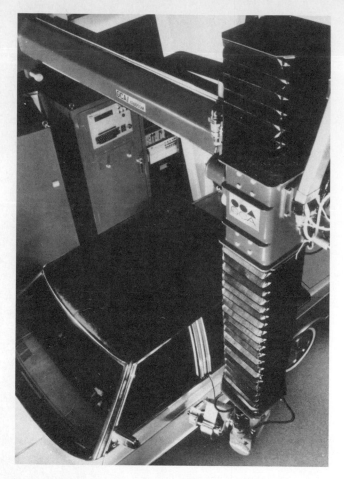

Fig. 15-6 Robot applying pinstripes to an automobile. *(GCA Corp.)*

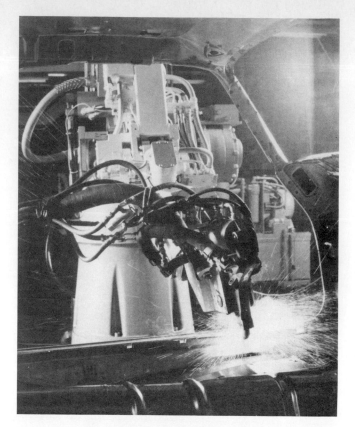

Fig. 15-7 Spot welding robot application. *(Cincinnati Milacron)*

even vision systems may be used. In addition to the robot itself, a robotic welding system can involve a lot of support equipment, such as the welding gun, gas and power sources, cooling water, and fixtures. There are two basic types of welding systems:

1. Spot welding. This is a common technique used to weld large mechanical assemblies, such as auto bodies (Fig. 15-7). It uses an electrical resistance gun to make multiple welds at points which will structurally bond the parts together. The robot's task is to squeeze the parts together, weld the spot, hold the parts, then turn off the gun and release the parts. Some spot welding operations can involve a large number of welds and complex maneuvers. To use a robot in such applications, one must consider a number of factors, such as:

☐ The path of the weld sequence

☐ The positioning of the gun

☐ The access of robot to the welding position

☐ The coordination of moving parts and multiple welding stations

2. Continuous welding. This type of welding is used when a complete bond or seal is required between two parts. A number of different welding technologies can be used. Most involve an electric arc which melts a rod of

metal to form the weld. In some cases an inert gas is introduced to prevent oxidation; this is sometimes referred to as metal inert gas (MIG) welding. To use a robot in this type of application requires continuous path control, highly accurate positioning, and sensing (Fig. 15-8). Arc welding robot systems are more complex and expensive than most spot welding systems. Arc welding systems are used in applications that require high precision and quality control or when it is physically difficult for human workers to perform the task. Arc welding robot applications involve additional support equipment, such as wire feeders, gas supplies, and joint-tracking devices, which must all be controlled by the system (Fig. 15-9).

15-4 OTHER ROBOT TASKS

Some robots are designed to perform special tasks that involve more than just materials or tool handling. In many cases, these tasks require additional capabilities and support equipment. Following are the major categories.

ASSEMBLY TASKS

Applications for robots are expanding rapidly in manufacturing operations that involve assembly tasks. In general terms, these involve the bringing and fitting together of parts (Fig. 15-10). Such operations are usually labor-intensive and subject to error, and they often require precision. Many assembly tasks are relatively low-volume batch operations which normally do not lend themselves

Fig. 15-8 Metal inert gas (MIG) arc-welding robot with seam-tracking vision control system. *(American Cimflex Corp.)*

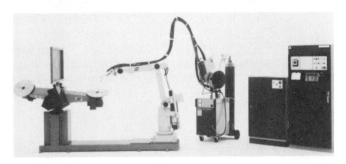

Fig. 15-9 Arc-welding robot system. *(ESAB North America Inc.)*

Fig. 15-10 Robot performing assembly task. *(International Business Machines Corp.)*

to automation. Programmable robots now make it practical and economical to automate those types of operations. Assembly robots are usually designed differently from those intended to perform materials or tool handling tasks. Some of their typical features include:

Small and light structure for handling small parts

High accuracy

Capability for high-speed movement

Capability for complex motion

Servo control with multiple sensing systems

In addition to the robot itself, assembly applications usually involve a variety of support equipment, such as fixturing, materials handling systems, assembly tools (e.g., power screwdriver), and inspection devices. The parts to be assembled will typically have to be oriented, fed, and presented to the robot for assembly. Some of these tasks may be performed by support equipment (e.g., hard automation) or by the robot itself. The actual assembly tasks may use compliance devices to ease the mating of parts. Assembly operations can also be integrated through the use of robotics. One robot workstation could be used to assemble a complete unit involving multiple assembly tasks and parts. As an alternative, multiple robot stations can be tied together into an automated assembly line, each robot performing a particular task. In either case, complex programming and control systems are required.

INSPECTION TASKS

Most manufacturing operations include some inspection tasks. They may be visual, mechanical, or electrical in nature, but they all involve controlling the process or quality of the product being manufactured. Traditionally, inspection tasks have been manual. Because of that, they become expensive and subject to error in high-volume or complex production operations. Robots can be used to automate some inspection tasks. The role of the inspection robot may be passive. That is, the robot may be used just to handle or sort parts. It can also take a more active role, where it handles inspection devices, manipulates the part, and collects data. Such automated systems can increase the throughput of inspection operations while assuring high repeatability.

Inspection robots can take a variety of forms, depending on the inspection technique and types of parts involved (Fig. 15-11). Some use contact sensor systems, such as tactile probes for mechanical measurements. Others use noncontact systems, such as optical or vision techniques, to inspect for mechanical features or defects. The inspection system may be a separate automated operation in a manufacturing line, or it may be integrated into a particular operation in the process. A robotic workstation may be designed to include inspection tasks with other manufacturing tasks it is programmed to perform, such as assembly operations.

Fig. 15-11 Robots performing automatic inspection of automobile body. *(Cincinnati Milacron)*

DATA COLLECTION

A task that is not visible and is often taken for granted in manufacturing operations is data collection. To control the flow of product, the operation of the process, and the quality of the product, data is collected throughout the manufacturing line. The data may include results of electrical or mechanical tests, inspection results, parts counts, serial or job numbers, equipment operating parameters, and so on. One of the key features of robots is that they are able to automate the collection and processing of data as well as the physical manufacturing operations. The use of sensors, such as optical readers or contact switches, provides a means to collect data about the parts or operations involved. When the robot control system is tied into the other tools and machines in the operation, it can be used as a monitoring or control system for the work cell. Data collected by a robot can be transmitted to a host system for control or reporting, just as in other computer control systems on the factory floor.

15-5 INTEGRATING THE ROBOT TASK AND THE WORKPLACE

In most applications, the tasks performed by the robot are interrelated with other activities and functions of the workplace. A robotics application must therefore be integrated with the workplace environment to perform its tasks. These other factors can take several different forms, as follows.

ORIENTATION OF PARTS

Parts that are not symmetrical must be oriented in a predetermined way so the robot can properly grasp or handle it. This may be accomplished by special fixturing or automated tooling, or even by sensory control of the robot itself. It is essential that the orientation of a part be controlled so that the robot can perform the programmed task without errors or damage. To assure this, the orienta-

tion of the part should be permanently established when it is initially picked up. The applications program should avoid any situations that would cause the robot to lose this orientation intentionally or accidentally, such as by dropping the part or putting it down in an uncontrolled manner.

CONTROL OF SEQUENCE OF OPERATIONS

To prevent damage or potentially dangerous movements, each step of an operation must be controlled. When multiple pieces of automated equipment or machines are involved, it is essential that their actions be coordinated. Robotic manipulators, machine tools, and materials handling equipment are each powerful and potentially dangerous devices. When they are all in motion, collisions or accidents can occur. Sequence control and system interlocks can be provided by such devices as limit switches, photo detectors, and pressure switches.

LAYOUT OF THE WORKPLACE

The efficiency and throughput of an automated operation depends on how well the workplace is laid out. Many applications involve the feeding of parts and multiple tools or machines. Where objects are placed relative to each other will influence the complexity of the robot's path and the speed of its operation. The work can be arranged around the robot such that it has easy access to all the parts and tools involved. Depending on the size and quantity of parts involved, it may be more practical to bring the work to the robot either directly or by moving it past the workstation. Another alternative is to move the robot past the work. Each approach has its merits and therefore should be considered for a particular application.

LINE TRACKING

In those applications where the robot and workpiece are moving in relation to each other, their positions must be tracked and coordinated. When a robot is in a fixed position relative to a moving conveyor line of parts, it must be programmed to perform its task on the basis of the position of the parts, independent of the motion of the line. This is referred to as "stationary-base line tracking." The robot controller adjusts the coordinates of the preprogrammed path by using position sensors to track the motion of the line. An alternative approach is to move the robot so that it is synchronized with the motion of the line ("moving-base line tracking"). This obviously requires an additional transport system as well as a control system. The control program for line-tracking applications must consider such factors as the working envelope of the robot relative to the line, the need to stop when the limits of the path are reached, and the ability to correct errors or malfunctions. This type of system can often be found in applications such as spot-welding auto bodies, spraypainting, and conveyorized assembly lines (Fig. 15-12).

Fig. 15-12 Stationary-base line tracking is used to permit robots to spot-weld car bodies as they move past on a conveyor line. *(Cincinnati Milacron)*

15-6 METALWORKING APPLICATIONS

The metalworking industry comprises a wide variety of manufacturing operations which shape and form the many different products made out of metals. In most cases, the manufacturing environment is characterized by heavy equipment, heat, dirt, and noise. Although this is not a pleasant environment for human workers, these operations have traditionally been labor-intensive. They are normally batch operations that require skill and control that can turn out to be very expensive. In recent years, robots have been applied sucessfully to many of these tasks to improve productivity and reduce the hazards to humans. Following are some of the most common types of metalworking operations that can use robots.

CASTING

Many metal parts, particularly large, heavy objects and those with thick cross sections, are formed from molten metal which has been poured into some kind of mold. Two of the most common manufacturing processes are:

1. Die casting. In this technique the metal is cast into a permanent mold in a machine. The casting must then be removed from the die-casting machine, quenched to cool it, and then trimmed of any flash. A typical robot application in a die-casting operation includes:

☐ Preparing the die for casting (i.e., lubrication, flash removal)

☐ Loading inserts into the die

☐ Ladling molten metal into the die

☐ Unloading castings from one or more die-casting machines

☐ Quenching the casting

☐ Placing the rough casting on a trimming press

☐ Removing trimmed material from the workstation

☐ Placing the trimmed casting on a conveyor or pallet

In a die-casting operation, the robot control system must be tied into the operation of the die-casting machine and support equipment, such as conveyors and trimming machines. It must assure that there are no objects which may interfere with the operation of the die-casting machine, such as the lubricating mechanism, casting material, or humans. The use of a robot in this type of operation can save money by increasing throughput and reducing labor costs, and it should also improve quality and save material and floor space. It will also obviously remove human workers from the hazards and undesirable nature of the environment.

2. Investment casting. In this technique an expendable pattern is used as the mold, which is formed by a refractory coating over wax. It is often used to manufacture more complex parts in larger quantities. A typical robot application in an investment-casting operation may include:

☐ Removing wax patterns from the die

☐ Assembling wax patterns into a cluster to be cast together

☐ Dipping the wax cluster into refractory material for coating

☐ Loading and unloading clusters for a sanding operation

☐ Loading and unloading clusters into a furnace to fire the mold and remove the wax

☐ Pouring molten metal into the mold

☐ Removing and quenching the casting

☐ Moving the casting to trimming and finishing operations

Since this technique is used to produce castings with smoother finishes and more complex shapes, the control of the process and materials is even more important than in die casting. Robots can provide careful and consistent handling of the shell molds and assure uniform coatings to produce high-quality castings with minimal losses. They can also increase the throughput of this complex operation, particularly when heavy loads are involved.

FORGING

Many heavy metal objects which are designed for high strength are formed by a forging process. This basically involves using hammers, presses, or dies to shape metal which has been heated until it is soft. The work involved is heavy, hot, dirty, and fast-paced. Several different types of forging techniques are used in manufacturing.

Typical tasks involved that a robot can perform are:

☐ Preparing the presses for the forging operation (e.g., lubrication, cleaning)

☐ Loading the forging press

☐ Actuating the press

☐ Removing the forged parts

☐ Moving the parts to a finishing operation

A robot can increase the productivity of this type of operation and can avoid the need to subject human workers to this undesirable and hazardous environment. The robot, however, must be rugged, and special provisions must be taken to protect it from the heat and shock involved.

HEAT TREATING

Some metal parts are subjected to special heat treatments to increase their hardness or strength. These processes can take a number of different forms, such as tempering, annealing, or surface hardening. Normally, the parts are placed in a furnace, perhaps in a special environment with other materials, for a prescribed time at a particular temperature to achieve the desired results. At the end of the treatment, the parts are quenched to cool them, usually at a controlled rate. Robots can be used to perform a variety of tasks in a heat-treating application (Fig. 15-13):

☐ Orienting the parts

☐ Placing the parts in the furnace

☐ Removing the parts from the furnace at the prescribed time

☐ Quenching the parts for the specified time

☐ Placing the parts on pallets or conveyors

These are relatively simple tasks that are tedious and undesirable for human workers. In most cases, a pick-and-place-type robot is adequate for such an application.

PRESSWORK

Many products are made out of sheet metal. The sheet metal is usually formed into its final shape by a series of press operations, such as bending, stamping, punching, and cutting. These operations are all dangerous. Robots are therefore used frequently to avoid exposing people to this hazardous, noisy, and monotonous operation. Robots are particularly well suited for handling the transfer and loading of large, heavy stampings (Fig. 15-14). They can also reduce the time it takes to set up and change over for batch runs.

MACHINE TOOL OPERATIONS

Aside from actually performing some machining operations, such as the tasks described earlier in this chapter, robots have been very useful in machine tool operations. Machine tools include lathes, drilling machines, grinders, milling machines, broaches, and multitool machining centers. In these operations robots are typically used for tasks such as (Fig. 15-15):

Loading and unloading machine tools

Transferring parts between tools

Palletizing and depalletizing parts

Transferring parts to and from materials handling systems

These are primarily pick-and-place-type tasks which normally require robots that have a large work envelope and a heavy payload capacity. The robot's control system usually must tie into the machine and materials handling controllers to coordinate the entire sequence of operation. A single robot may service several machine tools.

15-7 PLASTICS MANUFACTURING APPLICATIONS

Like metals, plastics are used to produce a wide variety of parts and products. They have become a substitute for metals by offering comparable function at lower cost. In addition, they are lighter and more easily formed and finished. There are number of different types of plastics manufacturing operations, each of which provides opportunities to use robots.

MOLDING

Molding is the most common process used to fabricate plastic parts. A variety of molding techniques are used,

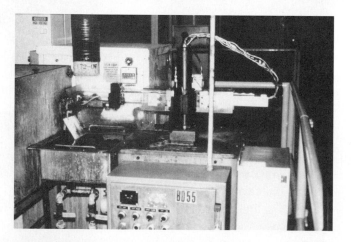

Fig. 15-13 Heat-treating application. (*International Business Machines Corp.*)

Fig. 15-14 Presswork application: robot unloading part from hot form press. (*copyright Unimation Inc., a Westinghouse Co., Shelter Rock Lane, Danbury, CT 06810, used with permission*)

Fig. 15-15 Machine tool operations: robot loading lathe. *(Cincinnati Milacron)*

such as compression, injection, and extrusion. Typical tasks involved that a robot can perform are (Fig. 15-16):

Loading inserts and charges into molds

Unloading molded parts from molding machines

Trimming flash from molded parts

Cleaning mold cavities

Transferring or packing molded parts

A robot can perform these tasks for several machines by controlling their sequence of operation and cycle times

Fig. 15-16 Plastic-molding application: robot arm unloading molded plastic assembly. *(Automated Assemblies Corp.)*

and coordinating the materials handling system. Automating and integrating an operation like this can increase throughput, improve quality, and reduce scrap. Using a robot to unload parts from a molding machine averts the safety exposure of a human worker reaching into a potentially hazardous area.

COMPOSITE MANUFACTURING

Special plastic materials are now being used to fabricate high strength, lightweight structures, such as parts of airframes (e.g., tail sections of airplanes). These materials are called "composites" because they combine materials such as graphite fibers with plastics. These materials are usually made in the form of sheets which can be laminated, cut, and formed into shapes. Robots can be used to perform a variety of tasks in composite manufacturing:

☐ Stacking sheets of composite material

☐ Cutting stacked sheets into patterns

☐ Applying adhesive and coating materials

☐ Transferring stacked pattern to laminating and forming presses

☐ Loading and unloading presses

☐ Stacking finished structures or transferring them to the materials handling system

This type of operation can be slow and expensive if it is not automated. An integrated process using robots, automated materials handling, and a computer-based design system can improve throughput and quality while minimizing waste of the expensive materials involved.

APPLYING ADHESIVES

Many plastic products are fabricated by bonding several pieces together with adhesives. This is even true of some large structural assemblies, such as those used in automobiles. This type of operation is not complex, but it requires control, repeatability, and materials handling. For human workers the tasks involved can be tedious and even physically taxing, particularly for large assemblies, and the operation is subject to error. A robot can be programmed to perform these tasks in much the same manner as one would program a continuous welding operation (Fig. 15-17).

COATING

Plastics are sometimes applied to parts and products as a spray coating for protection or finishing. This type of operation is much like painting and can be performed efficiently and safely by a robotic spray-finishing system.

CUTTING

Plastic parts must often be cut to size and form after they are molded. Since they cannot be machined like metal

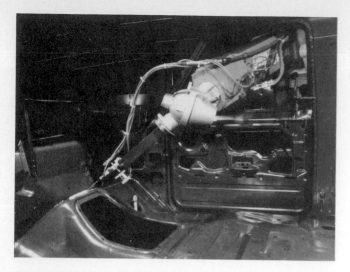

Fig. 15-17 Robot applying adhesive to interior of car body. (Cincinnati Milacron)

Fig. 15-18 Robot cutting plastic with water jet. (GCA Corp.)

parts, special cutting operations may be used. One new technique which has proved to be accurate and efficient is water jet cutting. A robot with continuous path control can cut complex and customized patterns in sheets of molded plastic or fabric (Fig. 15-18).

15-8 ASSEMBLY APPLICATIONS

The manufacture of many end products involves some type of assembly operation. The assembly tasks themselves may include mating, joining, and inserting steps, as well as support tasks such as materials handling, inspection, and finishing. Following are the major types of assembly operations which have used robots:

MECHANICAL ASSEMBLY

A wide variety of manufacturing operations involve assembling products out of mechanical parts. The products can range from small parts to large structures. When assembling small parts, control of orientation and automated parts feeding are important to the performance of a robot. In large assembly tasks, such factors as reach, work envelope, and payload may be the most important considerations. Perhaps the best example of the successful use of robots in a mechanical assembly application is automobile production (Fig. 15-19). The tasks that have been performed by robots include materials handling, welding, inspection, and finishing, as well as the assembly of parts. In many cases, robots are the key to automating the entire manufacturing process to reduce costs, increase throughput, and improve quality.

ELECTRICAL AND ELECTRONIC ASSEMBLY

Most electrical and electronic equipment is manufactured by assembling a large number of parts. The assem-

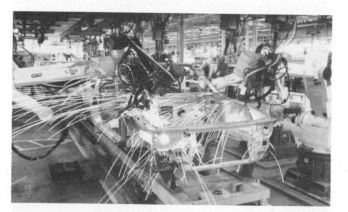

Fig. 15-19 Mechanical assembly application: automobile assembly line. (General Motors Corp.)

blies involved may be printed circuit boards, motors, cables, switches, and so on. The tasks required to produce such assemblies include component insertion, soldering, wiring, inspection, testing, and materials handling. In the past, much of this work was done manually. Today, many of these operations have been automated by using robots (Fig. 15-20). When all these operations are integrated into an automated assembly process, errors and defects can be eliminated while throughput is increased. Some electronic assembly operations are so complex or so delicate that manual approaches are impractical or undesirable.

ELECTROMECHANICAL ASSEMBLY

As the name implies, electromechanical assembly is a combination of electrical and mechanical assembly. Many high-technology products are assemblies of advanced electrical, electronic, and mechanical parts. Some of the best examples are computer-related equipment such as printers, disk drives, and tape drives. The assembly tasks often involve high precision and cleanliness to assure that the final product will operate properly and reliably. Many of the tasks involved lend themselves to automation with the use of robots (Fig. 15-21). These may include materials handling, inspection, and test operations, as well as the assembly of small parts. Some manufacturing operations automate the entire process, from the handling of small parts coming in to the packaging of the finished product. This provides consistent quality and high throughput with minimal space and labor. Automated assembly has become a popular approach to manufacturing some high-volume products, and it may also be applied to batch or customized production.

15-9 PROCESS OPERATIONS

Many manufacturing operations involve continuous processes rather than discrete fabrication or assembly steps.

Fig. 15-20 Electronic assembly application: printed circuit board assembly line. *(Universal Instruments Corp.)*

Fig. 15-21 Electromechanical assembly application: Computer equipment assembly line. *(International Business Machines Corp.)*

These processes are usually complex and controls on the materials, equipment, and environment are required. Products manufactured in a process environment include:

Semiconductors

Magnetic tape

Magnetic disks

Chemicals

Pharmaceuticals

Nuclear fuel

Although such processes are often advanced and sophisticated, they have also traditionally not been very automated. Recently, however, robots have been used successfully to perform tasks in such operations to improve throughput and quality as well as minimize the potential sources of contamination (Fig. 15-22). For ultra-clean manufacturing areas, such as a semiconductor clean room, robots must be specially designed to assure the cleanliness required. The robots used for these applications are usually small electric-drive units because the

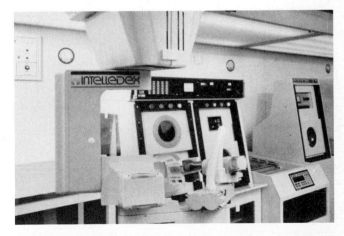

Fig. 15-22 Process manufacturing application: semiconductor clean-room operation. *(Intelledex Inc.)*

tasks typically involve light payloads, high speeds, and contamination-free operation. The robot control systems are often sensor-based, since their operation must be coordinated with manufacturing equipment and must tie into a process control system. The tasks typically involve materials handling, since they are usually related to loading and unloading process tools. However, they may also involve inspection, testing, and packaging. The ultimate manufacturing line for many of these process-intensive products may employ few, if any, human workers, relying solely on robots and automated process equipment.

15-10 SUMMARY

As robotics technology has advanced, the use of robots in manufacturing has expanded rapidly. Most robots are designed to perform certain types of general purpose tasks which can be applied to a variety of different industries. Robots are most frequently used for some form of materials handling task, but they can also use tools to do work. Some of these tasks place significant demands on the performance and control of the robot. Many require sensors, fixtures, and materials handling systems. In such applications, the robot's operation must be integrated with the entire workstation.

Robots are used in many different types of manufacturing operations. They are used to perform undesirable tasks as well as to improve productivity. In many cases, they perform multiple tasks, including the preparation of production machines, the handling of materials and tools, and the control of the process sequence. Robots may also be used to service several machines in a work cell. The use of robots in heavy industrial applications avoids the need to expose human workers to hazardous, noisy, and monotonous tasks. Robots have recently begun to be used in assembly and process operations. These applications usually require complex control systems that are integrated into high-speed, automated processes. Some also require contamination control for clean manufacturing environments.

REVIEW QUESTIONS

The answer to each question can be found in the section(s) indicated at the end of the question.

1. What are the basic types of tasks for which most general purpose robots are designed? [15-1 and 15-2]

2. Identify some of the advantages to using a robot for materials handling. [15-2]

3. Give some examples of tool handling tasks that can be performed by robots. [15-3]

4. What are some of the requirements of robots used in machining applications? [15-3]

5. Describe the major characteristics and advantages of a robot finishing application. [15-3]

6. Identify some of the different requirements that can be involved in welding applications. [15-3]

7. What are some of the typical features required for an assembly robot? [15-4]

8. Identify the factors which must be considered when integrating a robot into an automated workstation. [15-5]

9. Give examples of some of the major types of manufacturing operations that use robots. [15-6 through 15-9]

10. Describe some of the typical tasks performed by robots in manufacturing operations. [15-6 through 15-9]

11. Identify some of the common benefits to using robots in heavy industrial applications. [15-6 and 15-8]

12. What are some of the unique characteristics of automated assembly operations? [15-8]

16
IMPLEMENTING ROBOTICS IN MANUFACTURING

16-1 INTRODUCTION

Knowing how robots work and where they can be used is only the beginning of applying robots in manufacturing. A great deal of time and effort can be spent in developing and implementing a robot application. If robots have not been used in a manufacturing operation before, there may be concerns and problems that delay the implementation or perhaps even cause it to fail. The experience gained in using robots successfully in manufacturing applications is the best insurance against such problems. Without that experience, it is essential that a thorough job be done in preparing for the application.

This chapter will address the process involved in developing a robot application. It will not make the reader an expert, but it will cover the basic elements that should be pursued. It will start with how to decide where and when to use a robot. Then it will cover the process of implementing the application in manufacturing. In addition, there are special sections which deal with safety considerations and simulating robot applications.

16-2 DECIDING WHERE AND WHEN TO USE A ROBOT

BASIC CONSIDERATIONS

Robots are not placed in manufacturing applications just because people believe that they are the best way to automate or that they are the only answer to improving productivity. Developing and implementing a robotics application can cost a lot of money. Therefore, the decision should be made carefully, after thorough consideration of all the costs and alternatives as well as the potential benefits involved. The basic factors are:

1. Economics. The potential savings which can be achieved must be weighed against the costs of implementing and maintaining the robot application. All the tangible elements which can be identified should be quantified.

2. Intangibles. Potential benefits as well as disadvantages which cannot be quantified, but which may have a bearing on the success of the application, should be considered. These may include items such as safety, quality, human factors, and risks.

3. Alternatives. There are a number of different ways to implement most manufacturing operations. Before choosing a specific solution, at least the basic alternatives of manual and fixed automation should be evaluated along with a robotics solution. In many cases, the most practical solution uses a combination of these approaches.

4. Time. The amount of time it takes to develop and implement a solution to a manufacturing operation can vary significantly among alternatives. In some cases, it may be necessary to choose a solution that can be implemented within a specific time frame even though another alternative offers greater potential savings.

5. Robot selection. Even if the robotics alternative is chosen, there are many different robots that could be selected. The capabilities of and differences between all those that can perform the task should be considered.

ECONOMICS

When evaluating the economics of a potential robotics application, one of the primary factors to consider is the relative cost of manual labor. The cost of wages and benefits for skilled manufacturing workers has increased substantially in recent years (Fig. 16-1). However, the cost of installing and operating a robot has not increased very much. This trend has made the use of robots more attractive in many manufacturing applications where labor costs are high, such as the automotive industry. Manual labor, of course, is not the only alternative available for a production operation. Special purpose, fixed automation equipment could also be considered. When evaluating the relative costs of these alternative approaches, one finds that (Fig. 16-2):

1. Manual labor is the least expensive approach for low-volume production. This is because it avoids the expense of obtaining and installing special tools and equipment that would be required to automate an application. Such an investment cannot be justified if it would only be used to produce a relatively small number of parts.

2. Special purpose automation equipment is usually the

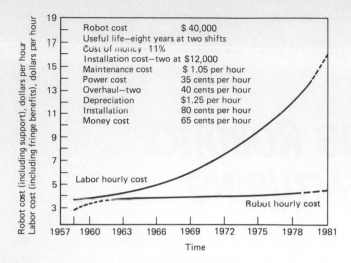

Fig. 16-1 Cost of labor versus cost of robot. *(Roger Allan, "Busy Robots Spur Productivity," IEEE Spectrum, Sept. 1979; © 1979 IEEE)*

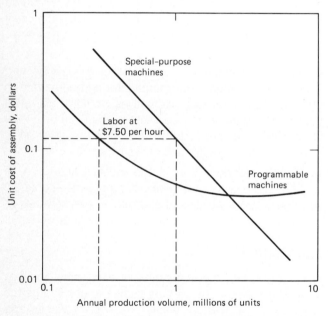

Fig. 16-2 Relative cost of fixed and programmable automation versus manual labor. *(Roger Allan, "Busy Robots Spur Productivity," IEEE Spectrum, Sept. 1979; © 1979 IEEE)*

least expensive approach for very high-volume production. In this case the investment can be applied to large quantities of parts. Fixed automation can be very efficient when it is designed to produce the same part over and over again at high speeds for a long period of time. This is the basic concept of mass production, which has been successful for many high-volume products.

3. Robots are a form of programmable automation, which is often the most economical approach for medium-volume production operations. Robots are flexible enough to adapt to changes in setup for batch production, which avoids the need to invest in expensive special purpose tools and equipment. At the same time, robots can often operate more efficiently than manual laborers.

When at least moderate production volumes are involved, the investment in a robotics approach may be the lowest-cost alternative.

When evaluating the economics of these alternatives, cost elements other than wages and investments must be considered. The available working times of a human, a robot, and special purpose automation equipment can differ significantly. Robots do not have to take breaks to avoid fatigue or have lunch. They do not take vacations and are never absent from work. They do not have to be replaced for every shift and they can work around the clock. Like all machines, however, they do have to be maintained, and they can fail. Robots often experience less "downtime" than either human workers or automated equipment. Industry data has shown that robots exhibit downtimes in the range of 2 percent, with a mean time between failures (MTBF) of over 400 hours. This can be compared to typical "downtimes" of 3 to 4 percent for humans and up to 10 percent for special purpose automation equipment.

Another important cost factor to be considered is quality. Errors can cause defects which may result in scrap or rework. The ultimate economic measure is the total cost of producing good products. Once programmed properly to perform a task, a robot can repeat it without error. In most applications, the quality of products manufactured by robots is better than that produced manually, and it may be at least as good as that produced by other automation techniques.

An indicator often used to evaluate the economics of alternatives is the return on investment (ROI). A relatively simple calculation determines the rate of return and payback period for an investment (Fig. 16-3). The rate of return is like the interest rate paid by a savings account or the yield on a common stock. It indicates how much the investment saves over another alternative. A good application should yield a greater rate of return than could be obtained from normal monetary investments. The payback period is the amount of time it will take to recover the cost of the investment from the savings that it generates. A typical guideline for a good application is for it to payback in two or three years or less.

SELECTING THE RIGHT APPLICATION

If a manufacturing operation currently does not use any automation, it may be difficult to decide where to start introducing robots. Even in operations that already use some robots, it may not be obvious which application to pursue next. Following are some basic guidelines that have worked sucessfully:

1. Start simple. That is, if robots have not been used before, pursue a simple application first. Complex tasks can be difficult and time-consuming to automate, even for those with prior experience. A robot can be applied to a simple task with relatively little effort in programming, debugging, and support equipment.

$$\text{PAYBACK PERIOD} = \frac{\text{Total investment}}{\text{Net annual savings}}$$

WHERE Total investment = Capital + Expense of application
= Robot + Application tooling + Application
development + Installation + Debugging

Net annual savings = Difference in operating costs between existing or alternative
approach and robot proposal, in terms of Labor + Benefits + Overhead expense
+ Support costs + Maintenance expense + Tools + Depreciation + Scrap.

Example:
Total investment in robot application = $80,000
Annual robot operating cost = $60,000
Existing annual operating cost = $100,000

$$\text{PAYBACK PERIOD} = \frac{80,000}{100,000 - 60,000} = 2 \text{ years}$$

$$\text{ANNUAL RATE OF RETURN} = \frac{\text{Annual savings}}{\text{Investment}}$$

$$= \frac{40,000}{80,000} = 50\%$$

Fig. 16-3 Simplified economic analysis of a robotic application.

2. Start where there are poor working conditions. This should include tasks which expose workers to hostile environments or potential hazards. The advantage to using robotics here is not just the labor savings, but also the removal of humans from undesirable tasks.

3. Select applications that have the highest potential cost leverage and short-term payback. These include operations which have productivity problems due to bottlenecks, low throughput, high scrap, or rework. Automating these operations should yield the greatest savings. Applications that have less leverage, but still have the potential for savings, can be pursued later.

4. Start with applications that have a high chance of success. If robots have not been introduced into the manufacturing operation before, it is important that the first application succeed. Choosing a complex and challenging application first can result in a failure that may discourage future applications.

5. Look for repetitive and tedious tasks. These are better suited for machines than for people. In most cases, a machine will be more productive in such an application. The tasks are also usually simple and relatively easy to implement.

6. Do not try to use the maximum capabilities of the robot. Select a task that is well within the normal range of the robot's specified capabilities. Attempts to challenge or stretch these capabilities often lead to failure.

7. Select an application that is easy to measure. The performance of the robot and the productivity of the operation need to be measured to determine whether the application was actually successful. Without a simple measurement, one may not really know that the robot paid off as expected.

NON-ECONOMIC FACTORS

In addition to the normal economic factors, other considerations may influence the decision on where and when to use a robot:

1. Benefits to workers. Robots can prevent the exposure of workers to hazardous and undesirable tasks. The benefits may not be easily quantifiable, but they certainly include the reduction of accidents and of worker dissatisfaction. Industrial environments can create a wide variety of hazards. These may include toxic materials, vapors, fumes, sprays, high temperature, radiation, noise, dust, vibration, shock, and risk of fire or explosion.

2. Benefits to products. Automating some manufacturing operations can improve the quality or reliability of a product. The cost savings may not always be quantifiable and may not even directly benefit manufacturing. Automation can, however, result in reduced field service (repair) cost and in improved customer satisfaction.

3. Benefits to manufacturing. In addition to the direct cost savings, the introduction of robotics may have strategic value to the manufacturing operation. Robotics can begin to prepare a factory for the future, when it must be able to compete in a world of advanced manufacturing technologies and high productivity. The sooner a manufacturing operation starts along that path, the faster it will get there.

APPLICATION DEVELOPMENT

After a robot application is selected and justified and a decision is made to implement it, the real work begins. Depending on the scope and complexity of the tasks involved, the process of developing an application can involve a long and sizable effort. Application development includes all the work required to go from a concept of the application to the actual installation in manufacturing. It includes the planning, design, and debugging efforts on all the hardware and software involved. The basic steps in this process are:

☐ Developing a plan of action and schedule of events

☐ Establishing a team to work on the project

☐ Becoming familiar with the existing operation (if there is one) by thoroughly analyzing each element of the tasks involved

☐ Becoming familiar with the capabilities and limitations of the equipment involved (including the robot)

☐ Obtaining support and advice from experienced experts, support groups, and suppliers

☐ Identifying constraints and potential problems

☐ Choosing a backup approach

☐ Laying out the operation

☐ Designing, selecting and ordering all the tools and equipment

☐ Choosing a software architecture and programming environment

☐ Preparing the manufacturing area for installation (e.g., services, rearrangements)

☐ Establishing a training program

☐ Installing the tools and equipment

☐ Programming and debugging the application

☐ Training operator and support personnel

☐ Testing the performance of the application

A couple of professionals may be able to develop a simple application which involves one robot and a single task in only a few months. However, a number of teams of people may have to work for several years before finishing the implementation of a large automation system which involves the integration of many manufacturing operations with a lot of application tooling.

GUIDELINES FOR IMPLEMENTATION

A robot application that pays off on paper is not an assured success. The process of implementation can be filled with problems and complications that need to be avoided or overcome. Some general guidelines can make this job easier:

☐ Be thorough in the analysis of the tasks and evaluation of the equipment involved.

☐ Have the full support of management before starting.

☐ Include manufacturing as well as support personnel in the project.

☐ Determine the real capabilities and limitations of the robot and other equipment involved. Do not rely solely on specifications and word of mouth.

☐ Take advantage of experts and the experience of others.

☐ Provide for thorough training of all operator and maintenance personnel.

☐ Establish a complete maintenance and spare-parts plan.

☐ Provide for adequate safety protection.

☐ Anticipate potential problems from all sources (e.g., hardware, software, parts, people).

☐ Have a backup plan in case the first does not work.

Once a few successful applications are in place, the process of introducing additional robots into a manufacturing operation will become easier. The people involved will have more experience and confidence in robots and will have learned from any mistakes they may have made in the past. Their familiarity and training can save time and help to avoid problems in subsequent applications.

HUMAN CONSIDERATIONS

The introduction of robots into a manufacturing operation can raise a number of concerns on the part of both workers and management. Workers often have a fear of losing their jobs, since robots are usually viewed as replacing people. They may also have concerns about their safety when working close to robots. People who have not worked with robots or computers before also tend to be concerned about how they may have to interact with a robot. Management personnel, too, are not without fears. They are usually concerned about how the workers will accept the installation of robots. Management will certainly have some fear of the risk of failure if the firm has not had any successful experience with robots before. Since the cost of implementation is usually high, it takes a long time before management will see the investment pay off. In the meantime, the robot may be viewed as a risk that is losing money. If workers are being displaced by robots, management must also be concerned about the problems involved in placing or retraining them.

A number of actions can be taken during the implementation of robot applications to avoid or minimize many of these concerns:

1. Start with hazardous and new operations that will not be viewed as a threat to existing workers.

2. Involve the workers and the support groups early in the process of developing the application. This will give them an opportunity to understand the objectives, become familiar with the operation of a robot, and perhaps even influence how the robot application is implemented.

3. Conduct education and training programs for those who will be involved in and affected by the robot application. This should eliminate some fears or concerns that come from a lack of knowledge or familiarity with robots.

4. Keep people informed about the plans and progress of the project. This may avoid speculation and concern caused by a lack of information or a perceived cloud of secrecy.

5. Encourage ideas and suggestions for improving the operation. This will not only make people feel that they are a part of the project—but it may reveal some better ways to implement the application or avoid problems.

16-4 ROBOT SAFETY

WHAT ARE THE POTENTIAL SAFETY HAZARDS?

Robots are often placed in applications to avoid exposing workers to potential safety hazards, but robots themselves have the potential for creating safety exposures. Robots are powerful mechanical devices that move in response to programmed instructions. Unless there are provisions for eliminating potential hazards, robots can unintentionally be involved in industrial accidents. Although the safety record of robots has been excellent to date, as their use expands rapidly it becomes even more important to assure their safe operation.

One revealing perspective on why a robot can be a safety exposure is the frank, but accurate, description of the common industrial robot as "a one-armed, blind idiot, with limited memory and feet nailed to the floor, that cannot speak, see, or hear without the use of peripheral devices. Further, it is unable to make independent decisions and thus it will obey incorrect instructions or fail due to malfunctions of the controller." (*Introduction to Robotics.* J. Rehg, instructor at Piedmont Technology College, 1984.)

From the time robots were first conceived, safety has been a concern. In his famous science fiction novel *I, Robot*, published in 1950, Isaac Asimov proposed the following three laws of robotics:

1. A robot may not injure a human being or, through inaction, allow a human being to come to harm

2. A robot must obey the orders given to it by human beings except where such orders would conflict with the First Law

3. A robot must protect its own existence as long as such protection does not conflict with the First or Second Law

A robot that could meet all three of these laws would probably be considered inherently safe. That is, no special provisions would need to be added to the application to protect people from the robot. Although that is still a goal in the real world of manufacturing, the state of the art of robotics today cannot provide inherently safe robots. Therefore, safety provisions must be made as part of the development of every robot application.

The major hazards in the use of a robot are:

☐ Being struck by it while it is moving

☐ Being struck by objects dropped or ejected by the robot

☐ Being trapped between the robot and some other object in the area

These exposures exist only when a person is inside the work envelope of a robot. This situation can arise from several circumstances:

☐ Walk-through or lead-through teaching

☐ Application development and debugging

☐ Maintenance work

☐ Production operations where humans work in close proximity to a robot

The accidents that have occurred were mostly during manual rather than automatic operation. More often than not, they involved manual intervention to fix a problem, such as a jamming or snagging of parts or equipment. A robot should not be assumed to be safe when it is stopped. For example, a robot that is holding a heavy object, could drop it. In some cases it may also not be safe to hold an object in a stationary position, such as when heat or motion are involved (e.g., an object may overheat if it is held too long in a furnace; it may prove hazardous to hold an object in a stationary position when it is on a moving conveyor with other objects). Some of the reasons accidents occur are:

☐ People are not aware of or forget the potential hazards involved with robots

☐ People are preoccupied or are not paying attention to what is going on around them

☐ People are not following procedures or are taking chances in their actions

These reasons are basically the same ones that cause many industrial accidents. Therefore, a complete safety plan for a robot application must include many of the common ingredients of industrial safety programs in general, as well as some special provisions for robotics.

WHAT IS REQUIRED TO MAKE A ROBOT APPLICATION SAFE?

A complete robot safety plan must address the following requirements at each level of the manufacturing operation:

1. The robot itself must be designed for safety. Its control and drive systems should have the safety provisions expected of any high-quality industrial machine. It is just as important to have a thorough understanding of the manipulator's capabilities and motion.

2. The workplace must also be designed for safety. Provision must be made to restrict access to the area and avoid potential exposures to moving- or trapping-type accidents. This may involve the way the workplace is laid out, as well as special safety equipment and controls. For example, operators can be isolated from the robot work space by physical barriers, and by transfer mechanisms between them and the robot (Fig. 16-4).

3. The entire manufacturing operation should be safety-oriented. Plantwide programs which include education, training, and maintenance activities can help to prevent accidents in all workplaces. Safety consciousness and a generally safe working environment are the most effective ways to avoid accidents.

The first step in developing a safety plan for a robot application is to thoroughly understand the robot's design and operation in terms of:

☐ The maximum working envelope

☐ Danger spaces outside the working envelope which may be exposed to hazards from trapping, dropping, or ejecting

☐ Hazards associated with other equipment adjacent to or part of the application

☐ Parameters of the application and equipment (e.g., weights, forces, speeds, distances)

☐ Potential sources of failure or error (e.g., hardware, software, drive systems, power)

SAFETY PROVISIONS

Safety can be built into a robot application by making provision for it in each of the following areas:

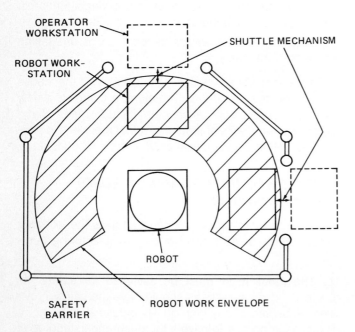

Fig. 16-4 Layout of a workplace to isolate the operator from the robot. (*Arnold D. Potter, "Safety for Robotics," reprinted with permission from the Dec. 1973 issue of Professional Safety, official publication of the American Society of Safety Engineers*)

1. Prevention. Much can be done to prevent accidents even before a robot application is implemented. Methods include:

Laying out the workplace to eliminate any points of potential entrapment, jamming, or snagging.
Training all operators and support personnel in safe operation of and potential hazards associated with robots in general and the specific application in particular.
Authorizing access to only those trained and qualified to work safely in the vicinity of the robot.
Developing procedures for the safe operation of the robot and the specific application involved.

2. Controls. A wide variety of techniques can be used to control the safety of a robot application. These usually take one of the following forms:

System operation. There are a number of ways to control the operation of the system to provide for safety. These include:
Emergency stop buttons to disconnect power in the event of an accident or potential hazard
Operator controls to restrict the motion of the robot during programming or debugging
System controls that detect errors or malfunctions

Physical access. The most visible and effective controls attempt to prevent access by humans to the working envelope and danger zones of the application. These can take several forms (Fig. 16-5):
Barriers, such as gates, fences, chains, or ropes
Detection schemes, such as pressure-sensitive mats, photoelectric cells, or ultrasonic sensors

Application environment. The tools, equipment, parts, and physical environment of the application, in addition to the robot itself, can also be a source of hazards. Some of the techniques used to control such exposures are:
Limit switches or stops to restrict the range of movements
Sensors on the manipulator and application tooling to detect errors or intrusions
Interlocks that cut power to all equipment when a failure or intrusion occurs
Environmental and equipment safety controls that minimize potential hazards from fire, explosion, or toxic materials

Warnings. Another essential type safety control, which should only be considered to be secondary, is some form of warning. Examples include:
Signs that are visible and clearly state the hazards and restrictions
Flashing lights that direct attention to the fact that the robot is operational
Painting of the robot and its restricted zone with bright colors to draw attention

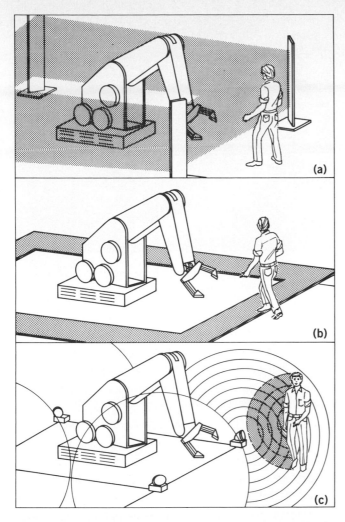

Fig. 16-5 Robot detection schemes: *(a)* photoelectric *(b) pressure sensitive, and (c)* sonic. *(Frank K. Welch, "Don't Neglect Safety in Your Automation Planning," Production Engineering, Nov. 1985)*

16-5 SIMULATING ROBOT APPLICATIONS

WHY USE SIMULATION FOR ROBOTICS?

Most of the effort required to develop and implement a robot application is involved in layout of the workstation, design of the equipment, and selection and programming of the robot. The development of computer graphics modeling and simulation tools has made it possible to perform those tasks more efficiently without using an actual robot. This avoids the need to use a robot that could be used more productively in performing manufacturing tasks. It can even avoid the need to obtain and install a robot before the application is developed and debugged. Robot simulations are used to:

☐ Reduce the time, labor, and expense required to develop a robot application

☐ Verify that the actual robot installation will perform the tasks required as intended

☐ Optimize the efficiency of the robot's operation

☐ Assure that the robot can make all the movements required

☐ Avoid collisions with obstacles or other equipment

☐ Determine cycle times of operation

☐ Evaluate alternative solutions

☐ Try out unusual manipulation movements

☐ Demonstrate how an application will look and perform

☐ Determine the effects of variables in the robot's performance (e.g., speed, torque, force, load, acceleration)

☐ Identify errors, inaccuracies, and problems in the robot's path

☐ Provide feedback to allow dynamic changes to the applications program

Interactive graphics simulation is similar to teaching a robot on-line with a pendant in a walk-through mode. It allows the applications developer to use actual spatial relationships in the workstation and determine the real capabilities of the system. Changes can be made interactively to fix problems and try alternatives. The design of the workstation, selection of the equipment, and programming of the robot can all be done off-line and then downloaded to the factory floor for execution. Simulation is particularly useful when complex motions and multiple pieces of equipment are involved. In such cases, it is not always obvious whether the motions or tasks required can actually be achieved.

The selection of a robot is not a simple task. Although many robots look alike, they vary in their specifications and capabilities. Simulation permits the application developer to evaluate a variety of different robots in the application without actually having to buy them and try them out.

ROBOT SIMULATION TOOLS

Robot simulation requires the use of several advanced computer functions:

Interactive graphics

3D graphics modeling

Dynamic simulation

Geometric databases

Kinematic analysis

Robot-level program translation

A typical robot simulation package contains the following features:

1. A library of geometric and parametric information. This may include robots, end effectors, workpieces, application equipment, basic workstation designs, and basic movements. Each must be described in terms of dimensions, axes of movement, speeds, and so on. The more complete the data, the better the model.

2. A description of workstation composition. This provides the capability to select and position the various components of the application equipment and tools, including the robot. An interactive system can allow the position of these components to be modified as the application is being debugged.

3. Task simulation. This uses geometric modeling and kinematic analysis to simulate the motions of the robot and other application equipment. This is the most visible and interactive part of the package (Fig. 16-6). It is used to develop motion sequences, check for limitations or interferences, and determine dynamic performance characteristics.

4. Robot programming. Once the application has been simulated successfully, the robot can be programmed from the motion sequences that have been established. Some packages use a general purpose language that can be translated automatically into the specific control language of the robot. Robots can be programmed off-line using interactive computer graphics and simulation tools (Fig. 16-7).

Some robot simulation packages include additional features, such as cycle-time evaluation, economic analysis, robot calibration, and process feedback. Simulations based on solid geometric models are more thorough and accurate, since they provide complete representations of all the equipment and objects and their spatial relationships (Fig. 16-8).

Many robot simulation packages are in use today, among them:

☐ GRASP (General Robot Arm Simulator Program), developed by RPI.

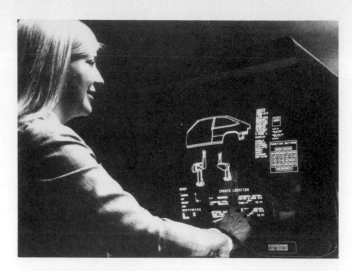

Fig. 16-7 Using graphics simulation to program a robot application off-line. *(General Motors Corp.)*

☐ PLACE (Position Layout and Cell Evaluator), developed by McDonnell-Douglas. PLACE may be used in conjunction with BUILD, COMMAND, and ADJUST.

☐ ROBOT-SIM, developed by GE Calma. Used in conjunction with ROBOT-PRO.

☐ ROBOGRAPHIX, developed by Computervision.

☐ I-GRIP (Interactive Graphics Robot Instruction Program), developed by DENAB.

☐ ROFAC (Robot Factor), developed by Scientific Management Corp.

☐ AUTOMOD, developed by Automation Simulations Inc. Used with AUTOGRAM.

☐ RCODE, developed by SRI International.

☐ ROBOTEACH, developed by General Motors.

☐ CATIA ROBOTICS, developed by Dassault Systèmes.

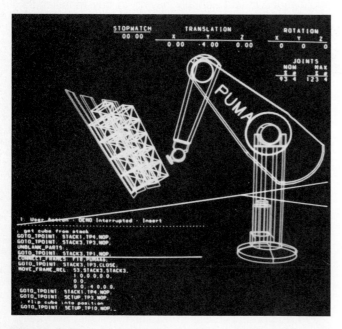

Fig. 16-6 Simulating a robot application with interactive computer graphics. *(McDonnell-Douglas Manufacturing Industry Systems Co.)*

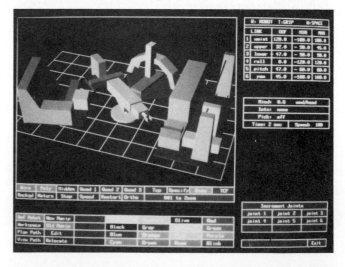

Fig. 16-8 Solid-model robot simulator. *(courtesy of W. E. Red, Systems Automation Laboratories, Brigham Young University, and AutoSimulations, Inc.)*

USING SIMULATION TO DEVELOP ROBOT APPLICATIONS

The typical steps involved in the simulation of a robot application are:

1. **Design the workstation.** This includes the selection or definition of all the tools and equipment in the application (e.g., robot, conveyor, fixtures, machines). Their positions may be modified during the simulation.

2. **Define the work points.** This describes the positions the robot must reach during the task (Fig. 16-9).

3. **Define the path of the robot.** This checks the movements required to reach the workpoints against the work envelope of the robot. It also involves checking for interferences and collisions.

4. **Run the simulation.** This allows the applications developer to visualize the movements of the robot to verify that the sequence and the path are as intended. The animation of motion can be smooth or incremental. It can be slowed down or interrupted at any point. Some simulation tools provide zooming capability for a closer look at critical positions. This may involve the tracking of moving objects and equipment, such as conveyors, as well as the simulation of product flow.

5. **Perform a dynamic analysis.** The effects of varying performance parameters can be determined by kinematic analysis techniques [e.g., speed, acceleration, torque, force, and cycle time (Fig. 16-10)]. Characteristics of the motion path and errors in it can also be determined, such as path accuracy, overshoot, and settling time.

6. **Modify the simulation model.** Changes can be made in positions, paths, and parameters to correct errors or optimize performance. Objects and obstacles can also be deleted or replaced.

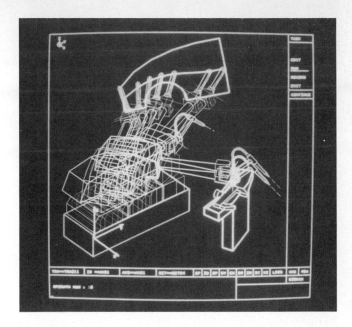

Fig. 16-10 Dynamic analysis of a robot application. *(International Business Machines Corp.)*

7. **Generate the robot program.** The output of the simulation sequence is translated into a program format. Some packages require a detailed programming effort at this point, while others can generate instructions directly in a robot-level language.

8. **Execute the robot program.** When the actual robot application is installed, the program is downloaded to the robot controller for execution. In most cases, a calibration routine is required to adjust the program for the actual positions in the workstation.

16-6 SUMMARY

The decision to use a robot in a manufacturing application should be based on a thorough assessment of all the alternatives, costs, and benefits involved. Robots often have an economic advantage over manual and fixed-automation approaches for medium-volume, batch-type operations. They can also provide noneconomic benefits to workers, products, and the manufacturing operation. Robot applications should be selected for both their potential benefits and their chance of success.

Developing robot applications can involve a great deal of time and effort. Doing a thorough job and relying on experience can avert implementation problems later. Although robots are often used to avoid safety hazards to human workers, robots themselves can present a potential safety problem if the application is not adequately designed and controlled. The process of developing and debugging a robot application can be made much more efficient by using computer simulation tools. Ideally, both the design and manufacture of a product can be automated with computer-based technologies.

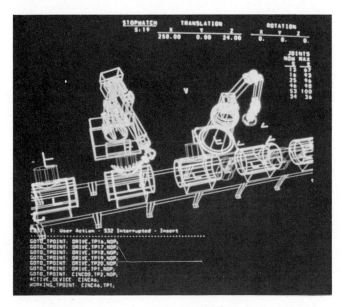

Fig. 16-9 Simulating a robot application to define the work points. *(McDonnell-Douglas Manufacturing Industry Systems Co.)*

REVIEW QUESTIONS

The answer to each question can be found in the section indicated at the end of the question.

1. Identify the basic factors which should be considered in deciding where and when to use a robot. [16-2]

2. Describe the economics of manual and fixed automation versus the use of robots. [16-2]

3. What are some of the basic economic advantages that robots have over human workers and special purpose automation? [16-2]

4. Define what ROI means. [16-2]

5. Identify some of the noneconomic benefits of using robots in manufacturing. [16-2]

6. Describe the process involved in developing a robot application. [16-3]

7. What are some of the human considerations involved in introducing robots into a manufacturing operation? [16-3]

8. Identify the major safety hazards associated with the use of robots. What can be done to control them? [16-4]

9. How can simulation be used in developing a robot application? [16-5]

10. Identify some of the advantages of simulating robot applications. [16-5]

PART

MANUFACTURING SYSTEMS

CHAPTER 17
SYSTEM ARCHITECTURE

CHAPTER 18
MANAGEMENT SYSTEMS

CHAPTER 19
INTEGRATED MANUFACTURING SYSTEMS

Parts 2 through 4 of this book covered the key technologies that enable us to achieve computer-automated manufacturing (CAM): computers, graphics, and robotics. If these technologies are to work together in an automated factory, they must be integrated into a common manufacturing system. A system, in the general sense, is a collection of parts that make a whole. This is more than true in the case of integrated manufacturing systems. When all the pieces of automated manufacturing technology are tied together, their capability is greater than the sum of the individual parts. An integrated system not only provides control and management of the manufacturing operations; it also makes possible the automation of some tasks that have traditionally been inefficient, time-consuming, and costly. The key to integration is a system that can move both data and material efficiently.

There are many aspects to the subject of manufacturing systems. Part 5 is organized into three chapters to deal with them. Chapter 17 addresses system architecture. This deals with how data processsing systems can be tied together to provide communication and control. Chapter 18 covers management systems. This is a broad subject that includes each of the principal types of data systems in a manufacturing operation (i.e., technical, logistical, and administrative data), as well as techniques to optimize manufacturing performance. The objective of Chap. 19, "Integrated Manufacturing Systems," is to explain how all the elements of CAM can be tied together into an integrated system. The chapter deals with the integration of the movement of both data and materials as well as the use of all the key computer tools which can improve manufacturing productivity.

Part 5 is not intended to be comprehensive enough to make experts in system integration. However, it explains the relationships and interdependencies between the various computer-based automation tools which were covered in previous parts of the book. It also illustrates how they can be tied together into an integrated system to achieve computer-automated manufacturing.

SYSTEM ARCHITECTURE

17-1 INTRODUCTION

We have seen in previous chapters that computers are the basis for all the key technologies which enable us to automate manufacturing operations. Systems tie computers to other devices and other computers in order to provide for the transfer of information between them. Such information systems are an integral part of manufacturing automation. They are the means to integrate all the hardware and software into an automated manufacturing system. Although individual tools or groups of tools may be automated, the information they generate may not be shared with other parts of the manufacturing operation that could use it. A system which permits the efficient transmission of information between machines,

workstations, and management is a key to improving the productivity of the factory. Manufacturing data systems can permit access to information which was not previously available or was too costly to obtain. This can help managers make better decisions and run the factory more efficiently.

An automated manufacturing operation involves many computers and systems to handle technical, logistical, and administrative data. To make it possible for this data to be shared and to enable these systems to communicate with each other, an overall, plantwide information system must be established. This system ties all the tools, processes, and administrative activities together (Fig. 17-1). It serves as a manufacturing control system by tying data collection systems on the factory floor to tool

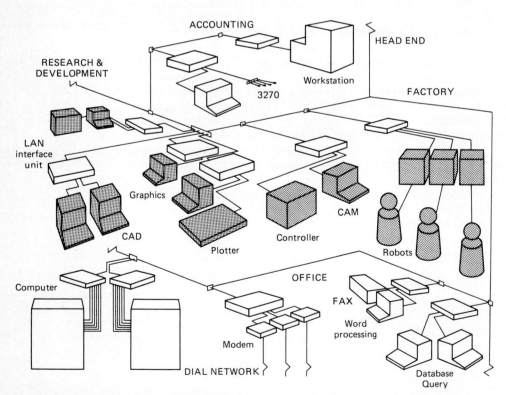

Fig. 17-1 A manufacturing system: computer networks permit information to be shared between functions. *(Tom Dixon, "Networks Bridge the Gap Between Islands of Automation," Electronic Packaging and Production, Dec. 1984)*

control, process control, and floor control programs. It also provides a link between the development and manufacturing functions through the exchange of design, performance, and cost data.

In order to establish such a plantwide information system, a system architecture must be adopted to provide the communication links with all the pieces. The architecture is the scheme which permits all the hardware and software to communicate with each other. It involves a structure, a physical method of interconnecting hardware and interfaces for software systems. A number of different approaches can be taken to establish such a system. The selection of a particular approach depends on the size, complexity, and performance requirements involved. This chapter will cover the basics of system architecture and how it is applied in a manufacturing environment. In particular, it will address:

Features of a manufacturing system

Hierarchical systems

Local area networks

Control systems

Trends

17-2 WHAT IS A MANUFACTURING SYSTEM?

TYPES OF MANUFACTURING SYSTEMS

Three principal types of data need to be handled in a manufacturing operation:

1. **Technical data.** This includes design, tool control (e.g., numerical control—NC) and test data.
2. **Logistical data.** This includes a variety of production control and planning information related to such subjects as demand, capacity, schedules, and inventory.
3. **Administrative data.** This includes data collected on the factory floor relating to manufacturing performance (e.g., labor, throughput, equipment utilization, cycle time, quality) as well as information derived from this data (e.g., cost).

Traditionally, separate computer systems have been developed to handle each of these types of data. In some cases involving complex operations, several systems may have been developed for each type of data. It is not unusual to find, even in factories that use a lot of computers and have automated operations, that these systems are separate and independent. They may have been developed at different times by different organizations to serve their specific needs. The result is usually that the specific tasks intended for the system are automated, but the data cannot be exchanged with other systems or used for some higher-level function. A typical factory has systems for applications such as:

Tool control

Shop floor control

Process control

Engineering design

Shop floor data collection

Production planning

Tool planning

All these systems and more are necessary to run a large manufacturing operation. The ability to exchange information between systems and work together will affect the efficiency and productivity of the entire operation. The nature and functions of these systems will be addressed in Chap. 18.

SYSTEM REQUIREMENTS

The primary high-level functions of a plantwide information system include:

Database management

Data collection

Data communication

Monitoring and control

Management reporting

Large manufacturing systems are made up of a number of smaller subsystems with many users. For the factory to run efficiently, performance demands must be placed on the system, such as:

☐ High availability (e.g., it may need to support manufacturing operations 24 hours a day, 7 days a week)

☐ Access to multiple databases (for data exchange between functions)

☐ Accurate, timely, and complete data (to assure efficiency and integrity)

☐ Traceability of materials, products, and tools (to control quality and reliability)

☐ Measurements of performance (of tools, processes, and people)

☐ Fast response (for quick access to data)

☐ Friendly interfaces (for ease of use)

INHIBITORS

Although few people would argue with the objectives of establishing plantwide manufacturing systems, a number of factors have inhibited their development or success in the past:

☐ The systems can be very complex

☐ The development effort can be costly and time-consuming

☐ There has been a lack of a standard architecture and interfaces

☐ The technologies involved are still advancing and changing

These factors are all real and should not be ignored. However, they can each be overcome with a sound approach to system design and implementation. The alternative—not establishing an integrated manufacturing system—will lead to suboptimization and inefficiency, which, in the long run, can be much more costly.

17-3 HIERARCHICAL SYSTEMS

MANUFACTURING ORGANIZATION

Most business organizations have a hierarchical structure. Such an organization structure defines which individuals report to whom, and separates functions and tasks into manageable units. In addition to separating the functional activities at the lowest level of the organization, the structure provides for different forms of these activities at higher levels. For example, the manufacturing organization is usually separated from the engineering organization because the nature of their activities and skills are different. The lowest level in each organization is involved primarily in activities which affect its own day-to-day operations. Moving up the organizational hierarchy, one finds more involvement in planning and control activities as well as greater interaction between functions (Fig. 17-2). Even operational control occurs at several levels, depending on the size and complexity of the manufacturing process:

Equipment and tool control. This is the lowest level of the process, where each individual operation (i.e., process step, machine, or workstation) must be controlled.

Floor or line control. All the steps or tools in a particular manufacturing process must be tied together in some type of control scheme.

Plant or manufacturing control. All the processes in the plant must somehow be tied together by plant-level management.

Since, from a management viewpoint, this is a logical way to organize the plant, it makes sense for the information system to be designed with a similar structure. Most manufacturing control systems, therefore, are hierarchical and have a structure which tends to follow that of the management organization. In concept, the technical, logistical, and administrative data systems each operate at several levels which, in themselves, have interdependencies (Fig. 17-3). At the lowest level are the activities involved in executing operational tasks (e.g., controlling the operation of a tool). Moving up the system hierarchy, the activities are more supervisory in nature, such as monitoring, reporting, and planning. The architecture of the manufacturing system must provide for the communication and exchange of data both vertically and horizontally in this hierarchical structure.

TASKS OF THE MANUFACTURING CONTROL SYSTEM

A hierarchical manufacturing control system must perform different types of tasks to satisfy the control needs at each level in the structure. Information required at the operational/execution level includes:

What job to work on next

Where the job was last

Where the job goes next

What tasks must be performed on the job

What limits or controls are on the tasks

Information required at the monitoring and control levels includes:

Priorities and schedules of jobs

Location of jobs in the process

Process and tool control limits

Status of tools and workstations

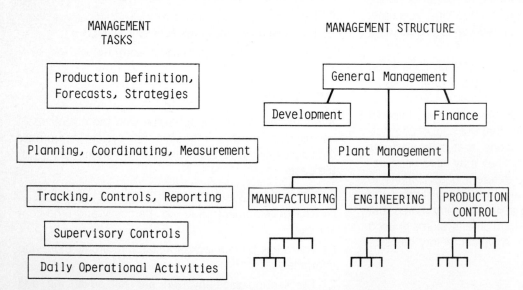

Fig. 17-2 A typical manufacturing organization.

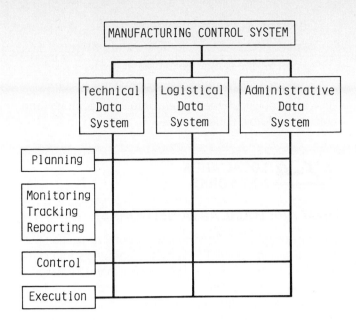

Fig. 17-3 Structure of a manufacturing control system: basic concept.

Availability of materials and tools

Information required at the management and planning levels includes:

Overall schedules and demand

Performance measurements and reports

History records

Product and process definitions

In a hierarchical control system, the computers or controllers at each level take direction only from one computer at a higher level and give direction to the computers tied to them at a lower level. The overall manufacturing tasks, in terms of long-range plans and objectives, are entered into the system at the highest levels (e.g., product definition and demand). These tasks are then separated into more specific and detailed tasks to be executed at the next-lower level in the system (e.g., production schedules). This process of breaking down the tasks continues to the lowest level of the hierarchy, where the individual manufacturing operations are performed. Information on actual results then flows up through the levels of the system to provide feedback on performance. This type of system structure merely serves to automate what a typical management structure would do to control the operations of a plant.

HIERARCHICAL SYSTEM STRUCTURE

The hierarchical structure of a system controlling a large manufacturing operation typically involves four or five levels, as shown in Fig. 17-4. The scope of information and the response time required to operate is generally less at each lower level. This is compatible with the way the organization operates at those levels of management, and it is a means to minimize the amount of data handling

System level	Organization level	System response time
Plant level	Plant management	Days
Production line	Mfg management	Hours
Process center (mfg cell)	Supervision	Minutes
Process step (workstation)	Operator	Seconds
Equipment (tool)		Nanoseconds

Fig. 17-4 Hierarchical structure of a manufacturing control system.

and processing required throughout the system. The functions at each of these levels are summarized below. Actual manufacturing control systems may have more or fewer levels with different designations, but this provides a generalized example of the concept.

1. Plant level. This is the highest level of control, where the technical, logistical, and administrative data systems must come together. It deals with plans, objectives, and measurements for the entire manufacturing operation. Plant-level control must also provide database management and interfaces to related functions, such as development and customer service.

2. Production line. Large plants are usually made up of several production lines for different products or processes. At this level in the system, the focus is on operational management of those lines. This includes scheduling jobs, allocating resources, establishing priorities, and measuring performance.

3. Process center (or manufacturing "cell"). This level may or may not exist in a control system, depending on the complexity of the process. Many production lines are divided into process centers which each perform a unique set of tasks using related equipment and processes. The controls at this level deal with the sequencing of jobs, the scheduling of equipment, and the handling of materials.

4. Process step (or workstation). At this level the system is dealing with the real-time control of individual processes. It may involve several pieces of equipment, depending on the nature of the process. This is typically the level that must interface directly with the operator.

5. Tool (or equipment). The system at this level is tied directly to individual pieces of manufacturing equipment. This is where tool control programs operate by providing NC data and monitoring actual performance.

DISTRIBUTED SYSTEMS

As discussed in Part 2, "Computer Technologies," there are several ways to design a hierarchical computer system. One could have a central host with several levels of satellite computers and controllers that are tightly coupled to each other in a hierarchical structure. This satisfies the division of tasks and the overall control objectives for the system, but it also has some limitations and drawbacks. In such a system, each computer can communicate only with the ones tied to it, directly above or below it in the hierarchy. In this configuration, therefore, each computer is totally dependent on the one above it for information and data processing support. Data from other parts of the structure must first flow up before it can be shared at the same level. In addition, if one computer goes down, it will interrupt the operations of the satellites below it.

An alternative is some form of distributed hierarchical system (Fig. 17-5). Such a system provides more computer power at lower levels in the structure as well as a means for peer-to-peer communications. This permits computers to operate in small "clusters" or groups to control segments of the manufacturing operation without being totally dependent on a higher-level control system. Some of the advantages of this type of system include:

☐ Higher perceived availability of computer power

☐ Reduced risk of system interruption

☐ Support for specialized hardware and software

☐ Support for the exchange of data at the peer level

☐ Easier modification and expansion of the system

☐ A breakdown of the system into less complex and more manageable subsystems

17-4 LOCAL AREA NETWORKS

WHAT ARE LOCAL AREA NETWORKS?

To make distributed systems work, a communication network must interconnect the group of terminals and computers in such a way that they can exchange data on a peer-to-peer basis. There are a number of different approaches to designing such networks. Since the objective is to provide data communication that is efficient (i.e., high-speed and low-cost) between a closely clustered group of devices, these networks are usually designed to serve restricted geographical areas. Therefore, they are commonly referred to as "local area networks" (LANs).

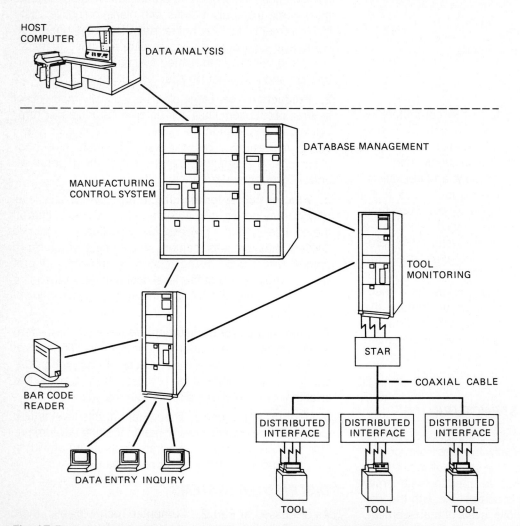

Fig. 17-5 A distributed hierarchical system. *(International Business Machines Corp.)*

Since the distances involved are limited, LANs can use techniques which provide high-speed data communication over relatively inexpensive transmission lines; this is not possible for a wide area or long-distance network. LANs are an effective way to satisfy the objectives of distributed systems because they:

☐ Reduce the demand on the central host system

☐ Provide for an exchange of data between local clusters of peer devices

☐ Reduce the dependency on the central host system to maintain operation

☐ Increase the utilization of the total computer power in the system

☐ Minimize the overall cost of the system

A simple alternative to a LAN would be to tie all the devices directly to a common computer. This would be similar to a local telephone network which uses a private branch exchange (PBX). This is often done for individual stand-alone terminals. However, when a large number of devices need to exchange data, this approach becomes impractical and expensive because:

☐ Each device must have its own line to the central computer

☐ All communication between peers must flow through the central computer

☐ Data communication is limited by the telephone transmission rate

Advantages of a LAN include:

☐ All devices share a common transmission line

☐ Communication can be peer-to-peer

☐ High-speed transmission techniques can be used (as high as 10 Mbps versus 9200 bps for telephone)

The use of LANs in manufacturing is a key to improving both productivity and control. With direct links between tools and workstations, data can be exchanged more efficiently than if it were collected centrally or even manually and then retransmitted or reentered into the system. LANs also have some additional features which offer advantages to manufacturing:

1. LANs permit terminals to share common computer resources, such as high-speed printers and large-capacity disk drives

2. LANs allow incompatible equipment to be tied together with a common interface

3. LANs can provide a convenient means for communication between workstations (e.g., electronic mail)

To provide all these advantages is not a simple task. LANs need sophisticated and reliable data transmission and access control schemes to make them work. Several different approaches have been developed to implement LANs. Since this is a relatively new field, researchers are continuing to improve the capabilities of LANs.

TYPES OF LANs

LANs are described in terms of three basic design characteristics:

1. Topology. This is the physical configuration, the way the devices in the network are connected together. The most common configurations are the bus and the ring, which are shown in their simplest form in Fig. 17-6. In the bus configuration, each device shares access to the communication network. In the ring, all communication passes through every device in the network.

2. Control mechanism. Each device must be able to gain access to the communications on the network. Communication on LANs is done in the form of message "packets," which contain routing and control information as well as data. Three major types of mechanisms are used today to control the communication of these message packets on a network:

Contention control. This scheme, used in some bus configurations, is based on relatively simple rules of communication which avoid conflict or contention between the devices using the network. A device wanting to use the network waits until the line is clear. The standard scheme which has been developed has a complicated name: "Carrier Sense Multiple Access With Collision Detection" (CSMA/CD). It was adopted by Xerox under the trademark Ethernet.

Message slots. This is an alternative method of controlling communications on a bus. Data must be transmitted in a fixed-length format called a "slot." Each device in the network is entitled to transmit one slot at a time when the line is not busy. During the transmission cycle from one end of the bus to the other, each device has an opportunity to use at least one message slot. This scheme was adopted by AT&T as Fasnet.

Token ring. This is a common scheme used to control communications on a ring network. It provides access for devices to use the network by passing a "token" around the ring (Fig. 17-7). The token is a bit in the data structure which is detected by each device as it is transmitted around the ring. It is initially "free." When a particular device puts a message on the line, it changes the token to a "busy" indicator. The message continues around the ring, being retransmitted by the other devices until it

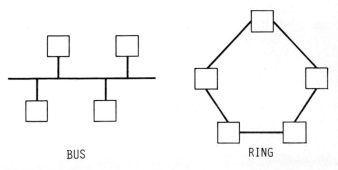

BUS   RING

Fig. 17-6 Basic network configurations: bus and ring networks.

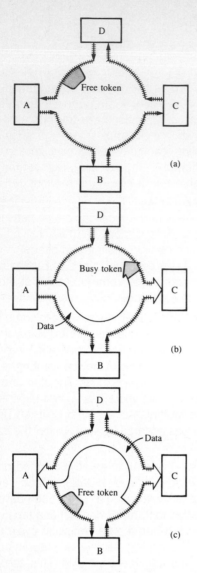

(a) Sender (node A) looks for a free token, then changes free token to busy and appends data. (b) Receiver (node C) copies data addressed to it; token and data continue around the ring. (c) Sender (node A) generates free token upon receipt of physical header and completion of transmission, and continues to remove data until receipt of the physical trailer.

Fig. 17-7 Token ring access control protocol. *(Norman C. Strole, "A Local Communications Network Based on Interconnected Token-Access Rings: A Tutorial," IBM Journal of Research and Development, Sept. 1983; © 1983 by International Business Machines Corp., reprinted with permission)*

reaches its destination. At that point the token again becomes free, allowing access to the network. This scheme has been adopted by some computer manufacturers, such as IBM, as well as by users, such as General Motors.

3. Transmission mode. LANs use serial rather than parallel data transmission to minimize costs while still making it possible to operate at high speeds. Two different modes of transmission are used:

Baseband. This is the simplest approach, involving only one stream of data on the line which connects the net-

work. Either twisted-pair wiring or coaxial cable can be used.

Broadband. This allows multiple streams of data to be transmitted simultaneously, using a system similar to cable TV. Although more complicated, broadband transmission makes it possible to use one network for a variety of applications (e.g., video, voice, and data).

DIFFERENCES BETWEEN LANs

Each of the approaches to a LAN can have advantages for specific applications. No one is best in all cases. There are differences, however, which should be understood before selecting which type of LAN to use. These differences can be expressed in terms of the following factors:

1. Accessibility. This is determined by the control mechanism used.

☐ CSMA/CD provides equal access to the network for all devices. This also means that no priority schemes can be used.

☐ Message slots can provide equal access and can adopt a priority scheme.

☐ Token rings normally provide access based on the position of the device in the transmission route. However, priority schemes can also be used.

2. Performance. The performance of token rings is less affected by communication work load than that of buses. Token rings can also achieve high levels of utilization by using variable-length messages.

3. Reliability. The physical reliability of a LAN is influenced by the topology of the network. A standard ring is subject to interruption if one of the devices in the ring fails. However, there are hybrid ring designs which use relays to bypass a failed device. A bus does not have the same exposure since each device only monitors the line and does not have to regenerate the signal on it.

4. Expandability. Both bus and ring networks can be expanded. Facilities can even be prewired to provide for future additions to the network. Adding a station to a ring will cause an interruption unless a hybrid bypass design is used. Additional devices will not affect the transmission in a ring since the signal is regenerated by each device. A bus will have a limit to the number of devices that can be added before the signal gets too weak.

17-5 CONTROL SYSTEMS

TYPES OF CONTROL SYSTEMS

The type of system used to control a manufacturing operation depends on the nature of the process involved. There are two basic types of manufacturing processes:

1. Discrete. This is where individual parts move in batches through a series of separate steps, such as in machining or assembly operations. The steps involved are usually some form of fabrication, assembly, inspection, or test. Quality control for such processes is performed by specialized equipment, such as instruments or testers, or even by manual inspection. In most cases, the product can be repaired or reworked if errors occur. Even when a part or product must be scrapped, the error usually affects only one or a few units before it is corrected.

2. Continuous. A continuous process involves the transformation of materials as they flow through a series of interdependent steps (although this varies in degree). Such processes include the manufacture of chemicals, semiconductors, glass, and metals. There is usually some "yield loss" associated with such a process due to the many variables which can affect the product. When errors occur, they can also affect a lot of material before they are corrected. In this type of environment controls are usually performed by sophisticated equipment which must be constantly monitored.

Modern manufacturing processes often have some characteristics of both discrete and continuous processes. They are often complex, involving many steps. Once a process which can successfully produce a product has been developed, the variations in the process must be controlled in order to assure that a functional and economical product can be reproduced. A complex process may be subject to thousands of variations. There are two basic types of process variables that a control system has to deal with:

1. Uncontrollable. These are variations caused by factors outside of the process, such as the physical characteristics of the materials involved. The manufacturing process must be designed to accept such variations within some expected limits.

2. Controllable. These are variations caused by factors within the process, such as the manufacturing equipment and environment (e.g., time, temperature, rates). They are usually parameters that are established as part of the process setup or measured as a result of the process.

Control systems monitor process variables and act on them when they exceed specified limits. The variables are monitored by computers collecting and processing data derived from signals which are generated by instruments and equipment controllers. These signals may come in several different forms:

☐ Continuous signals from analog measurements, such as temperature or pressure

☐ Binary signals, such as those from switches

☐ Pulselike signals, such as those generated by a stepping motor

For a control system to deal with many different types of signals, it must have interfaces that can convert signals into a usable form of data. The I/O channels of control systems therefore use devices such as analog-to-digital convertors, counters, signal conditioners, and transducers for this purpose.

Several types of control systems may be associated with a manufacturing operation. One or more may be involved in controlling the actual process steps. Another, such as computer numerical control (CNC), may control the operation of the tools. Another may control test equipment. There may also be a system for indirect controls, such as quality inspections. They may each be independent, or they may be integrated into a total process control system. In a large, complex manufacturing process, the control system must be very sophisticated. The entire process may be under the constant monitoring and direction of a centralized computer control center (Fig. 17-8).

FUNCTIONS OF A CONTROL SYSTEM

The major tasks performed by a process control system are:

1. Data collection. Using the special I/O interfaces, the control system obtains data from the various signals which are generated by the process variables. Since this may involve a great deal of data, it is only necessary to collect enough data to detect changes in the process. In most control systems it is sufficient to scan or sample the signals at a fixed frequency rather than continuously monitoring them.

2. Data communication. The collection of the data by itself does not help control the process; the data must be used. One use is to communicate to operators or other, higher-level control systems which can respond to changes. This can take the form of displays, messages, warnings, or alarms.

3. Process control. The most important function of the system is to actually help control the process variations. This involves computational processes, including comparisons against limits and algorithms that determine process corrections.

4. Process operation. The start-up and shutdown of a complex manufacturing process may require automatic control of the procedures.

Fig. 17-8 Computer control center for monitoring and controlling a complex manufacturing process. (*photo courtesy of Allen-Bradley Co., Inc. a Rockwell International Co.*)

5. Diagnostics. Control systems may include programs that can analyze unexpected process variations or failures in order to determine possible causes.

6. Supervisory control. Large process control systems usually include some higher-level management functions, such as optimizing the performance of the process or scheduling and sequencing operations.

Typical requirements which are placed on process control systems include:

High availability and reliability

High accuracy

Fast response

Large number of variables and I/O interfaces

Backup safety controls

The benefits of using process control systems can be significant, particularly with complex processes. In some cases, it would not be possible to produce large quantities of functional, high-quality product without a sophisticated computer control system. Additional benefits typically include:

☐ Increased production throughput (by optimizing the process)

☐ Improved quality (from consistency)

☐ Reduced indirect labor (e.g., for inspections and checking)

☐ Reduced materials costs (fewer losses and less waste)

☐ Lower energy costs (due to tighter controls)

PROCESS CONTROL TECHNIQUES

The basic objective of most process control techniques is to detect variations in the process and respond to them with adjustments which will keep the process within prescribed control limits. This requires that several fundamental conditions be satisfied:

1. It is essential to identify process variables that are critical to the manufacturability of the product (i.e., variables that can have a significant effect on the functionality, quality, or cost of the product)

2. Control limits must be established for each of these variables to specify nominal values and allowable deviations that will assure the manufacturability of the product

3. A system must be put in place to monitor, measure, and collect data on these process variables (e.g., sensors, instruments, signal converters, and data processing equipment)

4. A technique must be adopted to take the observed data and generate corrections in the process when any of the variables exceed their control limits

A number of techniques are used in industry to perform this process control function. They have similarities,

but they vary in complexity, sensitivity, and precision. The approach adopted for a particular application depends on the control needs of the process and the economics involved. All use data processing functions (i.e., data acquisition, data communication, computation). Three of the major types are:

1. Feedback. A control device is used to perform the following functions (Fig. 17-9):

☐ Compare a parameter measured at the output of a process step to a specified control value

☐ Generate a control signal if any deviation is detected

☐ Activate some mechanism which can adjust the input of the process to correct the deviation

This technique is called "feedback" simply because the error is detected on the output of a process step and the correction is made on the input.

2. Feedforward. In this case, the control device functions in a similar manner, but the control signal is sent ahead in the process sequence to compensate for the deviation that was detected. This is not practical in all processes, but it can be effective and economical (since it can avoid losses).

3. Adaptive control. This is a more powerful and complex technique. It responds to unexpected changes in the operating conditions which may be due to factors outside of the process itself—for example, changes in the environment (e.g., temperature), disturbances (e.g., vibrations), or variations in the materials (e.g., density). An adaptive control system must have sensors to detect these changes as well as techniques to compensate for them in the control parameters of the process. This approach can be used for both discrete and continuous process applications. It can be important in an application where the cost of the product is high—that is, too much is already invested in it to scrap. It can also be effective when the cost of running the operation is relatively high (e.g., it is expensive to interrupt a highly automated line).

In a critical machining operation, for example (Fig. 17-10), the control system can compensate for variations in the tool and the workpiece by overriding the programmed instructions in response to inputs from sensors. This can improve the quality of the finished product as

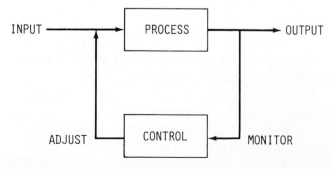

Fig. 17-9 Feedback control.

Fig. 17-10 An adaptive control system: machining application involving complex cuts and difficult-to-machine materials on large workpieces. *(Cincinnati Milacron)*

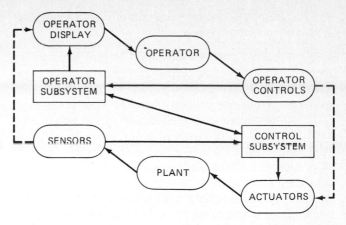

Fig. 17-11 Control process: operator and plant are controlled by hardware and software subsystems. *(Bruce W. Weide, et al., "Process Control: Integration and Design Methodology Support," Computer, Feb. 1984; © 1984 IEEE)*

well as increase the throughput of the operation. In some cases where very large, complex workpieces are involved, adaptive process control might even be required to make the job practical at all. Adaptive control systems require more inputs than other control techniques. In a machining operation, the NC program may have to include information on the material of the workpiece, the shape of the cutting tool, and allowable variations in the workpiece and the cutting tool.

ARCHITECTURE OF CONTROL SYSTEMS

The design of a control system for a manufacturing operation should consider such factors as:

The nature of the process (i.e., discrete or continuous)

The number of individual operations under computer-based process control techniques

The effects of the control of each operation on other process steps

The amount of data involved in the process control tasks

The response times required for the control mechanisms

These factors will influence the architecture of the control system as well as the choice of hardware and software to be used. The system must tie all the elements of the control process together (Fig. 17-11): operator, data from sensors, actuators on machines, and the control hardware and software. In the past, process control systems were usually designed for individual operations or machines using analog control techniques. Digital techniques later provided faster response and lower costs. Direct digital control (DDC) systems, however, usually involved separate, dedicated computers for each tool. This approach became too expensive and unreliable for plants with complex interactive processes. Advanced computer technologies, such as microprocessors, minicomputers, and distributed processsing, have made hierarchical control systems practical (Fig. 17-12). The modern approach to controlling the process of a complex

manufacturing operation is based on:

Decentralizing the control process

Distributing the acquisition of data

Distributing the data processing

Distributed hierarchical control systems provide advantages over centralized or dedicated control systems:

1. Distributed processing allows faster response than a shared central processor

2. Data can be shared between controllers by communication schemes outside the central host (e.g., LANs)

3. Failures in the system usually affect only one operation and not the entire process

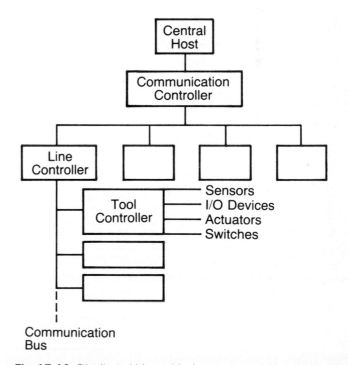

Fig. 17-12 Distributed hierarchical process control system.

4. The distributed hierarchical system can be modified and expanded more easily

Most process control systems run in real time and are event-driven. They must respond immediately to unscheduled events that change the state of the process—for example, a variation in one of the process parameters (e.g., temperature) or a malfunction of the equipment. Such a control system uses what is referred to as "interrupt logic." Signals are received about process "events" from sources such as sensors, timers, or even the operator. They interrupt the control program so that it can switch to an appropriate subroutine to respond to the signal. The computer then returns to the execution of the program that was interrupted. In order to deal with the large number of interrupt signals that may be involved in a complex process, the computer assigns priorities on the basis of urgency and importance.

The software for process control systems often has to provide a number of special functions, such as:

☐ On-line, real-time task execution

☐ Interrupt logic

☐ Shared communication channels

☐ Management of the central processing unit (CPU) and memory resources for multiple tasks

☐ Maintenance of a system database

☐ Multiple or parallel processing

☐ Multiple I/O interfaces

☐ Fault tolerance

The typical structure of a process control program includes several basic functions (Fig. 17-13):

A set of control limits or goals for the system

A model of the process and control parameters

An input from sensors to the model

An output to generate commands to actuators

Such a control program typically samples the input signals at a certain frequency, compares them to the control limits, and generates output signals to actuators which can adjust process parameters based on the process model. Once a general process control program has been developed, it can be modified and adapted to specific applications. For example, interactive computer graphics techniques can be used to make it easy to input or change process control points and parameter limits (Fig. 17-14).

A true process control system provides a model of the entire manufacturing operation. The overall performance of the operation can be affected by each of the process steps. In addition, the control of one step can affect others. A process control system also involves many of the plant's activities and personnel. It must be designed to deal with engineers, scientists, programmers, and technicians as well as operators. Modern computer graphics and easy-to-read color displays can provide operators with a quick overview of the state of the entire process (Fig. 17-15). They can also display specific data from process inputs and alert operators to deviations from control limits (Fig. 17-16). Computer control systems also make it easy to store and display historical data and analyze process trends (Fig. 17-17).

RELIABILITY OF CONTROL SYSTEMS

The reliability of process control systems is more important than that of most other application programs. An interruption or failure in a complex manufacturing process can cost substantially more than the system itself. The reliability objective for process control systems is to have a very high "availability" (Fig. 17-18). That is, there

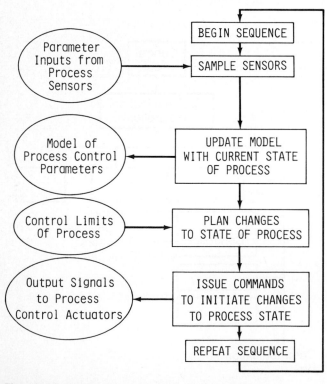

Fig. 17-13 Structure of a process control program.

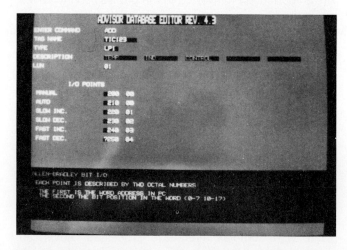

Fig. 17-14 Interactive computer graphics techniques for inputting or changing process control points or parameter limits. *(photo courtesy Allen-Bradley Co., Inc., a Rockwell International Co.)*

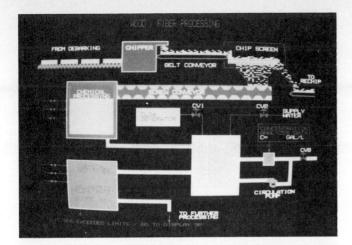

Fig. 17-15 Color graphics display of the status of a manufacturing process. *(photo courtesy of Allen-Bradley Co., Inc., a Rockwell International Co.)*

Fig. 17-16 Color graphics display of process status at control points. *(photo courtesy of Allen-Bradley Co., Inc., a Rockwell International Co.)*

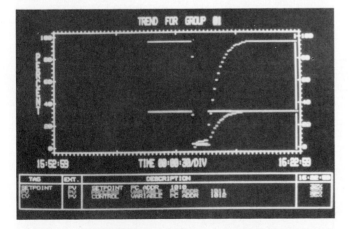

Fig. 17-17 Color graphics display of historical trends from process control data. *(photo courtesy of Allen-Bradley Co., Inc., a Rockwell International Co.)*

must be a long mean time between failures (MTBF) and the failures must only require a short mean time to repair (MTTR). Process control systems can be interrupted by a number of causes, such as:

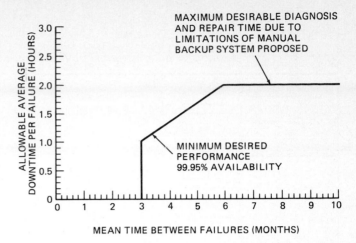

Fig. 17-18 Reliability requirements for process control systems. *(Theodore J. Williams, "The Development of Reliability in Industrial Control Systems," IEEE Micro, Dec. 1984; © 1984 IEEE).*

Machine malfunctions

Data errors

Control signal errors

Data communication errors

Timing errors

Computer failures

Power failures or disturbances

Operator errors

For a complex control system to perform its functions without interruption for very long periods, it must be able to tolerate such errors and failures. Internal failures must not be perceived externally to the system. The system must be "fault-tolerant" and must be able to quickly diagnose and repair internal failures. There are a number of approaches to providing such fault tolerance:

☐ Distributed hierarchical control permits processes and tools to operate when failures occur in higher-level systems, since the control programs and data processing tasks reside in the local controller.

☐ Communication channels can be protected from disturbances, such as electrical interference from motors, by shielded cables. Error detection schemes can also be used. These schemes include a redundant bit in each message. If the bit is not detected, the message may have to be repeated.

☐ Redundancy can be built into the computer control system itself. This can take a variety of different forms (Fig. 17-19). In a "duplex system," two computer systems compare data to detect errors. One is the primary or "master" system. If it fails, the "slave" system takes over the control process. In a "triplex system," three computers are used to "vote" if their data does not agree (i.e., majority rules).

Another important aspect of the reliability of a process control system is that the programs must be designed to function properly and to be error-free. Unlike most appli-

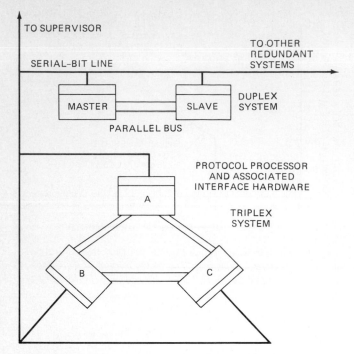

Fig. 17-19 Redundancy schemes for process control systems: comparison of duplex and triplex schemes. *(Theodore J. Williams, "The Development of Reliability in Industrial Control Systems," IEEE Micro, Dec. 1984; © 1984 IEEE)*

cation programs, complex real-time process control systems usually cannot be tested and debugged on a real process. This could cause substantial interruption and losses. Instead, they must be simulated to thoroughly check out all their functions.

STANDARDS

Large manufacturing operations involve many different processes and activities that use computer systems. To provide efficient process control, data communication, and management reporting, it is desirable to tie these systems together. Since typically many different types of manufacturing and computer equipment are involved, this is not a simple task. It requires that the architecture of the overall system be designed so that it can provide a structure and set of interfaces which are compatible with such an environment. Industry has spent a great deal of time and effort to develop standards. The most fundamental standard is the model of the system architecture. The International Standards Organization (ISO) has adopted an open system interconnection model that provides a framework for the development of specific data communications standards (Fig. 17-20). It separates data communications considerations into seven "layers" of basic tasks which are relatively independent of each other. Standard communication protocols can then be developed for each of these. The major elements of this structure are discussed below:

Layer 1 is called the "physical layer" since it provides the actual connections between the machines at the lowest level of hardware in the system (e.g., tool controllers and workstations). The protocol for exchanging data at this level is unique to the particular network adopted (e.g.,

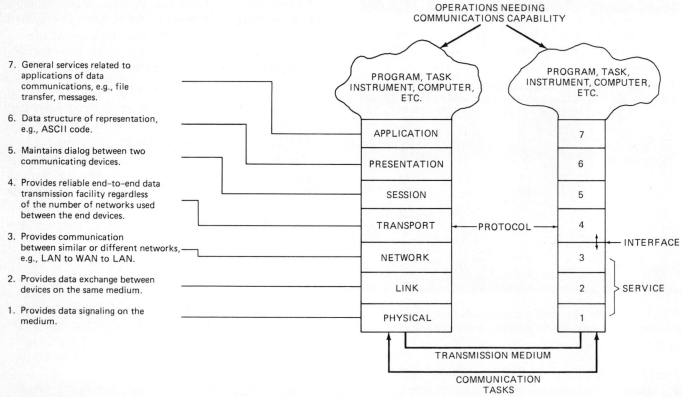

7. General services related to applications of data communications, e.g., file transfer, messages.

6. Data structure of representation, e.g., ASCII code.

5. Maintains dialog between two communicating devices.

4. Provides reliable end-to-end data transmission facility regardless of the number of networks used between the end devices.

3. Provides communication between similar or different networks, e.g., LAN to WAN to LAN.

2. Provides data exchange between devices on the same medium.

1. Provides data signaling on the medium.

Fig. 17-20 ISO open-system interconnection model. *(Maris Graube and Michael C. Mulder, "Local Area Networks," Computer, Oct. 1984; © 1984 IEEE)*

token ring, token bus, or baseband bus).

Layers 2 through 4 are relatively independent of the applications involved and provide compatibility between networks. **Layer 2** is called the "data-link layer" since it puts the data into a standard format so that it can be communicated between networks.

Layer 3 is called the "network layer." It provides the control of the routing and paths of messages.

Layer 4 is called the "transport layer." It assures the reliability of communications by providing feedback that transmissions were received.

Layers 5 through 7 are application-dependent. **Layer 5** is called the "session layer" since it initiates and terminates communications between nodes in the system.

Layer 6 is called the "presentation layer." It assures that the languages used in the communications between nodes are compatible.

Layer 7 is the last and highest one, since it deals with the "application" itself. It provides the interface between the user and the network.

A number of standards have been developed for most of these levels. They are being used by industry in the design of manufacturing equipment and communication systems (Fig. 17-21). Some standards efforts have been aimed specifically at the computer-automated manufacturing (CAM) environment, such as:

1. **Manufacturing automation protocol (MAP).** This is a token ring-based protocol compatible with the ISO model. MAP was developed by General Motors; it has been adopted by many other suppliers and users of computers for manufacturing applications.

2. **Product definition data interface (PDDI).** This standard for the exchange of design data was developed by the U.S. Air Force ICAM (integrated computer-aided manufacturing) project. It can provide the vital link between automated design systems and automated manufacturing systems.

3. **Process data highway (PROWAY).** This is a token bus standard developed for industrial use by the U.S. committee of the International Electro-Technical Commission.

COMMUNICATIONS TECHNOLOGY

The first link in a communication system is the cabling that connects the equipment. Although it may seem to be simple, cabling can have a significant influence on the cost and performance of a data communications system. Recent developments in cabling technology have focused in two areas:

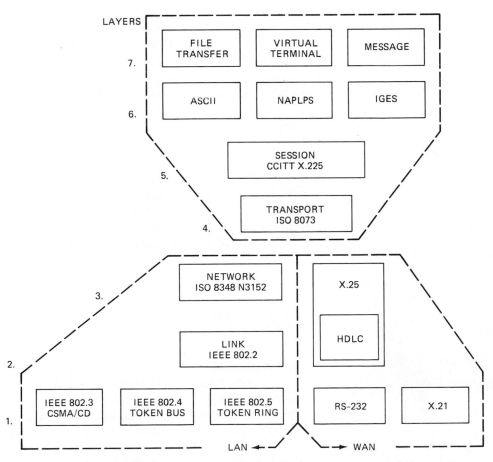

Fig. 17-21 Industry standards compatible with ISO architecture: protocol standards for local area networks (LANs) and wide area networks (WANs). *(Maris Graube and Michael C. Mulder, "Local Area Networks," Computer, Oct. 1984; © 1984 IEEE)*

1. A single broadband coaxial cable strung throughout a factory can carry all the different types of information required to run the operation (i.e., data, voice, video, facsimile).

2. A fiber-optic cable uses light to transmit data. It has a much higher capacity than wire cable and is immune to the electrical disturbances that may be found in an industrial environment. Special electronics have been developed to convert optical signals into digital signals at high data rates (Fig. 17-22). Pulses of light are generated by lasers and transmitted through thin cables of glass fiber which connect computers to peripherial equipment.

Communication networks are also becoming more efficient. Special telephone services can now provide high-speed, high-volume data transmission through a common central line and switching system for multiple applications in the same facility. Long-distance data communication networks can tie factory systems together through satellite relay links or even FM radio transmissions.

COMPUTER TECHNOLOGY

The advances in semiconductor technology have made it possible to provide special data communication functions on integrated circuit (IC) chips. Digital signal processors, for example, can be used to replace the more expensive and less reliable analog filters that have traditionally been used in process controllers. Customized ICs have also been developed to provide standard interfaces for LANs. Such high-density devices make it possible to build standard data communication features directly into manufacturing equipment.

Special communication processors (sometimes called "front ends") have been developed to relieve central host computers of data communication tasks and to provide standard interfaces to user networks.

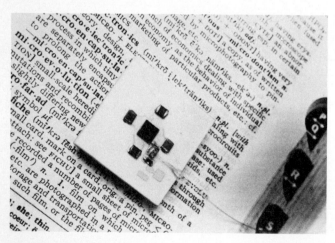

Fig. 17-22 Optical communications IC chip capable of receiving data from I/O devices via fiber-optic cable at the rate of 400 Mbps. At that rate the entire text of a 20-volume encyclopedia could be transmitted in less than 3 seconds. *(International Business Machines Corp.)*

In the area of software, a number of developments offer improved capabilities for manufacturing systems:

1. High-level real-time process control languages are easier to use and provide more function than the traditional methods of programming controllers (e.g., relay ladder–type logic).

2. Simulation tools can be used to design a process and model its behavior.

3. Expert systems can be used to automate tasks involved in both the design and operation of a process control system. Intelligent control systems offer the potential not only to control a process, but to optimize it. Such systems can explore alternatives and even learn by modifying their knowledge base from experience.

As the cost of computer power has come down so dramatically, it has become almost negligible compared to some sophisticated types of manufacturing equipment. It is therefore practical to use a large amount of computer power for tasks such as monitoring and controlling processes.

APPLICATIONS

Several general factors in the evolution of the industrial environment will influence the application of manufacturing systems:

1. The complexity of products, processes, and equipment will make it essential to use computer control, even in small factory operations.

2. High-technology products will change the nature of many factories. Job shop type operations, which have been in a labor-intensive, discrete manufacturing environment, will evolve into a more continous flow–type process environment, which is capital-intensive.

3. To manufacture products in high volumes at low cost, factories will have to operate around the clock. This will improve the utilization of their facilities and keep the inventory moving. To keep the operation running without interruption, however, will require computer systems that can control automated processes and respond to unexpected events.

17-7 SUMMARY

An automated manufacturing operation involves computer-based systems for tools, processes, and management activities. In most factories, the technical, logistical, and administrative data systems were developed separately. To establish a plantwide manufacturing control system, an architecture which links all the hardware and software together is required. Such plantwide systems can be complex and difficult to implement.

A hierarchical approach to system design closely parallels the traditional organization of a plant. The control system performs different information handling tasks at each level of the system structure. Complex manufacturing operations often use a distributed hierarchical architecture to optimize the overall efficiency of the system. Local area networks (LANs) can be used to provide peer-to-peer communications in such a system. There are a number of types of LANs, which have different design and performance characteristics.

The design of a control system must be tailored to the nature and needs of the particular manufacturing process involved. Control systems must deal with different types of process variables and data signals. Computer process control may not only help run a manufacturing operation more efficiently; it may even make some complex processes practical. The reliability of such systems is critical to the entire manufacturing operation. Therefore, they must be designed for high availability and fault tolerance.

Standards are being developed for the architecture and interfaces of manufacturing systems. Advances in computer and communications technology have also improved the capabilities of manufacturing control systems.

REVIEW QUESTIONS

The answer to each question can be found in the section indicated at the end of the question.

1. What is system architecture? [17-1]

2. Identify the principal types of data which need to be handled in a manufacturing operation. [17-2]

3. Describe the primary functions of a manufacturing control system. [17-2]

4. Identify some of the performance requirements for large manufacturing systems. [17-2]

5. Describe the basic architecture of a hierarchical manufacturing system. [17-3]

6. Identify some of the typical tasks which must be performed at each level of a hierarchical manufacturing control system. [17-3]

7. What are the advantages to using a distributed architecture for a hierarchical manufacturing system? [17-3]

8. Define a local area network. [17-4]

9. Identify some of the advantages of using LANs. [17-4]

10. What are the basic characteristics which describe the design of a LAN? [17-4]

11. Identify some of the major differences between types of LANs. [17-4]

12. Describe the basic functions of a control system. [17-5]

13. What are some of the benefits of using process control systems? [17-5]

14. Identify the major types of process control techniques. [17-5]

15. What is interrupt logic? [17-5]

16. Describe the basic approaches to providing fault tolerance in a computer system. [17-5]

17. Why is there a need for system architecture standards? [17-6]

18. Identify some of the developments in computer and communications technology which have improved the capabilities of control systems. [17-6]

MANAGEMENT SYSTEMS

18-1 INTRODUCTION

A system is not just a computer or network of computers and peripheral equipment. It is also a management process that has been automated by the use of computers. Most of the tasks involved in managing a production operation can be done by or with the support of computer systems. The management process involves processing and communicating three types of manufacturing data: technical, logistical, and administrative. This chapter will describe the major management systems used in manufacturing to handle each of these types of data.

Technical data systems include the generation of process plans and numerical control (NC) programs from

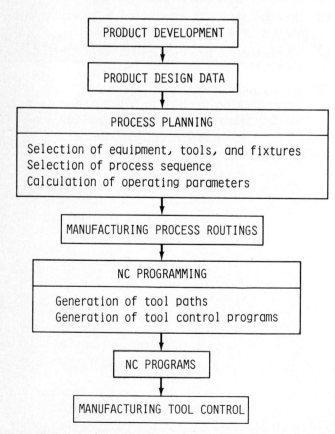

Fig. 18-1 Technical planning process.

design data. Logistical data systems are involved in the planning and scheduling of production. They can be limited to the production control of materials or cover the entire scope of the resource planning process. Administrative data systems include a variety of support functions. Most of these relate to measurements of manufacturing performance, such as quality, labor, equipment, and cost data.

Computers can also be used to design the architecture of a manufacturing line—both the physical layout of the tools and the flow of the materials and product. In addition, computers can be a valuable tool to help management optimize manufacturing operations. This may include balancing the line, reducing inventory, improving cycle time, or using statistical techniques to control the process.

Computer systems have become an integral part of the management process. Many complex production operations could not be managed or run without them. Even most small manufacturing operations benefit from the use of computer systems. This chapter will describe how computer-automated manufacturing (CAM) technologies are used to make the management of manufacturing operations more efficient.

18-2 TECHNICAL DATA SYSTEMS

THE TECHNICAL PLANNING PROCESS

The principal technical planning task for manufacturing is to convert design information into data that can be used to manufacture the product. This process typically involves a sequence of steps which starts with the product design and ends with programmed instructions that control the operation of manufacturing equipment (Fig. 18-1). The technical planning process is the link between the development and manufacturing organizations. Product design data is the prime source of technical information that manufacturing uses to determine how to make the end product. This design or engineering data typically includes information about the product and its parts, such as:

Bills of materials (i.e., a complete list of all the parts that make up the end product)

Geometry (the physical shapes of the product and its parts)

Dimensions

Tolerances

Materials

Special requirements (e.g., surface finishes or treatments)

Traditionally, all this data was contained in an engineering drawing which manufacturing engineers would use to obtain the information they needed to plan the manufacturing process. Today, this information is often found in a computer-automated design (CAD) system. It may be stored in the form of engineering drawings or it may be incorporated in a computer model of the product. Manufacturing uses a technical data system to extract and process the information it needs in order to plan and operate the manufacturing process.

The first step of the technical planning process is called "process planning." The manufacturing engineer or process planner uses the design information which describes the product to select processes and machines that can be used to fabricate and assemble the parts. The planner works out the details of the specific tools and fixtures that will be required as well as the critical parameters that will be used to control the operation of the machines. The end result of this activity usually takes the form of a manufacturing "routing." This describes the entire manufacturing process in detail, including the sequence of operations and the settings or control limits on each tool.

The next step is NC programming. The programming activity involves taking the process definition and parameters and developing specific instructions for all the computer-controlled manufacturing equipment. The result is a set of computer programs that will be used to operate the tools. In machining operations this step involves NC part programs. In other types of manufacturing processes the type of data may differ, but the function is basically the same. In the manufacturing of electrical or electronic products, test programs must be developed in the same manner from electrical design data.

GROUP TECHNOLOGY

A technique that can be very useful to manufacturing in the technical planning process is group technology (GT). GT is a method of grouping parts together to take advantage of their similarities when they are designed and manufactured (Fig. 18-2). Families of similar parts can share the same process plans. This can reduce the effort involved in process planning as well as increase the efficiency of the manufacturing operation. There are several benefits to having more parts manufactured using the same process:

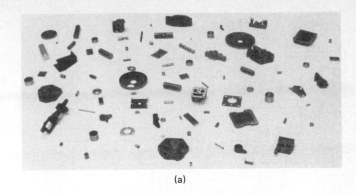

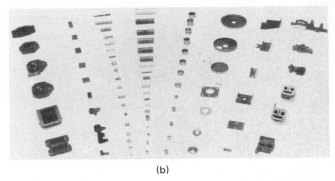

Fig. 18-2 Grouping parts together into families on the basis of their similarities: (*a*) mixed parts, (*b*) parts families. *(OIR/Organization for Industrial Research)*

☐ The number of machine setups should be reduced

☐ The work-in-process (WIP) inventory can be lower

☐ The production scheduling job should be simpler

☐ Standard process plans can be optimized for lower cycle time and higher machine utilization

GT is, in itself, a technical planning process. It involves a disciplined approach to the following steps:

1. Identifying the attributes of parts in terms of their physical or mechanical characteristics (e.g., shape, size, materials)

2. Identifying similar attributes between parts and grouping them together using a classification and coding scheme

3. Generating or selecting process plans on the basis of the similarities that have been identified

Parts may be classified by similarities in either of two different types of attributes (Fig. 18-3):

1. Design attributes (e.g., size or shape)

2. Manufacturing attributes (e.g., process requirements)

This classification process is not always simple and straightforward. Parts that have the same design attributes may not have similar manufacturing attributes and vice versa. For example, parts may look similar but differences in materials, dimensions, and tolerances may prevent them from being manufactured the same way. There are also many cases where parts that are not similar in

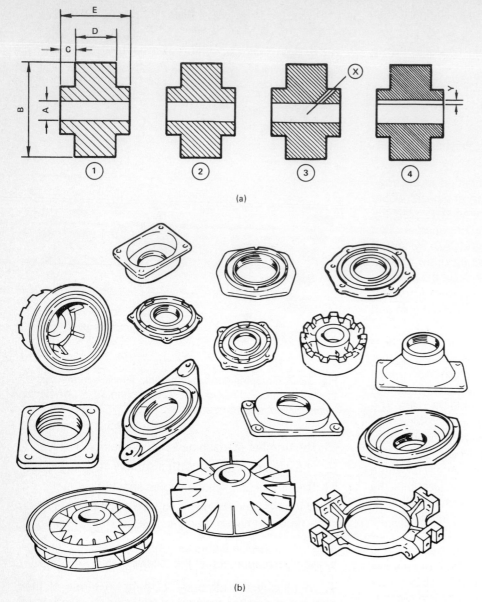

(a)

(b)

Fig. 18-3 Classification of part attributes: similarities based on (*a*) shape and (*b*) process requirements. *(OIR/Organization for Industrial Research)*

appearance are manufactured by the same process steps.

For the classification of attributes to be useful in either the design or manufacture of the parts, the groupings must be organized into a standard coding system so that the information can be stored and retrieved efficiently. A number of GT coding schemes are in use. Most involve a numerical code with a series of digits that categorize each of the attributes either quantitatively or qualitatively (Fig. 18-4). To provide all the information that might be required for both manufacturing and design use, a code may have 20 to 30 digits. Coding schemes classify attributes such as shapes, dimensions, tolerances, materials, and machine operations (e.g., machining steps, machinability). Some coding schemes use a long series of digits to identify all of the part attributes. This is referred to as a "chain" structure or "polycode." Another approach is a hierarchical or "tree" structure, which is also referred to

as "monocode." These codes are more compact since each digit is a subset of the preceding one. Some of the most commonly used coding schemes are:

1. OPITZ (developed by H. Opitz of Germany)

2. MICLASS (which stands for "Metal Institute Classification System"; developed by the Netherlands Organization for Applied Research)

3. CODE (Manufacturing Data Systems)

4. GTTC (which stands for a generic "group technology characterization code" being developed by the U.S. Air Force ICAM project)

Parts classification and coding makes it possible to take advantage of standardization in both design and manufacturing. Some of the benefits that can be derived from using such a system include:

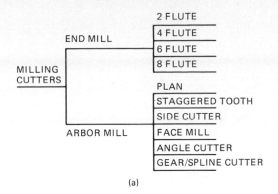

EXAMPLE OF MONOCODE

(a)

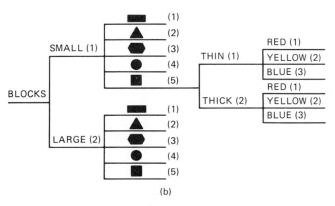

EXAMPLE OF POLYCODE

(b)

Fig. 18-4 Group technology coding schemes. *(OIR/Organization for Industrial Research)*

☐ Reducing the number of unique part designs and nonstandard features

☐ Minimizing the number of unique manufacturing process plans

☐ Increasing lot sizes in manufacturing

☐ Automating the exchange of design and manufacturing data

Parts classification and coding can result in lower costs by increasing the efficiency of manufacturing. This can be in the form of reduced effort, less investment, or greater utilization of assets. As we will discuss further, GT can be used for several purposes:

1. Design. To select standard parts and avoid the need to design and manufacture new ones.

2. Process planning. To automate the generation of manufacturing process plans based on standard process sequences.

3. Line layout. To design the placement of equipment and flow of materials on the basis of standard process plans.

COMPUTER-AUTOMATED PROCESS PLANNING

The job of process planning is to derive manufacturing routings from engineering design data. This involves in-terpreting the engineering information and making manufacturing decisions, such as:

☐ What operations should be performed (e.g., specific machining steps)

☐ How those operations should be performed (e.g., direction and depth of cuts)

☐ What equipment should be used to perform those operations (e.g., selection of machine tools)

☐ What tools and fixtures are required for that equipment

☐ What sequence of operations should be used (i.e., process steps)

Process planners base their decisions on their experience and their interpretation of the data. This can result in inconsistency between planners, variations in plans over time, and even errors. To do this job well requires a lot of knowledge about manufacturing as well as practical experience. It can be a tedious and time-consuming task and will not necessarily result in the most efficient use of manufacturing resources.

Computer-automated process planning (CAPP) techniques were developed to improve the efficiency of the task as well as the manufacturing operations. There are two basic approaches to automated process planning:

1. Variant or retrieval. This involves using or modifying standard process plans that have been developed for families of parts. It uses GT techniques that classify these standard plans by family groupings on the basis of design and manufacturing attributes. When a new part is designed, a standard plan which has similar characteristics can be retrieved from the system (Fig. 18-5). The plan can then be modified to adjust for any unique requirements (therefore "variant"). A number of GT-based variant process planning systems are available, such as:

☐ MIPLAN, which was developed by the Netherlands Organization for Applied Research, the same group that developed the MICLASS GT system

☐ CAPP, which was developed by Computer Aided Manufacturing-International, Inc. (CAM-I)

2. Generative. This approach involves the automatic creation of a new unique process plan based on an analysis of the engineering data. Such a system uses algorithms or rules to make process decisions and select from possible alternatives. Some may use a solid model of the part stored in the design system to extract geometric data about features and to identify surfaces which require machining operations. This data may then be matched to standard tools and process sequences to develop a manufacturing routing. Such process plans may be modified and rerun several times in order to optimize their cost and performance. Generative process planning systems may use an artificial intelligence (AI) technique to make the manufacturing decisions. Rules are established about how to perform certain operations given certain information about the design. Automating generative process plans is a relatively new and complex approach. One

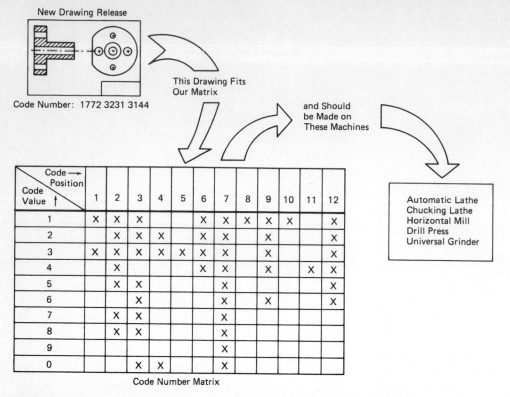

New Drawing Release

Code Number: 1772 3231 3144

This Drawing Fits Our Matrix

and Should be Made on These Machines

Automatic Lathe
Chucking Lathe
Horizontal Mill
Drill Press
Universal Grinder

Code Value ↑ / Code Position →	1	2	3	4	5	6	7	8	9	10	11	12
1	X	X	X			X	X	X	X	X		X
2		X	X	X		X	X		X			X
3	X	X	X	X	X	X	X		X			X
4		X				X	X		X		X	X
5		X	X				X					X
6		X					X		X			X
7		X	X				X					
8		X	X				X					
9							X					
0			X	X			X					

Code Number Matrix

Fig. 18-5 Variant or retrieval-type process planning based on group technology. (OIR/Organization for Industrial Research)

such system is:

GENPLAN, developed by the Lockheed-Georgia Co.

The benefits of automated process planning systems include:

☐ Reduced workload for manufacturing engineers

☐ Accurate and consistent process plans

☐ Optimization of process plans in terms of economics and utilization of equipment

☐ Reduced time required to prepare process plans

☐ Capability for integration with other manufacturing systems (e.g., GT classification and coding, NC part programming)

COMPUTER-AUTOMATED PART PROGRAMMING

After a process plan or manufacturing routing is established, programs must be developed for the numerically controlled tools. In machining operations this includes the determination of:

Tool selection

Machining sequence

Tool paths

Cutting conditions

Traditionally, this was a tedious manual effort that relied on the experience and judgment of a part programmer and the use of reference material, such as machining handbooks. Computer-automated NC programming sys-

tems use minicomputer workstations with color graphics and geometric models to perform this task (Fig. 18-6). To select the optimum cutting conditions, some systems use a database compiled from experience, while others use mathematical equations. Expert systems have also been developed to perform this task by using a rule-based program to select the optimum cutting sequence and conditions. Most systems allow the operator to use simple commands, function keys, or a menu screen to select the cutting conditions. Using a geometric model of the part design, tool paths can be displayed on the screen and altered to optimize the process. Once a final set of conditions has been selected, the system will generate the NC program by post-processing the data.

Some of the advantages of this approach are:

☐ Reduced effort and skill required of the part programmer

☐ An optimized set of operating conditions

☐ Reduced lead time between the design and manufacture of a product

☐ Visualization of tool paths

☐ Alternative cutting conditions can be tried before actually implementing them in manufacturing

The technical data systems in manufacturing are all involved in converting design information into manufacturing programs. The process is basically the same whether the product is a machined part, a mechanical assembly, or an electronic component. Technical data

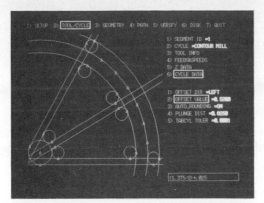

Creating the design of the part by describing its geometry
in a computer model

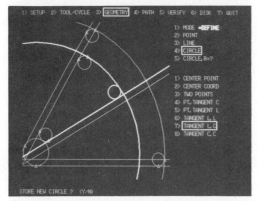

Selecting the machining process and tool parameters to
fabricate the part

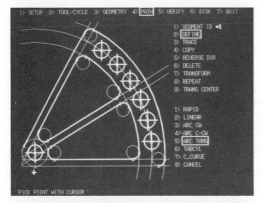

Defining the tool paths for the machining process

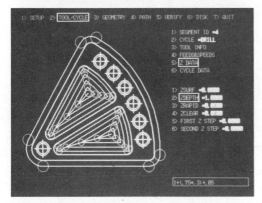

Verifying the cutter paths for proper operation

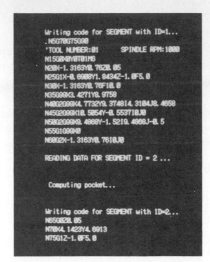

Generating the NC part program for the machine tool

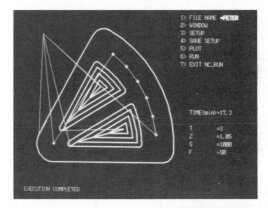

Simulating the machine tool operation and performance
(e.g. run times)

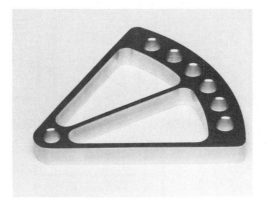

The finished part

Fig. 18-6 Computer-automated part programming. *(Bridgeport Machines Division of Textron Inc.)*

systems may involve physical fabrication data or logical testing data. The process starts with a design model from a CAD system and yields NC programs that control the operation of production equipment. Computers have changed a time-consuming and error-prone process into one that is fast, efficient, and economical. Automated technical data systems are essential to make the manufacture of complex products even practical.

18-3 LOGISTICAL DATA SYSTEMS

WHAT ARE LOGISTICAL DATA SYSTEMS?

Logistics, in general terms, means the science of transport and supply. In a manufacturing operation, logistics systems control the scheduling and movement of materials and products. The total logistics process involves three major phases of activity, each having several functions (Fig. 18-7). These functions all take some form of planning, scheduling, or controlling activity. All are involved in answering the basic questions of production management: What has to be made, when, and how many?

1. Master plan. This phase identifies the requirements for the end products. Production demands are derived from long-range marketing forecasts as well as actual customer orders. These demands, together with bills of materials for the products, are used to create master plans for manufacturing the parts and assemblies required.

2. Operating plan. This phase deals with relatively short-term planning of the production operations. It involves developing plans for manufacturing capacity, inventory, costs, purchases, and production schedules. These plans influence the operation of production lines and need to be updated frequently to respond to changes. For example, changes may occur in demand, product mix, bills of materials, costs, manufacturing capacity, or the availability of materials; each can affect the operating plans.

3. Plan execution. This is the phase of activity that is directly involved with the daily operation of production lines. It includes releasing and tracking all the job or shop orders currently in manufacturing. To assure that customer requirements are met, this phase must be capable of real-time responses to changes in manufacturing performance as well as demand.

The job of managing this logistics process can be complex for large manufacturing operations. It involves enormous amounts of data that can be dynamic. One measure of success is that the customer receives the right quantity of the right product on time. An effective logistics process, however, must also accomplish that task at minimum cost. Some of the problems that such a system must prevent or cope with include:

Excessive inventories

Inadequate capacities

Long manufacturing lead times

Poor utilization of equipment

Errors in data

Errors in production

Computers can provide the ability to efficiently handle the large amounts of data involved in the logistics pro-

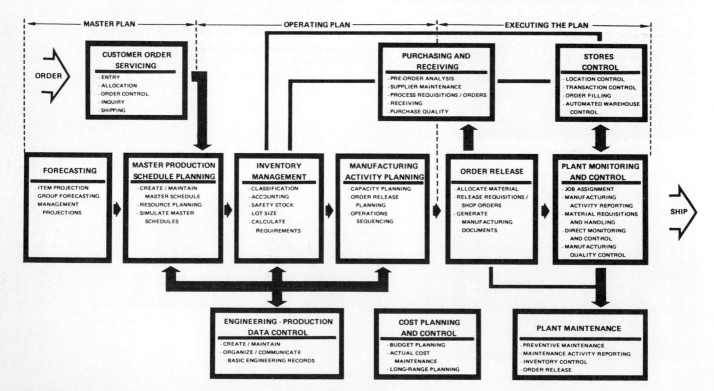

Fig. 18-7 The logistics process. *(International Business Machines Corp.)*

cess. They can also provide analytical tools to help make decisions and optimize the scheduling of the manufacturing operation. The type of computer-based logistics system used depends a great deal on the nature of the manufacturing operation. Some of the factors that can influence the design and capabilities of a logistics system are:

☐ Complexity of the manufacturing process
☐ Number of products and parts involved
☐ Quantities involved in the demand
☐ Schedules
☐ Frequency of changes to the parts and demand
☐ Costs of the parts and production equipment
☐ Interdependencies between different manufacturing operations
☐ Lead times for production orders and processing

Over the years, many computer systems have been developed to support the logistics process. As manufacturing operations became more complex and the capabilities of computer systems increased, logistics systems became more sophisticated. There have been three basic stages in the development of manufacturing logistics systems. Production operations today may be found at any stage, depending on their needs and level of sophistication in the use of the computer.

1. Basic production control systems. These involve relatively simple techniques for the scheduling and control of production operations. They may provide only a method of gross requirements planning to identify how many items must be produced. Since this approach is not sensitive to changes, inventories are used to protect against shortages. A further enhancement to this approach is to build in some form of "order point" control. This will trigger a need to order more parts or material based on planned requirements with some contingency. Such systems are still not responsive to dynamic changes in requirements.

2. Materials requirements planning (MRP). This is a modern and comprehensive approach to planning, scheduling, and controlling production operations. The production schedules are based on actual manufacturing status [e.g., inventory, WIP, open orders] as well as planned requirements. An enhancement to this approach is referred to as "closed loop MRP." It adds capacity planning functions to adjust schedules or manufacturing resources in response to over- or under-scheduling conditions.

3. Manufacturing resources planning (MRP II). This is a total manufacturing management system which ties in the resource and financial implications of production decisions. It involves the control of factors which cause schedule changes to occur, such as product design changes, production problems, or changes in customer requirements.

MATERIALS REQUIREMENTS PLANNING

An MRP system provides priority scheduling of the production of materials on the basis of requirements derived from customer orders, production plans, and product forecasts (Fig. 18-8). The demand for end products is broken down into requirements for parts and assemblies using bills-of-materials files. These requirements must then be netted against the inventory on hand to determine the requirements for purchasing and manufacturing. Specific production orders are created after adjusting these requirements for the appropriate lead times and desired lot sizes. For an MRP system to be effective, it must perform these tasks quickly and efficiently. It must also be responsive to changes and exceptions and cannot cause errors. The results must be readily accessible to production planners and manufacturing management in a timely manner and simple form.

These production planning tasks are typically accomplished by interactive display systems which permit changes to be entered on-line. In Fig. 18-9 the requirements for an assembly (A) are calculated by the system in a multistep process:

1. Requirements of parent assemblies (X and Y) and service requirements (independent demand) are identified

2. Requirements for the assembly (A) are then based on scheduled receipts and inventory on hand

3. Shop orders based on lead times and lot sizes are scheduled for these requirements

4. Requirements for component parts (B and C) are then "exploded" using the bills of materials of the assembly

MRP systems can involve large data handling, storage, and computation tasks. The manufacture of some products (such as automobiles or computers) deals with a hierarchy of thousands of parts and assemblies supplied by many different sources. Production planning and control systems must be designed to deal with that kind of product complexity as well as a number of basic manufacturing factors, such as:

1. Inventory. This comes in several different forms, including raw materials, operating supplies (e.g., spare parts, tools), purchased parts, WIP, and finished goods.

2. Manufacturing processes. These may involve not only a series of complex operations, but also a hierarchy of manufacturing operations that must be completed to produce a finished product. The flow may start with the manufacture of materials, followed by the fabrication of parts and then several levels of assembly operations.

3. Demand. The simple concept of product requirements can take several forms and become complicated. Some demands for parts are "independent," since their requirements are not related to other parts. Others are "dependent," like parts for a common assembly. The flow of the demand can also vary. Requirements for most products are incremental or periodic, not continuous.

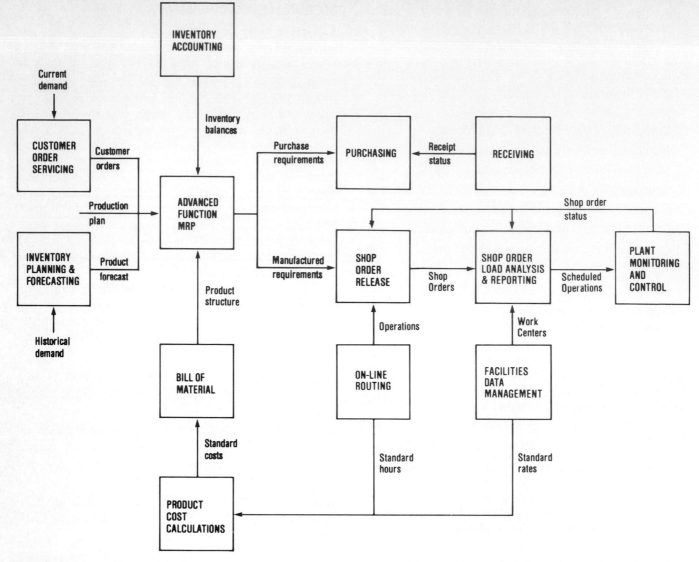

Fig. 18-8 Information flow in a materials requirements planning (MRP) system. *(International Business Machines Corp.)*

4. Lead times. These are amounts of time required for events to occur. The overall lead time in a production planning system refers to the time that elapses from when an order is placed to when a finished product is delivered. This lead time may include a number of smaller lead-time elements, such as for order processing, part manufacturing, final assembly, packaging, and delivery. The lead or "cycle time" within a manufacturing process also includes several smaller elements that must all be known and managed. For example, a typical production operation involves times for waiting (queueing), setup, running, and moving or transferring.

Once orders are released to production, a system must be in place to monitor and control the shop floor. The objectives of such shop floor control systems are to:

☐ Ensure that delivery dates are met
☐ Control the level of WIP
☐ Control manufacturing lead times
☐ Control queue lengths at production operations

☐ Prevent production bottlenecks
☐ Optimize the utilization of production equipment and labor
☐ Establish the priorities of production jobs

To meet these objectives, shop floor control systems perform three basic functions:

1. Releasing. This formally establishes a "job" or "shop order" for manufacturing to produce. The system must provide all the information necessary for manufacturing to execute the order. This typically includes a process routing and materials list as well as any documentation that may be used by manufacturing to keep track of the job and record the performance of operations. An on-line system can provide all of this information without the use of paper by making it available to operators on a display terminal at the workstation (Fig. 18-10).

2. Scheduling. This assigns each shop order to a specific sequence of production machines. Priorities must be established on the basis of delivery schedules, which may

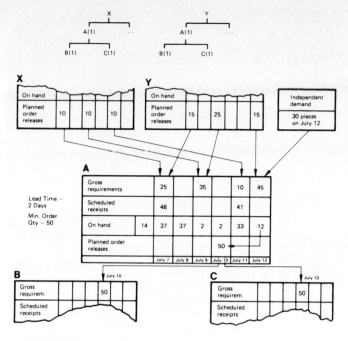

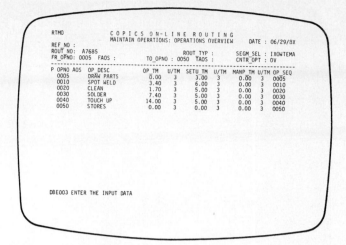

Fig. 18-10 An on-line process-routing application. *(International Business Machines Corp.)*

```
RPAD
FN:                    TIME-PHASED MATERIAL PLAN              PAGE:   1
ITEM NO: A                      DRILL
ORDER NO:              ORD POL: A  CTL: 1  SAFETY STK:        DATE: 06/29/8X
                                                             SHRINK:  0 %

DUE   *****  REQUIREMENTS  *****  *********  O R D E R S  *********  PROJECTED
DATE  TYPE  QUANTITY SOURCE  REF.   REL. TYPE S   NUMBER    F  QUANTITY  BALANCE

                                                        ON-HAND:            14
07/07 DEP      15 Y             PEG F  1404999   M
07/07 DEP      10 X             PEG F  1404999   M
07/07                 06/29 M   F  0404999   M        48              37
07/09 DEP      25 Y             PEG F  1405999   M
07/09 DEP      10 X             PEG F  1405999   M                     2
07/11 DEP      10 X             PEG F  1407999   M
07/11                 06/29 M   F  0400999   M        41              33
07/12 DEP      15 Y             PEG F  0407999   M
07/12 IND      30 INDEPENDENT DEMAND      P      M
07/12                 07/10 M   F  0406999   M        50              38



----- RPAD40: ITEM IN NEED OF REPLANNING                    -----
```

Fig. 18-9 How an MRP system works. *(International Business Machines Corp.)*

```
PMWP                    WORK CENTER PRIORITY
WORK CENTER 1102          STATUS  10 TO 50

SEQ ST SHOP ORDER OPER  DESCRIPTION      SCHED  SETUP RUN-REM QTY-REM
--  -- ---------- ----  -----------      -----  ----- ------- -------
 1 10  2380 0085        GRIND            04198X  0.1   4.5     350
 2 10   740 0040        GRIND & POLISH   04238X  0.2  13.4    1100
 3 30   830 0005        GRIND                4 L  0.1   7.1     600
 4 30  1230 0040        POLISH               5    0.2   2.7     165
 5 50  2310 0060        GRIND               12 L  0.1   4.9     250
 6 50   790 0030        FINISH GRIND         7 L  0.1   3.8     180

125I THERE ARE NO MORE OPERATIONS FOR THIS WORK CENTER
```

Fig. 18-11 A job-scheduling application. *(International Business Machines Corp.)*

change frequently. The work load of each operation should be balanced by considering capacity, utilization, queues, and potential bottlenecks. This information should be available on-line for each operation (Fig. 18-11).

3. Reporting. To provide management information on how manufacturing is doing relative to the schedules, the system must collect and report data about the status and performance of production operations. This may take the form of graphic displays which provide analyses of such factors as line loading (Fig. 18-12). It may also take the form of data summaries, such as performance reports (Fig. 18-13). Typical types of data that are provided by such systems include:

☐ Shop order schedules

☐ Machine loading profiles

☐ Work load outlooks

☐ Cycle-time performance

☐ Inventory status

☐ Serviceability performance (i.e., shipments versus schedules)

Shop floor control can be a complex process. It can involve a lot of data and many variables. Large production operations involve many parts, jobs, and machines. The production schedules are subject to changes in demand, priority, and manufacturing performance (e.g., losses, downtime, bottlenecks). In some complex processes it would be an accomplishment to just be able to keep track of where the material is, much less optimize the throughput, if computer-based MRP systems were not available. An MRP system can be as effective at increasing manufacturing productivity as the automation of the physical operations.

MANUFACTURING RESOURCES PLANNING

Manufacturing resources planning is also referred to as MRP II to distinguish it from materials requirements planning (MRP). MRP is really a subset of MRP II. MRP II

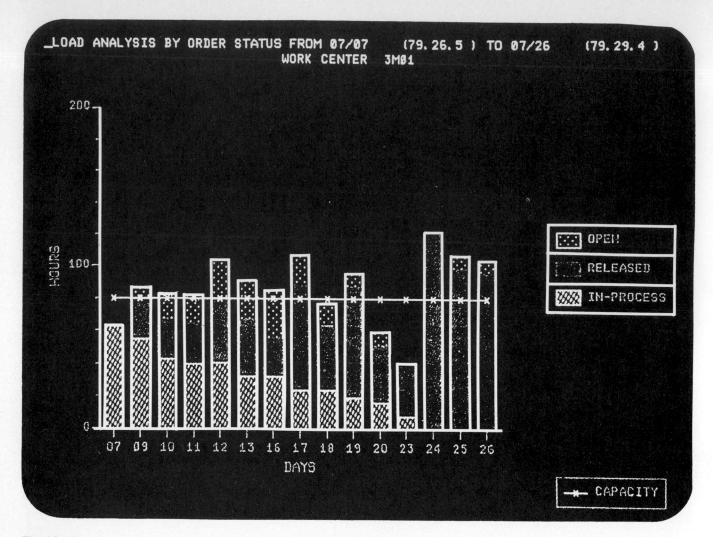

Fig. 18-12 A line-loading analysis. *(International Business Machines Corp.)*

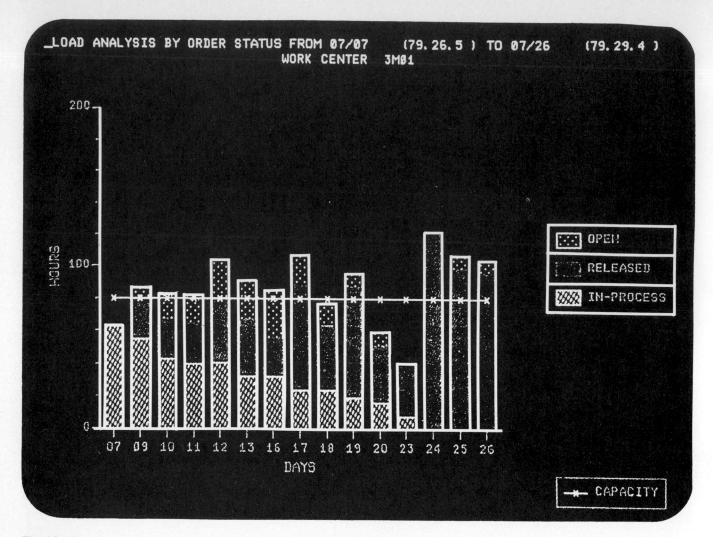

Fig. 18-13 A manufacturing performance report. *(International Business Machines Corp.)*

goes beyond production scheduling to tie in the implications of manufacturing resources, such as capacity and costs. Its objective is to optimize both the performance of manufacturing (e.g., serviceability) and the cost of operation (e.g., capital investment). Such a system encom-

passes all the management processes from high-level business planning through the execution and measurement of manufacturing performance (Fig. 18-14). MRP II adds steps to the production planning process to consider the resource implications of demand and performance changes. This allows manufacturing management to determine the feasiblity of executing a plan ahead of time. Potential problems and constraints can be anticipated and possibly avoided. MRP II may allow time to add or reduce resources in response to those changes so that orders are not lost or costs increased.

An MRP II system must be able to simulate the capabilities of the production operations involved. This allows management to evaluate the effects of changes or analyze "what-if" questions. It provides a higher-level function than the basic MRP system which controls the execution process. MRP II can predict key resource factors for management to consider in making planning decisions (e.g., inventory, labor, capital). The management of these resources can have a significant effect on the cash flow and profitability of a manufacturing operation. MRP II systems provide a means for management to keep the operational activities of production compatible with corporate financial plans and goals.

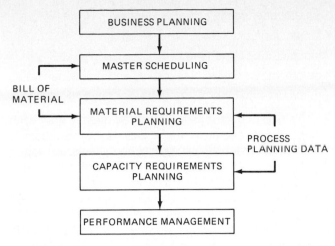

Fig. 18-14 Manufacturing resources planning process (MRP II). *(OIR/Organization for Industrial Research)*

18-4 ADMINISTRATIVE DATA SYSTEMS

In addition to the technical data that controls the manufacturing process and the logistical data that control the flow of the product, manufacturing needs a lot of other information to run the factory. Most of that information can be thought of as administrative data, since it does not fall into either of the other categories and it deals primarily with the administration or business control of the plant. The data is often collected from the factory floor operations for a variety of administrative organizations to use. The various methods of data collection will be addressed in Chap. 19. Following are some of the typical types of administrative data systems that are found in a manufacturing operation.

QUALITY CONTROL

The purpose of quality control activity is to assure that the manufacturing process is not producing defective product. It can be argued that this involves technical data, but the information is usually obtained outside of the tool or floor control systems. Most manufacturing operations have a quality control organization whose job is to acquire and analyze data which will determine whether the process is under control and the product being manufactured will meet its specifications. The data can be obtained from a variety of sources, such as:

Receiving or incoming inspections

In-process or in-line manufacturing inspections or tests

Final inspections and tests

Customer returns or failures

Failure analyses

In the past, quality control data was often taken and recorded manually, then either analyzed manually or entered into a computer system which could perform the necessary computations and data summaries. Today, more often than not, quality data is either recorded automatically by a computer-controlled instrument or entered directly into a computer system by the inspector. This eliminates the time and effort required to manually record and reenter data and also avoids a source of data errors. Quality control data can be used in a variety of ways, but it can be most effective as a means to control and optimize the manufacturing process to prevent defects. Statistical quality control techniques look at trends and deviations in the data in relation to predetermined control limits (Fig. 18-15). Some sophisticated quality control systems take greater advantage of the capabilities of computer systems

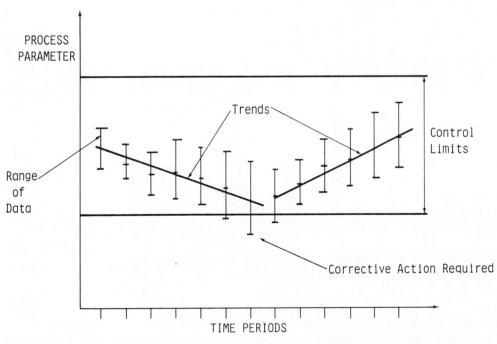

Fig. 18-15 Statistical quality control chart.

by including advanced features such as:

☐ Operator aids and alerts to prevent defects

☐ Feedback to operators on problems found

☐ Automatic sampling techniques which adjust the frequency and sample size for inspections based on actual history

☐ Advanced statistical analyses of process control data

☐ High-level management reports of the major problems

☐ Real-time monitoring and analysis of process data

MANUFACTURING PERFORMANCE MEASUREMENTS

Manufacturing management needs to have measurements which tell how well the plant is running. Management usually needs several different types of administrative information in addition to the logistical data which tells whether customer deliveries are being met. The performance of the people as well as the equipment must be measured. Typical parameters are:

Operator throughput (quantities of product processed)

Operator attendance and time working

Manufacturing losses (scrap or yield)

Manufacturing cycle time

Equipment downtime (breakdowns and maintenance)

Equipment throughput

This type of data must be collected directly from each manufacturing operation. In the past, operators recorded much of this manually on "production sheets" or "logs." To minimize the nonproductive time spent obtaining such data, automatic data recording or direct system input approaches must be used. Management reports can then be prepared from this data to summarize the performance of each factor being tracked in comparison to targets which have been established. This permits managers to quickly identify problems that may require corrective action. Computer systems can be used to collect such data and to prepare routine analyses and reports for management (Fig. 18-16).

PURCHASING

The purchasing organization is a key part of most manufacturing operations. Purchasing is responsible for ob-

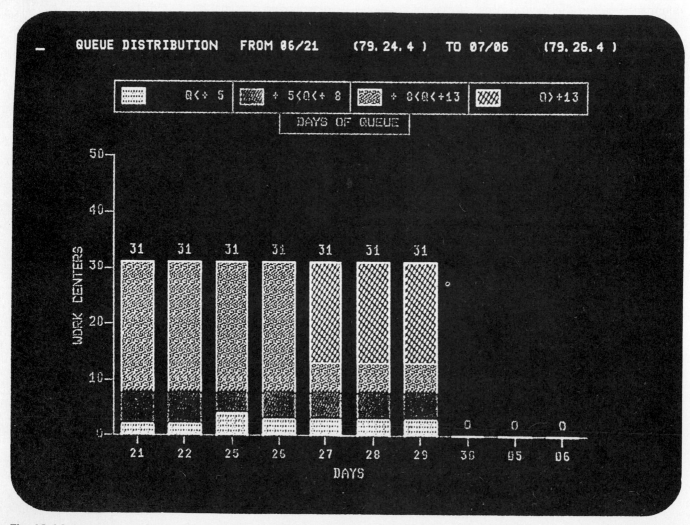

Fig. 18-16 Management report on manufacturing performance: queue distribution. *(International Business Machines Corp.)*

taining all the items that go into the manufacturing operation. These typically include raw materials, operating supplies, parts, and capital equipment. Manufacturing cannot perform its job if purchasing does not do its work. In large plants, the purchasing job can be very complex. It may involve large quantities of a wide variety of items from many different suppliers. To keep all this under control requires a computer-based purchasing system. The types of computer applications in purchasing usually take one of three forms:

1. **Document preparation.** The two main documents issued by a purchasing organization are purchase orders and payments to suppliers. These are the basic administrative input and output functions. In an organization with a lot of these transactions, computer systems can be used to help prepare the documents (Fig. 18-17).

2. **Status reports.** Buyers may have to keep track of thousands of transactions at a time. This would involve a lot of time and paperwork and might not even be possible if it were done manually. If the data involved is entered automatically or directly into a computer system rather than on a piece of paper, the system can prepare routine reports on the status of these transactions, which are easy to review. Some of the types of reports that might be used are:

Current open orders

Purchase order history

Receipts

Billings received

Payments made

3. **Performance reports.** Like a manufacturing operation, purchasing must measure its performance. Using data available to the system from purchasing transactions, reports can be prepared which summarize key indicators of the performance of the purchasing operation, such as:

Order delivery times

Receipts past due

Rejected materials

Payments due

FINANCE

Finance is a key administrative function in managing a manufacturing operation from a business viewpoint. The finance organization is responsible for financial planning, cost accounting, cash management, and business controls activities. To perform these functions, data must be obtained from the manufacturing operations, such as:

Direct labor claiming

Indirect labor charges

Quantities of parts and products started and completed

Materials consumed

Manufacturing losses

Overhead charges

The finance organization uses this data to determine the costs involved in the production operations and to compute the unit costs of the products that are manufactured. Since this involves a great deal of data from a variety of sources, it requires the help of a computer-based financial system (Fig. 18-18). Finance will also typically generate management reports which provide measurements of the financial performance of the plant versus targets. This may include such indicators as:

☐ Spending (direct and indirect charges)

☐ Unit costs (versus targets, trends, and comparisons)

☐ Direct labor productivity (performance versus labor standards)

☐ Inventory (cost and turnover rate)

☐ Capital (spending and depreciation)

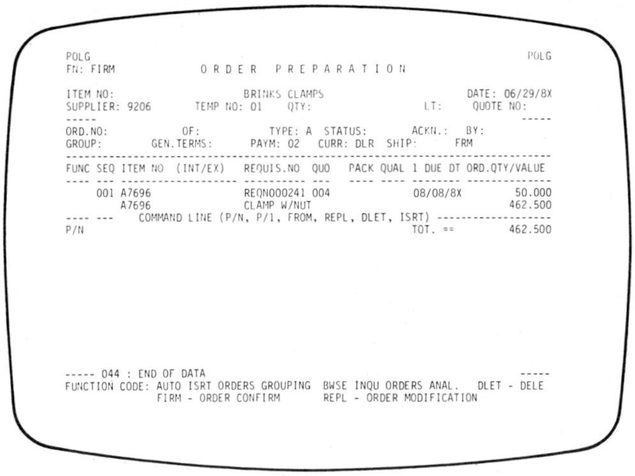

Fig. 18-17 Purchasing application: preparation of orders. (International Business Machines Corp.)

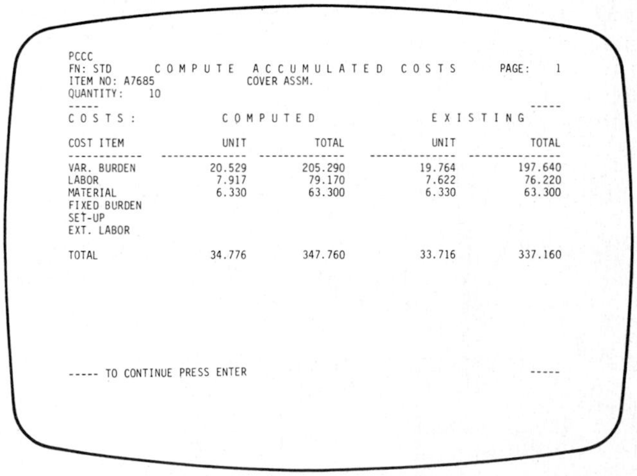

Fig. 18-18 Financial application: Product cost analysis. (International Business Machines Corp.)

Manufacturing management ultimately is responsible for deciding how to manufacture the products that have been designed for it to produce. Its challenge is to find the most efficient way to do it. The design of the product as well as the design of the manufacturing facility can have a significant effect on that efficiency. Manufacturing needs a systematic approach to influencing those designs before they are implemented and become too costly to change. Following are two major areas on which manufacturing should focus attention prior to the introduction of a new product or new production line.

LINE ARCHITECTURE

A manufacturing line is a system in much the same way that a computer network is a system. One involves moving and transforming physical objects, while the other involves moving and transforming data. Both should be designed for optimum efficiency. The architecture of a manufacturing line must be designed, just as one designs the architecture of the data systems that support it. Several important factors should be considered in designing the architecture of a manufacturing line:

1. Process versus product layout. The traditional approach to laying out a large manufacturing line is to group together similar equipment and processes which perform related functions. This creates a process-oriented line architecture made up of functional work centers or process "sectors" (Fig. 18-19). This is typical of an operation like a job shop, which must produce a variety of different products that have no standard process flow. This architecture allows machines to be shared and results in high utilization. However, it also creates a rather complex

product flow with many setup changes that can result in long manufacturing cycle times.

An alternative approach is to have a product-oriented layout. This groups machines into a work cell in a sequence to produce similar products. GT can be used to identify families of parts that can be produced by a flow through the same machines (Fig. 18-20). This "cellular" line architecture has a number of advantages over the process-oriented layout:

☐ Fewer setups required

☐ Reduced manufacturing cycle time

☐ More efficient materials handling

☐ More efficient utilization of operators and supervisors

☐ Reduced WIP inventory

This approach has often been adopted by plants that manufacture one standard product in large quantities. However, it has recently been found to be an efficient design for job shop–type operations as well, if they use GT to group their products on the basis of similar manufacturing process requirements.

2. Designing a facility for automation. In addition to the cellular layout, which lends itself to automation, a number of other considerations should influence the design of an automated manufacturing facility. The facility should:

☐ Provide for materials handling systems to move the product between machines and processes (see Chap. 19).

☐ Build flexibility into the layout for potential changes in processes and tools. This can minimize the cost and time required for future rearrangements.

☐ Leave room for adding automation features to equipment that is not currently automated (e.g., robot loading and unloading stations).

☐ In clean-room environments, move most of the people

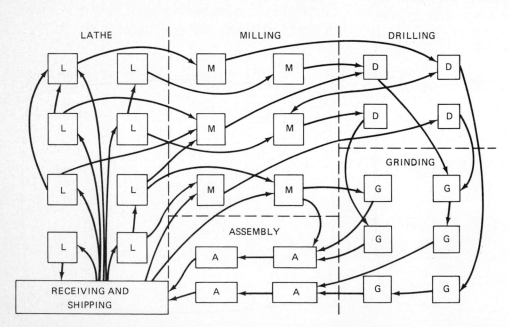

Fig. 18-19 Process-oriented manufacturing-line architecture. *(Mikell P. Groover, "Streamlining Process Routes with Group Technology," IEEE Spectrum, May 1983; © 1983 IEEE)*

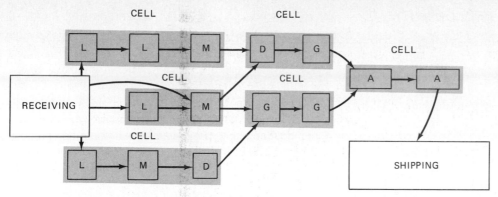

Fig. 18-20 Product-oriented manufacturing-line architecture. *(Mikell P. Groover, "Streamlining Process Routes With Group Technology," IEEE Spectrum, May 1983; © 1983 IEEE)*

and the bulk of the tools into peripheral support space to reduce sources of contamination.

Computer design tools can also be used to design the manufacturing facility (Fig. 18-21). This can include the physical structure of the facility, the layout of the equipment, and the flow of the product. By designing the line on a computer system, changes can be made easily and alternatives tried before actual construction. It can also significantly reduce the time required to design or rearrange a production facility.

3. Designing a product for automation. If automated production of a product is intended, manufacturing must influence the design of that product to assure that it lends itself to automated processes. This will not happen by itself. Product development and design activities have historically not provided for manufacturing automation considerations because of factors such as:

☐ Lack of manufacturing experience and skills in the development organization

☐ Primary focus on product function rather than cost or manufacturability

☐ Lack of an interface between the design and manufacturing systems

Fig. 18-21 Computer-aided facilities design application. *(International Business Machines Corp.)*

☐ Limited involvement of the manufacturing organization in the early stages of product development

Manufacturing must therefore take the initiative by developing guidelines for the design of products to assure that they are compatible with automated manufacturing methods. For example:

☐ Avoid the use of screws and bolts and use snap-on methods to fasten parts together

☐ Minimize the number of different features and variations in product models

☐ Use interchangeable parts

☐ Provide alignment and orientation guides

☐ Use stiff and rigid parts to allow for automatic handling

☐ Avoid features that have the potential to jam in automatic equipment (e.g., sharp edges, protrusions)

Manufacturing should also adopt a technique for evaluating product designs to determine how well they are suited for automation. A number of quantitative approaches have been developed to measure factors that can affect the productivity and quality of the manufacturing process. Typically values are assigned to each component of the product and the process on the basis of its impact on manufacturing. This can offer a systematic and objective approach to evaluating product designs before they are introduced into manufacturing. It can be used as a quantitative tool to feed information back to the development organization to use in improving designs and can lead to reduced costs and improved product quality.

18-6 OPTIMIZING MANUFACTURING OPERATIONS

A manufacturing plant that uses automated equipment and computer systems can still be inefficient. Manufacturing management must learn how to use these tools most effectively to increase the productivity of the production operations. There may be as many theories about how to run manufacturing as there are manufacturing managers, but there are also some fundamental techniques that have proved to be very successful.

WHAT IS MANUFACTURING PRODUCTIVITY?

In the beginning of this book, the basic objectives of manufacturing were said to be cost, quality, and schedule. The various measures of productivity can all be related in some way to these objectives. Typical indicators that are used by manufacturing management to evaluate the productivity of operations include:

High utilization of equipment

Low WIP inventory

High customer serviceability

Low scrap and rework

High inventory turnover

These and other general factors are difficult to argue with, but they are not easy to achieve and may even be in conflict at times. In order to focus on improving the productivity of manufacturing, management must have some measure of how it is doing today as well as what areas have the most leverage for overall improvement. For example, an objective look at how effectively manufacturing uses its resources may reveal some surprises. Figure 18-22 illustrates, in the form of a diagram, how much of the available equipment resource of a typical capital-intensive manufacturing operation might actually be used effectively to ship usable product. Some of the basic observations that can be made from this example are:

☐ The manufacturing equipment sits on the floor 24 hours a day, 7 days a week, but it is usually only scheduled to operate during part of that time. Even a manufacturing line which is scheduled to operate 5 days a week with 3 shifts a day starts out with a maximum utilization of its capital of only a little more than 70 percent.

☐ Most manufacturing operations do not schedule the equipment for all of that time. Some provision is usually made for meal times, breaks, and shift changes, which will reduce the scheduled time further.

☐ Not all the scheduled equipment time is available to manufacturing. The amount of down time due to equipment malfunctions, preventative maintenance, process problems, engineering time, and setup time can reduce that significantly (in the example in Fig. 18-22, down to less than 50 percent).

☐ Even when the equipment is available, manufacturing does not use it continually. Idle time can be significant, depending on how well the capacity and loading of the line is balanced. In Fig. 18-22, a utilization of 75 percent was assummed, which is not unusual. This leaves manufacturing with less than 40 percent of the total time to actually use the equipment.

☐ Once manufacturing personnel finally get to use the equipment, they do not always use it productively. Depending on the type of product and process involved, a large amount of time may be spent producing scrap, rework, or product that is not shippable (e.g., parts for engineering or quality control).

In Fig. 18-22, the net productivity of the equipment was less than 20 percent! That certainly leaves room for improvement. If each of the major factors cited could be improved by just 1 percentage point (for a total improvement of 5 percentage points), it would increase the overall net productivity of the operation by 25 percent! The point of the example is that there are opportunities that can improve the productivity of any manufacturing operation.

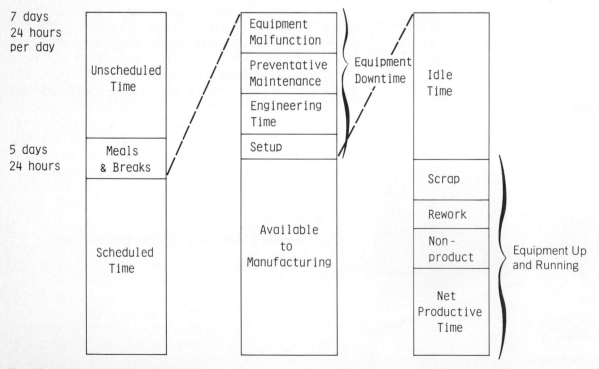

Fig. 18-22 Manufacturing productivity.

OPTIMIZING THE FLOW OF THE MANUFACTURING PROCESS

The way in which product flows through the manufacturing process can have a significant effect on productivity and cost. Many manufacturing operations have characteristics such as:

Significant fluctuations in production schedules

Bottlenecks at certain operations

Inconsistent utilization of equipment and labor

Large amounts of WIP

These all create inefficiencies that can result in the need for more labor and equipment than should be necessary to produce the amount of product required. If ways could be found to have the product flow smoothly and continuously, the process would operate more efficiently. Some of the methods which have been successful include:

1. WIP control. If work is allowed to move to the next process step regardless of the conditions at that step, WIP will eventually build up in front of the steps that have the slowest throughput or are experiencing problems (Fig. 18-23). This is sometimes referred to as a "push" system, since the product flow is based on pushing the work through the line without regard to the situation ahead of it. On the other hand, if the work is not allowed to move until it is needed at the next step (this could be based on some minimum "buffer" requirement), then excessive buildups

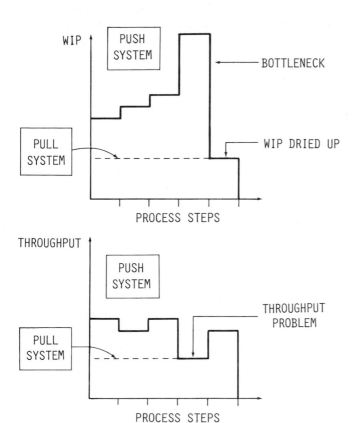

Fig. 18-23 WIP management: "push" versus "pull" systems.

can be prevented. This can reduce inventory, save space, and avoid production bottlenecks. Such an approach to controlling WIP is sometimes referred to as a "pull" system, since the needs of the next operation cause the product to flow.

2. Lot-size control. A line can be scheduled to produce one product at a time to minimize the number of setups required and optimize the utilization of the equipment. To meet a demand which has a mix of products on such a line, however, large inventories must be carried. If the setups can be performed quickly and simply, a mix of products can be run in smaller lot sizes. This results in faster cycle times and lower inventory levels.

3. Materials handling. If small lot sizes can be achieved, it is important to consider the physical movements involved in the flow of the products through the process. The methods and economics of transporting materials between workstations can be very different for large and small lots. The layout of the line may have to be modified to improve the use of space and minimize the amount of movement required.

4. Utilization of equipment and operators. In a small-lot-size, fast-cycle-time operation, it is important to prevent bottlenecks. Flexibility can be built into the manufacturing process in the way in which equipment and operators are used. It is easier to handle variations in quantities, mix, and performance in a process in which equipment can be set up rapidly and operators can run multiple machines.

5. Production scheduling. The easiest way to run a line with small lots is to schedule a consistent quantity and mix of product to be produced every day. This will result in a relatively constant work load and a minimum of delays, bottlenecks, and expediting (e.g., moving priority jobs faster). To create a smooth, continuous flow of product, the parts and material must get to the proper workstation when they are needed—not too early and not too late. This is popularly referred to as "just-in-time" manufacturing. It requires a disciplined approach to production control. One early technique that was used in Japan, even before computer systems, used a reorder card for every workpiece or batch of parts. This system, called "kanban," was a simple way to identify when more parts were needed. Such a manual system can be replaced with a more efficient computer-based scheduling system with the same results. However, another system is required to deliver small quantities of parts rapidly and frequently. Here again, an automated approach to materials handling can be the most efficient solution.

6. Quality control. If there are defects in the parts on the line, there will be rework, scrap, and bottlenecks. It will also be necessary to put in tests and inspections to find the defects. This increases costs and reduces throughput. A quality control system must be put in place to assure that all the parts meet the specifications, with a "zero defects" approach at the source of the parts. It is more effective to prevent the defects from occurring than to inspect them out of the process. However, to assure a foolproof, smooth-running production operation, automated techniques may be used to detect any problems as soon as they occur.

Although these concepts may all sound basic and simple, implementing them takes a lot of work. Inventory has traditionally been used to "protect" a manufacturing operation from running short of supply. It was an "insurance policy" against unplanned events. When inventory levels are first reduced, they will reveal problems in the efficiency of the production operation which will have to be fixed. Problems may include the quality of parts, the performance of equipment, or materials handling techniques. In these cases, changes are usually needed in the management systems used to run the line.

COMPUTER TOOLS FOR IMPROVING PRODUCTION EFFICIENCY

Optimizing a complex manufacturing operation is not easy. Even if the right concepts are adopted, manual methods are usually not practical. There are a number of analytical tools which can be used to solve or prevent production scheduling problems, all of which use computers to perform the computations and handle the data:

1. Linear programming. This is an iterative method of problem solving. The problem is described in mathematical form and different solutions are tried until the one that yields the best results is found. It can be a slow and tedious process, but a computer can do most of the work.

2. Algorithms. These are sets of mathematical equations based on some theoretical solution to a production problem. For example, some algorithms are intended to optimize the scheduling of production operations to minimize the impact of bottlenecks or balance the loading on the line for smoother flow. They may be based on experience, experiments, or theoretical models of production lines.

3. Simulation models. These are computer programs that simulate the operation of a manufacturing line. Each process step is described in terms of parameters that represent its performance (e.g., capacity, throughput, cycle time). Such models can be used to try out different production situations so that their effects on the overall performance of the line can be assessed. Simulations can highlight potential problems or constraints to avoid in production. Simulation techniques are often more practical than algorithms for complex manufacturing operations. It may be easier to describe the process than to develop a theoretical solution. Models also allow different solutions to be compared and sensitivity analyses to be conducted.

4. Decision support systems. This is a form of AI which uses knowledge representation to model the process. Decision support systems are rule-based, expert systems which can be used to either predict or react to the performance of a production operation. Decision support systems are newer tools than the others. Although they can be very useful and efficient, they are only as effective as the "expert" knowledge and decision processes programmed into them.

18-7 SUMMARY

Computer systems are used to automate the management process as well as the physical manufacturing process. Management systems handle the processing and communication of technical, logistical, and administrative data. The technical planning process is the link between development and manufacturing. This process translates product design data into manufacturing process routings and numerical control (NC) tool control programs. Group technology (GT) can be used to reduce the effort involved in process planning as well as to improve the efficiency of manufacturing operations. Computer-automated process planning (CAPP) is an efficient and consistent method of providing the technical data needed to drive the manufacturing process. NC part programming can also be automated. This can reduce the effort and time involved, and it can also allow alternatives to be tried before they are implemented in manufacturing.

Manufacturing logistics systems control the scheduling and movement of products and materials. This is a complex, data-intensive job for large manufacturing operations. Without computer-based systems, this process could not be managed. Materials requirements planning (MRP) systems provide a comprehensive approach to planning, scheduling, and controlling production operations. Shop floor control systems must be able to deal with changes in demand, priorities, and manufacturing performance. Manufacturing resources planning (MRP II) goes beyond the task of scheduling production to tie in the implications of capacity, labor, and costs.

Administrative data systems collect and report information which is required to monitor the performance of the manufacturing operation. Traditionally, such data was recorded manually or entered into separate computer systems. Much information of this type can be obtained as part of a computer-based shop floor data collection system.

Products and processes must be designed to be compatible with automated manufacturing so that the operation can run efficiently. Computer design tools and techniques can be used to do this. Just automating manufacturing is not enough to assure optimum efficiency and productivity. There are a number of management techniques which must also be used to optimize the operation of the factory. In most production operations, there are significant opportunities to increase productivity. Computer tools are available which can help to determine optimum production schedules.

REVIEW QUESTIONS

The answer to each question can be found in the section indicated at the end of the question.

1. Describe the technical planning process for manufacturing. [18-2]

2. Identify some of the benefits of using group technology. [18-2]

3. What are the principal applications of group technology? [18-2]

4. Describe the two basic approaches to automated process planning. [18-2]

5. What are some of the advantages of automated NC part programming? [18-2]

6. Describe the major phases of activity in the logistics process. [18-3]

7. Identify some of the factors that can influence the design and capabilities of a logistics system. [18-3]

8. Define materials requirements planning (MRP) and describe how it works. [18-3]

9. Identify the basic functions of a shop floor control system. [18-3]

10. Define manufacturing resources planning (MRP II) and explain how it differs from MRP. [18-3]

11. Identify some of the typical types of administrative data found in a manufacturing operation. [18-4]

12. Identify some of the factors that should be considered when designing products and processes for automated manufacturing. [18-5]

13. What are some of the measures of manufacturing productivity? [18-6]

14. Describe some of the techniques which can be used to optimize manufacturing productivity. [18-6]

15. Identify some of the computer tools which can be used to optimize production schedules. [18-6]

INTEGRATED MANUFACTURING SYSTEMS

19-1 INTRODUCTION

"Integration" means combining the pieces to make the whole. To automate manufacturing, we need to integrate all the "pieces" of the manufacturing system. Automation without integration may not reduce the cost of manufacturing significantly. If we automate only the operations performed by direct labor, we affect only a small portion of the total product cost. The largest part of product cost is driven by indirect activities which show up as overhead expenses (Fig. 19-1). These are the areas where technical, logistical, and administrative data systems can improve productivity. Unless these systems are tied together, along with the physical automation of the manufacturing process, the effectiveness of the overall operation will be limited, and it may even be destined to fail.

If computer-automated manufacturing (CAM) is to succeed, we need to integrate the key elements of the system (Fig. 19-2). This means tying the automated manufacturing process together with the automated materials handling system and the automated shop floor control system. This approach to the full automation of production operations can lead to the ultimate factory of the future. A "peopleless," "lights-out" factory is only possible if all the automated systems are integrated.

This chapter will address the key elements in the process of integrating manufacturing systems. We must start with the data that drives an automated manufacturing operation. An integrated database management system must be established as the basic element which ties the pieces together. An automated data collection system is then required to provide the data to that database. An automated materials handling system is needed to provide the materials and parts that keep an automated production operation moving. With these elements, we can begin to develop an integrated systems approach to automated manufacturing. These basic ingredients can be found in a "flexible manufacturing system" (FMS), which is a type of relatively small automated manufacturing operation initially developed and implemented by the machining industry.

To achieve CAM or computer-integrated manufactur-

Fig. 19-1 Typical elements of manufacturing cost.

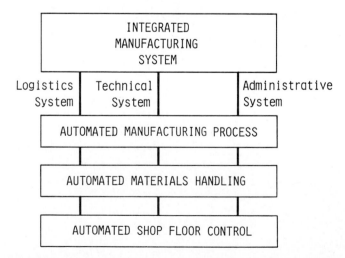

Fig. 19-2 Integrating the key elements of the manufacturing system.

ing (CIM), we need to use all the computer tools that have been discussed in this book. There are enough available today to do the job. They just need to be put to work—together. The integration process can be evolutionary. It starts with tying development to manufacturing. This requires a common database for the exchange of information. Then computer tools can be added to improve the productivity of activities associated with the manufacturing operation.

19-2 INTEGRATED DATABASE SYSTEMS

WHY DO WE NEED INTEGRATED DATABASES?

The key to integrating an automated design system with an automated manufacturing process is a well-structured common database. Both development and manufacturing rely on the frequent use of large amounts of information throughout the process, from the initial design of a product to its production (Fig. 19-3). Without a common database to share information, data must be recreated and reinterpreted at every stage of this process. Such duplicated effort can be avoided by organizing and storing the information needed by development and manufacturing in a common database. This may include information about:

Product design (e.g., geometry and dimensions)

Bills of materials (e.g., part numbers and parts lists)

Manufacturing (e.g., routings and schedules)

A major portion of the activity of development and manufacturing personnel is involved in searching for and communicating such information. If data is shared and readily accessible, the productivity of those people can be increased significantly as a result of the following benefits:

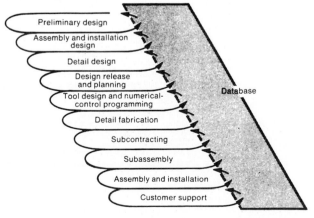

A master concept for a database encompasses manufacturing activities from preliminary design to customer support. Each activity can draw from the database and also contribute to and modify it.

- Preliminary design
- Assembly and installation design
- Detail design
- Design release and planning
- Tool design and numerical-control programming
- Detail fabrication
- Subcontracting
- Subassembly
- Assembly and installation
- Customer support

Database

Fig. 19-3 Providing a common database for development and manufacturing. *(William O. Beeby, "The Heart of Integration: A Sound Data Base," IEEE Spectrum, May 1983; © 1983 IEEE)*

☐ Eliminating duplicate and unnecessary efforts

☐ Reducing errors

☐ Increasing the speed of information exchange

☐ Providing more information than was previously available

☐ Improving the control of data (e.g., providing a single source, controlling changes, controlling data security)

☐ Preventing the use of outdated information

There are obviously many good reasons for having an integrated database, but it is not so easy to implement one. Establishing a database requires a substantial effort and may involve as large an investment as the physical automation of the manufacturing process.

REQUIREMENTS OF AN INTEGRATED DATABASE SYSTEM

The basic functions involved in a database system, relative to the information required by its users, are:

1. Defining the data
2. Inputting the data
3. Locating the data
4. Accessing the data
5. Communicating the data
6. Revising the data

Requirements that should be built into the design of such a system include:

Control of data integrity. Changes to the data must be controlled to ensure that unauthorized or inadvertent alterations or deletions are prevented.

Control of data security. Access to critical data must be controlled to prevent unauthorized use or loss.

Data storage protection. Provision must be made to protect the data from being lost in the event of system interruptions or failures.

System performance. The response time must be fast enough to make the system useful and convenient.

System functions. User requirements such as text handling, graphics, computation, and friendly interfaces need to be satisfied. To serve the needs of a broad base of users, the system must provide a variety of functions that are easy to use. Inexperienced users, in particular, need graphics and menu-based, natural language–like interfaces.

System compatibility. The system should be able to deal with a variety of different types of hardware, software, data, and users.

These requirements are not easy to satisfy. In a large and complex manufacturing operation, such a database may itself become very large and complex. The challenge is to make such a complex system easy to use and reliable.

HOW DO DATABASE SYSTEMS WORK?

The design of a database system depends on the nature of the computer environment it must operate in as well as the types of data and applications involved. The simplest environment has only compatible hardware and software in the system. This typically means that all the hardware comes from one manufacturer and uses a single operating system software package. In most large operations, this is not practical or even desirable. The greatest capability and flexibility exists in systems which use a variety of different hardware and software. To operate in an incompatible environment, the database system must be designed in such a way that individual systems can operate in their own environments and yet communicate between each other. Such a system can be made up of a number of subsystems for each major application area, each subsystem having its own database (Fig. 19-4). Subsystems can communicate through a central system which translates the data into a neutral format that can then be exchanged with other systems.

Integrating multiple systems requires a central system architecture with the following characteristics (Fig. 19-5):

1. A common communication network
2. A common communication language
3. A common database management system
4. Standard interfaces

Figure 19-5 illustrates an example of a database system which communicates over a broadcast network. The computer system at each node includes communication control, program control, and data administration functions. A number of manufacturing control processes can be linked to each node.

The network physically ties the systems together so that they can communicate. Each system has hardware and software associated with a particular application, such as process control, which communicates with the network through a communication controller. The data processing activities on each node are managed by a

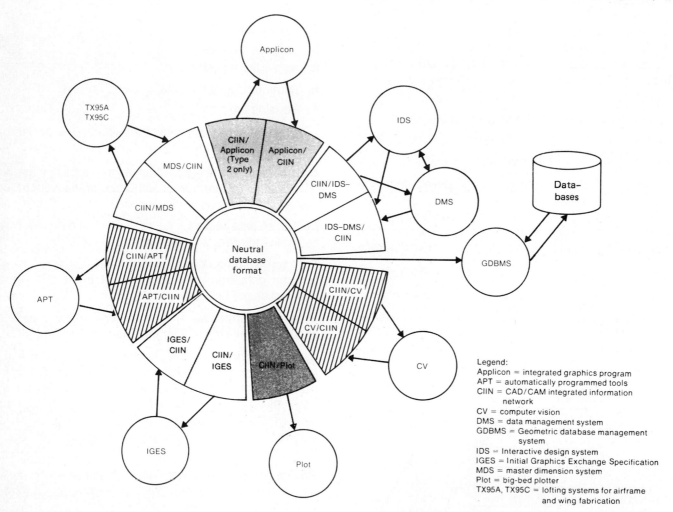

CAD/CAM integrated information network (CIIN) in use at the Boeing Commercial Airplane Co. consists of a variety of design and manufacturing hardware (outer circles) that communicates via translators (wedges) with a neutral database (center circle). Data entering the neutral base is in the native format of the source system; data leaving the neutral base is in the format of the destination system. The neutral database has access to other databases—design, manufacturing, marketing, and so forth—through a geometric database management system.

Fig. 19-4 Structure of an integrated database system. *(William O. Beeby, "The Heart of Integration: A Sound Data Base," IEEE Spectrum, May 1983; © 1983 IEEE)*

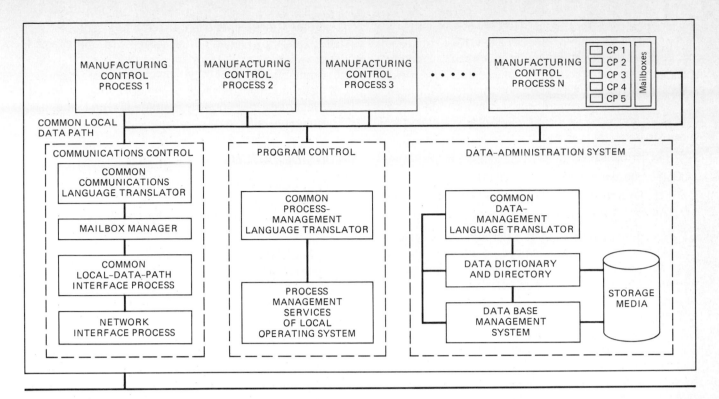

Fig. 19-5 Architecture of a database system. *(Charles McLean et al., "A Computer Architecture for Small-Batch Manufacturing," IEEE Spectrum, May 1983; © 1983 IEEE)*

program control function. The task of organizing data into a structured format so that it can be exchanged in a database system is called "data administration." The system requires several features to perform this function:

1. Data dictionary and directory system. This provides a standard structure for defining and indexing data.

2. Language translator. This performs the data manipulations necessary to transform data into a standard format which can be communicated between systems.

3. Database management system. This performs the actual allocation, storage, and retrieval of data by managing the computer resources in the system. It must also interpret the structure of the data being handled.

A typical transaction may involve the following steps:

☐ A request for information is received by the data administration system on the local node

☐ The language translator uses the local data directory to determine whether the request can be satisfied from the local database

☐ If the request can be satisfied, it is translated and forwarded to the local database management system

☐ If the request cannot be satisfied, it is communicated on the network to another node where it is handled by its data administration system

Two basic types of database management systems can be used:

1. Hierarchical system. This is an efficient approach in a structured environment. If the nature of the requests for information is known ahead of time, the system can be designed to sort the data hierarchically. This lends itself to applications such as financial or production control systems.

2. Relational system. This approach is more efficient in an unstructured environment, such as an engineering system. Since user requests are not always predictable and may change, the system must search for information on the basis of the relationships of the data to the information requested.

IMPLEMENTING AN INTEGRATED DATABASE SYSTEM

Three key elements are involved in implementing an integrated database system:

1. An organized and disciplined approach to storing and retrieving information. Group technology (GT), for example, offers a way to classify and code parts that can provide an information link between many applications in development and manufacturing. GT also makes it easier to adopt a data structuring scheme because decisions can be made for entire families of parts that have already been classified together. A GT database can be a useful tool for reducing costs by promoting the use of standard parts and processes.

2. A standard format for data. It is necessary to structure data into a standard or neutral format so that it can be communicated and exchanged between systems. The Initial Graphics Exchange Specification (IGES), for example, was developed by the National Bureau of Standards for the exchange of product definition data (e.g., design data, bills of materials, product documentation). It uses a standard data-file structure which is independent of the hardware and software being used. A typical data file is made up of the following sections:

START (for messages)

GLOBAL (for identification information)

DIRECTORY and PARAMETER (for design data)

TERMINATE (to signal the end of the file)

3. Procedures which support automated data systems. Establishing these procedures can be a very difficult job even though it does not directly involve the database system itself. All of the routine activities involved in the design and manufacture of a product have to be reviewed to determine how they may affect the implementation of an integrated database system. In many cases, these procedures will have to be changed to make them compatible with the needs of the system. If the controls and discipline required are not built into the management process, the system will not work.

19-3 DATA COLLECTION SYSTEMS

DATA COLLECTION REQUIREMENTS

The management systems covered in Chap. 18 were all dependent on data obtained from the factory floor. In order to control a manufacturing operation, accurate and prompt feedback on actual performance is needed. Data collection systems must therefore be put in place to provide information about the activities and results of production operations. There are two primary reasons for collecting manufacturing data:

1. To identify the status of manufacturing resources (e.g., equipment, people, or product)

2. To track manufacturing activities (e.g., production operations or interruptions)

Such data collection systems must perform the following basic functions:

1. Obtain the data from the appropriate source

2. Organize the data into useful forms (e.g., by sorting, combining, or summarizing)

3. Integrate the data into an accessible database

4. Communicate the data to users

In performing these functions, a good data collection system must be reliable, accurate, efficient, and easy to use. In the past, manual techniques were used; they were time-consuming, inefficient, and subject to errors. They relied on such things as time cards and production logbooks to record the data. The data was then collected from these documents and entered into a computer system (usually by key entry) for compilation and analysis. Reports would then be formatted and generated for management (often as hard copy). In an automated manufacturing environment, this type of process cannot be tolerated. Data must travel quickly from the source to the user. This requires automated techniques for data collection.

CONSIDERATIONS

Several factors must be considered in determining the type of data collection system to use. They may influence the requirements or constraints for the system, depending on the nature of the manufacturing operation:

1. Manufacturing environment. The characteristics of the production operation involved will have a significant bearing on what data gets collected and how it can be collected. These characteristics include:

☐ The quantities and production rates

☐ The variety of products and materials

☐ The complexity of the manufacturing process

☐ The physical layout of the production floor

☐ The organizational responsibilities in manufacturing

2. Data relationships. What data gets collected and how it is structured depends on the relationships that are important to the ultimate user of the data. For example, it may be necessary to be able to identify the specific machine and operator which processed a production job.

3. Physical environment. The conditions in manufacturing may determine requirements or limitations on the type of data collection system that is used. For example:

☐ Clean rooms may not permit sources of contamination, such as paper and pencils

☐ Dirty processes that generate grease, dirt, or soot may not be suitable for systems that rely on readable or legible data

☐ Particulates, such as smoke, carbon, or metal shards, can affect the operation of some equipment

☐ Shock and vibration will require rugged equipment

☐ Electromagnetic fields from electrical equipment can interfere with the operation of some types of equipment

TYPES OF DATA COLLECTION SYSTEMS

A number of different techniques can be used to collect data from the factory floor. Any automated or even semi-automated approach will have advantages over manual methods. Automated methods tend to be faster and more accurate, and they require less effort. The choice between different types of data collection systems depends upon the requirements and constraints of the application, such as accuracy, frequency of data collection, and cost, as well as the considerations discussed above. Following are the most common types of systems used in manufacturing:

1. **Key entry.** This is a simple way to eliminate paper as the intermediate form of recording data before it is entered into a computer system. Terminals, keyboards, or special keypads can be placed at workstations on the factory floor for data entry by operators. Key entry eliminates one step in the data collection process and provides a great deal of flexibility in the format in which data is collected. However, it still requires operator time and is susceptible to errors.

2. **Optical scanners.** These are machines that can read forms that are filled out by operators. They may be optical character readers (OCRs), which can read numbers and letters that are neatly written in a predetermined format (which also has the advantage of being readable by humans). A mark or dot form that can be filled in with pencil to represent numerical data can also be used. Such a form can be scanned quickly by a machine to record the data. Both approaches still require manual entry and may not be practical in some manufacturing environments (e.g., clean rooms or dirty processes).

3. **Bar-code readers.** Information can be coded on a label in the form of bars and spaces which can be read by a machine. A standard code is used throughout industry to represent numbers, which are usually also printed on the labels (Fig. 19-6). The coded labels may be applied to products or workstations as a means of identification. Operators can use a hand-held scanner to read the label and enter data automatically into a computer system. Scanners can also be installed on materials handling systems to automatically read the labels as products flow past them. These are very efficient and accurate methods to collect data with minimal operator effort. However, the data is limited to the fixed field of the bar code.

4. **Magnetic-stripe readers.** Data can also be recorded on a thin stripe of magnetic tape, in much the same manner as tape drives are used to store large quantities of information in computer systems. These magnetic stripes can then be applied to products, workstations, or even the identification badges of operators and then used in a similar manner to the bar-code labels. Hand-held magnetic wands can then be used to scan and read the stripes and automatically enter the data into the system (Fig. 19-7). Cards or badges with magnetic stripes can also be read by passing them through a slot in a machine with a magnetic reader. This approach has the same advantages in speed and efficiency as the bar code. It also has the capability to store more information and is more reliable in a dirty environment. The stripes, however, do not provide the added feature of being readable by humans.

5. **Voice recognition systems.** A relatively new technique allows operators to record data by talking directly to the computer (Fig. 19-8). Voice recognition systems can translate a limited vocabulary of numbers and words into digital signals that can be transmitted to a computer. Such systems can be used in applications where operators need to use their hands to perform other tasks while they are recording data (e.g., inspection, materials handling, testing, assembly). This can be an efficient approach for such applications, but it is limited in its capabilities and is still relatively expensive.

6. **Direct data collection.** The most efficient and accurate method of collecting data in manufacturing is having the machines and instruments on the factory floor communicate directly with the computer system. Not only can the operation of tools and processes be controlled by computers; their performance can be monitored as well. By using sensors and digital and analog recording techniques, data can be obtained directly and immediately from the process. This avoids all the potential cost, time, and errors involved in any intermediate recording technique. In a truly automated factory, this should be the principal method of data collection.

19-4 MATERIALS HANDLING SYSTEMS

WHY IS MATERIALS HANDLING IMPORTANT?

Materials handling is a major part of any manufacturing operation. The materials, parts, and products in a process

Fig. 19-6 Bar-code reading system. *(InstaRead Corp., a Rexnord Co.)*

Fig. 19-7 Magnetic-stripe reader. *(International Business Machines Corp.)*

Fig. 19-8 Voice recognition system. *(Westinghouse Electric Corp.)*

all have to be moved, loaded, unloaded, packed, and stored, often many times. These materials handling activities can consume a lot of labor, space, expense, and production time. The actual amount of time a product spends being manufactured is often only a small portion of the total cycle time of the process—in some cases, less than 10 percent. The rest of the time, the product is either moving or sitting idle. Work-in-process (WIP) inventory

can tie up a lot of assets. For manufacturing processes that have long cycle times and expensive products, the value of WIP may be as significant as the capital invested in the line. A reduction in WIP and an increase in the inventory turnover will reduce the amount of capital tied up in inventory. It will also reduce storage space.

Efficient materials handling systems can help to keep WIP moving and minimize the amount of time and space wasted when it is sitting idle. Even the earliest attempts to automate industrial processes recognized the importance of efficient materials handling (Fig. 19-9). When parts and materials are brought to the operator and the production machine, manufacturing time is not wasted by waiting. In some operations, materials handling and storage is the largest part of the product cost.

In an automated manufacturing operation, it is essential that the materials handling system also be automated. Otherwise, it would not be possible to keep the automated tools fully utilized and the product flowing smoothly. In addition, an automated process cannot be totally controlled unless there is a materials handling system that is integrated into the factory floor control system. An integrated materials handling system is made up of the following basic elements (Fig. 19-10):

1. Storage systems for large-scale bulk storage (e.g., warehouses) as well as small, in-line buffer storage

2. Transport systems to move parts and products from the storage systems to the production operations

3. Materials transfer systems to load and unload the work from production machines

These individual materials handling systems may each be automated, but they must be tied together by a factory control system to achieve an integrated manufacturing operation.

Fig. 19-9 Early approach to efficient materials handling in industrial automation: automobile manufacturing. *(Ford Motor Co.)*

Fig. 19-10 Integrated materials handling system. *(Joe Smith, "Our Flexible Systems Help Make a Better Product," Modern Materials Handling, July 1984; Copyright Cahners Publishing Co., Division of Reed Holdings Inc.)*

Using a computer-based production control system

Integrating the production control system and the materials handling system

Using the materials handling system to collect data on factory floor operations

Laying out the manufacturing process along with the materials handling system

Minimizing the storage of materials in process

Using the shortest paths between processes

Standardizing transportation and storage containers

Minimizing the frequency of loading and unloading operations

Improving the utilization of space by using "air space" for storage and handling (e.g., overhead conveyors, stacker cranes)

The overall effectiveness of a materials handling system should be judged on the basis of its contribution to the performance of the manufacturing operation. This can be measured in terms of cycle time, throughput, and WIP.

GENERAL CONSIDERATIONS

Many different approaches can be taken to automate a materials handling system. Before selecting one for a particular application, a number of factors should be considered:

1. Application considerations. The characteristics of the production operation involved will have a significant influence on the nature and requirements for a materials handling system. Some of the factors to be considered include:

The nature of the products and materials

The quantities and lot sizes

The cycle times of each operation

The throughput rates of each operation

The source of the materials

The layout of the facility and production line

2. Evaluation criteria. Alternative approaches should be considered and evaluated on the basis of their ability to meet the needs of the application. Some of the criteria which should be considered in comparing alternatives include:

The quantities and volumes that can be handled and stored

The time required to transport materials between operations

The method of transportation

The capacity of storage for WIP and inventory

The space consumed by materials handling and storage

The cost of operation and maintenance of the system

3. Optimizing the system. For any one production application, there may be several approaches to materials handling that will work. However, a number of techniques can be used to improve the efficiency of most systems:

AUTOMATED STORAGE SYSTEMS

Even with the most efficient manufacturing operation, there will always be a need for some inventory. This inventory must be stored at three different stages of the production operation:

1. Bulk storage of parts and materials to feed the manufacturing line

2. In-line storage to feed individual production operations

3. Bulk storage of finished goods at the end of the line

The general solution to automating these types of storage requirements is called an "automated storage and retrieval system," or ASRS. This includes several types of systems to satisfy storage applications of different sizes:

1. Stacker cranes. These are mechanisms which store and retrieve materials in high racks or "stacks" in a warehouse (Fig. 19-11). They use air space to provide higher-density storage than could be achieved by conventional warehousing techniques (e.g., forklift trucks).

2. Narrow-aisle equipment. This is a type of stacker-crane system that is designed to operate with narrow spacing between the rows of storage shelves to achieve high storage density (Fig. 19-12).

3. Miniload equipment. These relatively small automated systems are used to store and retrieve materials or small parts. They may be located in a warehouse or within a production line. Miniload equipment usually takes the form of a "carousel" of bins which store parts or subassemblies (Fig. 19-13). When the material is required, the system rotates the carousel to a position where a "picker" mechanism (such as a robot arm) retrieves it.

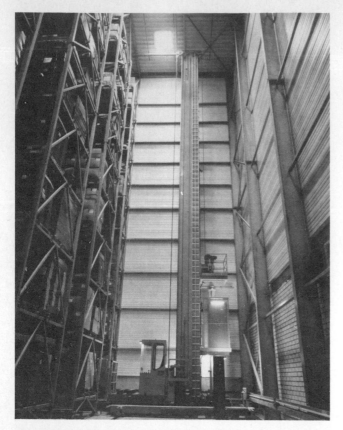

Fig. 19-11 Automated storage and retrieval system (ASRS) stacker crane. *(Clark Equipment Co.)*

Fig. 19-13 Mini-load equipment. *(Clark Equipment Co.)*

Fig. 19-12 Narrow-aisle equipment. *(Litton Industrial Automation Systems Inc.)*

ASRS systems can be scaled to a wide range of sizes to fit the needs of any production operation. They can be designed to interface with other automated materials handling systems which deliver the material directly to the

manufacturing line (Fig. 19-14). Some of the advantages of such systems include:

☐ Efficient use of storage space

☐ Secure control of the inventory

☐ Rapid delivery of materials

☐ Automatic location and tracking of inventory

☐ Elimination of errors in the storage and retrieval of materials

☐ Operation in environments which are undesirable for people (e.g., hot or cold warehouses, clean rooms, high altitudes, small spaces, heavy loads)

Fig. 19-14 ASRS equipment delivering parts directly to a manufacturing operation. *(Litton Industrial Automation Systems Inc.)*

Fig. 19-15 Conveyor transport system. *(SI Handling Systems Inc.)*

AUTOMATED TRANSPORT SYSTEMS

Materials can be automatically moved between storage systems and production operations by several different types of transport systems:

1. **Conveyors.** These were the earliest and still are the most common type of transport system. They provide a link between delivery points with a moving mechanism that the material is placed on (Fig. 19-15). Several types of conveyor systems are used:

☐ Continuous belt

☐ Roller

☐ Track

☐ Overhead trolley

☐ Air tables

2. **Carts.** These are vehicles that move between fixed positions on the factory floor. They are often used to transport heavy workpieces between machines (Fig. 19-16). Such carts usually move on rails and are pulled by a towline or continuous chain below the floor.

3. **Automatic guided vehicles (AGVs).** The AGV is a relatively new class of transport system. AGVs are vehicles or carts that can move in complicated and changing paths throughout the production floor (Fig. 19-17). The vehicles can be designed to handle a wide variety of materials and loads. They provide more flexibility than any of the other types of transport systems. The vehicles can be designed to include mechanisms for loading and unloading the materials, lifting heavy loads, and interfacing with production equipment. Three basic techniques are used to automatically guide such vehicles:

☐ Buried wires in the floor can be electrically detected by the vehicle

☐ Painted lines on the floor can be detected by a photoelectric sensor on the vehicle

☐ Computer-controlled guidance systems with wireless communication links to the AGVs

AGVs can also communicate with factory computer control systems, either by making a physical link at each

Fig. 19-16 Towline cart transporting heavy workpieces between machine tool stations. *(Hughes Aircraft Co.)*

Fig. 19-17 Automatic guided vehicles (AGVs) receiving chassis assemblies from automatic loading devices. *(Donald Downie, "Automatic Guided Vehicles Move Into the Assembly Line," Modern Materials Handling, Jan. 1985; Copyright Cahners Publishing Co., Division of Reed Holdings Inc.)*

workstation that the AGV stops at or by infrared or digital radio transmission. This provides a means to track and control the flow of materials, control the operation of production machines, and alter the instructions to the guided vehicle.

MATERIALS TRANSFER SYSTEMS

Once material is delivered to a workstation or production machine, it must still be moved into the proper position for that particular manufacturing operation. This is typically a loading and unloading task. Depending on the nature of the process and the machine involved, the transfer task may include buffer storage, lifting, and alignment or orientation requirements. The most common approaches to automating such tasks use either a fixed transfer mechanism (e.g., tables, shuttles, elevators) or robot arms (Fig. 19-18). Robots are more flexible in such applications for a number of reasons:

Fig. 19-18 Robot materials transfer system. *(Clark Equipment Co.)*

☐ Robots can handle a wide variety of loads

☐ Robots can perform accurate and delicate movements

☐ Robots can be reprogrammed easily

☐ Robots can be tied directly to the computer systems that control the process and the logistics

☐ Robots can perform multiple tasks

However, robots are usually limited in their capabilities by a relatively small work envelope and a lack of mobility. Some applications may therefore use a combination of hard and flexible automation techniques to perform these tasks.

The interface between the materials transfer system and the production operation can be very critical. It is therefore essential that the requirements of the application be identified and understood, such as:

Accuracy and repeatability

Orientation of workpiece and work position

Size, shape, and weight of workpiece

Relative movements between transfer system and workstation

Speed and throughput of operation

To assure efficient and trouble-free operation, several factors should be considered in the design of such a system:

☐ The workpiece carrier can include fixturing which will permit it to be worked on without being removed

☐ Orienting the workpiece on the carrier can avoid the need to orient it when transferring it to the production machine

☐ The movements of the materials transfer system must be synchronized with the timing of the manufacturing process to avoid delays and minimize queues

☐ Any critical surfaces (e.g., machined surfaces) should be protected during handling to avoid scratches and damage

☐ Flexibility should be built into the programming of the transfer actions to permit potential changes

19-5 FLEXIBLE MANUFACTURING SYSTEMS

WHAT IS AN FMS?

An FMS is a form of CAM. It is an integrated approach to automating a production operation. Although FMS's do not encompass an entire manufacturing line, they include a series of fabrication steps. Therefore, they are more complex than an automated manufacturing cell or module, which typically include between one and three machines. The primary characteristic of an FMS is that it is a computer-controlled manufacturing system that ties together automated production machines and materials handling equipment. The FMS is designed to be flexible so that it can fabricate a variety of different products in relatively low volumes. There are three major automated subsystems in an FMS:

1. Computer-controlled production equipment (e.g., numerically controlled machine tools)

2. Automated materials handling system (i.e., storage, transport, and transfer)

3. Manufacturing control system (including both tool and logistics control)

Some FMS's may have additional subsystems. For example, in a machining application there may also be systems for storing and retrieving tools, disposing of chips and cutting fluids, and inspecting workpieces. These subsystems must be tied together to achieve an integrated manufacturing operation.

Although FMS's were initially developed for machining applications, the concept has also been used in a variety of other manufacturing applications, such as:

Assembly

Semiconductor processing

Plastic molding

Sheet-metal fabrication

Welding

Such systems have proved to be practical and economical for applications with the following characteristics:

Similar types of equipment and processes

Families of similar parts

A moderate number of tools and process steps

Low to medium quantities of parts

Moderate precision requirements

FMS's are an automated approach to batch manufacturing operations. They are an alternative that fits in between the manual job shop and hard automation (Fig. 19-19). FMS's are best suited for applications that involve an intermediate level of flexibility and quantities. The job shop is best suited for very small quantities of many different types of parts. Since the job shop must provide the greatest degree of flexibility, most of its operations are

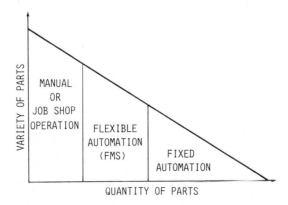

Fig. 19-19 Application of flexible manufacturing systems (FMS's).

manual. Hard automation with dedicated equipment is best suited for the production of very large quantities of identical parts. A large portion of the manufacturing industry involves the intermediate level of batch operations that lend themselves to the FMS approach.

ADVANTAGES OF AN FMS

FMS's are designed to provide a number of advantages over alternative approaches:

Reduced cycle time

Lower WIP inventory

Low direct labor costs

Ability to change over to different parts quickly

Improved product quality (due to consistency)

High utilization of equipment

Reduced space requirements

Ability to optimize loading and throughput of machines

Expandability for additional processes or added capacity

Reduced number of tools and machines required

Motivation for designers to limit variations and features

Some of these advantages can lead to significant savings. Direct labor can be eliminated almost entirely. Cycle time and WIP can be reduced to a fraction of what is normally experienced in a manual operation. An FMS is designed to have the production machines working most of the time rather than standing idle. An automated material handling system and a computer-based production scheduling system are needed to keep the machines fed with parts. FMS's use computer automation techniques to lower the overall cost of production operations.

FEATURES OF AN FMS

Each of the major subsystems in an FMS performs a number of functions and is dependent on the others to make the entire system work. The functions will vary, depending on the type of equipment and manufacturing operations involved. Following are some of the features of a typical FMS machining system (Fig. 19-20):

1. Production equipment. FMS's typically have a number of machining centers to provide general purpose machining capabilities. Machining centers offer the greatest flexibility, since they can perform many different machining operations (e.g., milling, drilling, boring). This is made possible by a tool-changing system that either is built into or supports the machining center. A part can therefore undergo multiple machining processes at a single workstation. Special purpose machines may also be included in the FMS to perform operations which are unique or require more efficiency (e.g., turning, grinding). The family of parts which the FMS is designed to produce will determine the capabilities required from the machine tools (e.g., accuracy, size, power).

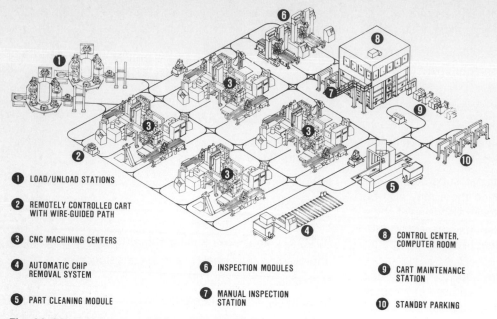

1. LOAD/UNLOAD STATIONS

2. REMOTELY CONTROLLED CART WITH WIRE-GUIDED PATH

3. CNC MACHINING CENTERS

4. AUTOMATIC CHIP REMOVAL SYSTEM

5. PART CLEANING MODULE

6. INSPECTION MODULES

7. MANUAL INSPECTION STATION

8. CONTROL CENTER, COMPUTER ROOM

9. CART MAINTENANCE STATION

10. STANDBY PARKING

Fig. 19-20 Features of an FMS machining system. *(Cincinnati Milacron)*

2. Support systems. Automated machine tools typically require several systems to support their operation. The tools required to perform the multiple processes of a machining center may be stored in magazines at each machine or in a central tool room. Local magazines provide fast access as well as backup capability (Fig. 19-21), but in a large FMS a central tool facility may be more efficient. Centralization not only permits the total number of tools to be minimized; it also provides the opportunity to perform additional functions automatically, such as:

☐ Measuring tools for wear

☐ Preprogramming for tool offsets on the basis of tool measurements

☐ Tool maintenance and repair

☐ Replacement of broken or worn tools

Many automated machine tools have built-in systems to monitor tool wear and detect tool breakage. They may use probes or noncontact techniques such as acoustic emission. When a tool needs replacement, the machine can signal the tool room for the delivery of a replacement. This may be performed by an AGV or cart.

Fig. 19-21 Large tool-changing magazine. *(Cincinnati Milacron)*

Automated machining operations also need to have the chips cleaned off the workstation and the workpiece. This may be performed by robots or special cleaning stations. Cleaning may involve turning the workpiece over, vacuuming, and washing.

3. Materials handling system. An FMS typically needs several types of materials handling systems to service the machines (Fig. 19-22):

☐ A transport system to move workpieces into and out of the FMS (e.g., overhead conveyors, AGVs)

☐ A transport system to move workpieces between machines within the FMS (e.g., conveyors, carts, AGVs)

☐ A buffer storage system for queues of workpieces at the machines (e.g., conveyor loops)

☐ A transfer system to load and unload the machines (e.g., robots)

For these systems to work effectively, they must be synchronized with the machine operations. The location and movement of workpieces must be tracked automatically. This is done by using sensors on the materials handling system and workstations. These may be either contact devices (e.g., switches) or noncontact devices (e.g., optical or proximity devices).

4. Computer control system. An FMS must be under the control of an integrated computer system that includes:

☐ Machine tool controllers

☐ Support system controllers

☐ Materials handling system controller

☐ Monitoring and sensing devices

☐ Data communication system

☐ Data collection system

☐ Central FMS computer

This control system must also tie into other computer systems in the factory. It needs access to data from other systems and provides data to them. In particular, the FMS system must communicate with the following systems:

☐ The design system which generates numerical control (NC) programs for the machine tools

☐ The shop floor control system which schedules loading and routing of the work

☐ The administrative system which provides management reports on the performance of the system

The FMS computer system and the hierarchical manufacturing control system integrate all the pieces of the FMS. Without the ability to tie the various controllers together and exchange information between systems, such an automated operation would not be possible.

APPLICATION CONSIDERATIONS

Following are some approaches which should be considered in order to optimize the overall efficiency and effectiveness of an FMS:

1. Minimize the process cycle time. The number of different jobs must be minimized in order to reduce changeovers.

2. Maximize the utilization of each machine. This can be done by balancing the work load in the system.

3. Use automated storage systems to keep work ready for machines to process. This can include loading the system for automated operation on off-shifts and weekends.

4. Provide for the detection of errors or problems. This includes the detection of the presence and absence of parts, jams, tool wear, machine failures, and so on.

5. Build in backup capabilities. The system should be able to run even when failures occur (e.g., use spare tools, isolate machines, have alternative materials transport paths, have additional machine capacity).

Fig. 19-22 Materials handling in an FMS machining system. *(Cincinnati Milacron)*

6. Include automatic measurement and inspection techniques. These control the process and assure product quality.

7. Use identification marking techniques. This permits automatic tracking of workpieces and tools.

A great deal of effort is required to implement FMS's. They are complex systems that require careful planning and thorough preparation. Some of the major tasks are:

☐ Selecting a family of parts that is both similar in design as well as business needs

☐ Specifying the capabilities and performance requirements of the system

☐ Evaluating the business case for the system

☐ Establishing an experienced team to develop the system

☐ Determining the size and complexity of the system

☐ Simulating the performance of the system

☐ Evaluating and selecting the equipment required

☐ Developing the control systems

☐ Preparing for the installation of the system

☐ Selecting the team to run the system

☐ Measuring the performance of the system

19-6 INTEGRATING MANUFACTURING SYSTEMS

WHY SHOULD WE INTEGRATE THE MANUFACTURING SYSTEMS?

We have seen that automation can be introduced into all phases of manufacturing activity. It can take the form of automating the operation of an individual tool, or it can be a complex FMS. Although these and other examples of automation can be productive and beneficial to manufacturing, they do not represent a computer-integrated factory. The concept of integrated manufacturing systems goes beyond implementing "islands of automation" in the factory. It ties all the design and manufacturing systems together into an overall factory management system. This is a step beyond the FMS: an integrated system is larger, more complex, and broader in scope. The question is not whether it is possible to do, but whether it is worth doing.

Most attempts at computer integration are limited in scope. They often involve linking only two major functions. An example of this would be tying a design system to a manufacturing system through the generation of NC programs. Although this represents an integration of systems, it is only a narrow slice of the manufacturing activities that can be automated. Many such systems can exist in a factory which would be considered by most people to be automated. Without the integration of those systems, however, the overall efficiency of the operation is suboptimum.

When systems are not integrated, they cannot exchange data automatically. If there is a need to exchange information between systems or tie the information from several systems together, it must be done manually. This requires data collection, reentry, analysis, and interpretation. Each of these steps is subject to inefficiencies and error. When paper is involved in this process, it limits the amount of information that can be transferred, as well as its accuracy and efficiency. When automated systems are tied together into an integrated manufacturing system, benefits to more than the individual automated functions can be realized. These typically include:

☐ Reduced labor requirements (for both direct and indirect activities)

☐ Reduced lead times in both the development and manufacture of products

☐ Increased flexibility in production capacity and scheduling

☐ Reduced levels of inventory in materials, WIP, and finished goods

☐ Increased utilization of resources (e.g., equipment, facilities, and labor)

☐ Improved ability to respond to changes (e.g., product design, demand, mix)

Obviously, these all sound too good to ignore. However, integrated manufacturing systems are not easy to implement. They can be extremely complex for large manufacturing operations, and developing them takes a substantial amount of time and effort. This is typically only feasible if it is an evolutionary process. One must start by understanding and planning the three basic ingredients to be integrated:

1. The tasks to be automated

2. The computer tools to be used to automate those tasks

3. The system architecture that will tie the computer tools together

MANUFACTURING TASKS TO BE AUTOMATED

An integrated manufacturing system must both automate and link together the basic tasks involved in the manufacturing process. These tasks fall into three major areas of activity (Fig. 19-23):

1. Product design. This includes defining the geometry and specifications for parts as well as the bills of materials for products. It is not only an activity at the beginning of the manufacturing process; design continues during the manufacturing life of a product. Engineering changes are made to correct, modify, and improve designs after they have been introduced into production. This design information is a key source of the data which drives the manufacturing process.

2. Manufacturing planning. Before a product can be manufactured, a great deal of activity is required to prepare manufacturing for it. This must start with planning

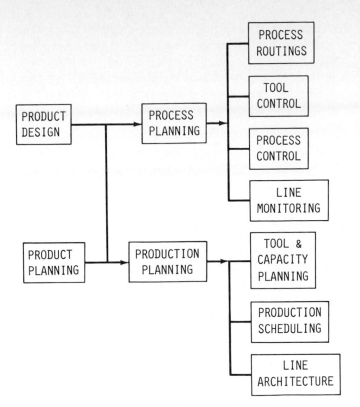

Fig. 19-23 Manufacturing tasks.

the manufacturing line itself, including defining the tools and capacity required as well as the architecture of the line. The process must then be planned, which results in process routings, operator instructions, and tool control programs. Like design, these are dynamic elements. They will continue to change as the product, process, and tools change.

3. Manufacturing execution. The most visible part of a manufacturing system is the actual production. The tasks involved here are primarily some form of control. The processes, tools, and production schedules must all be controlled by technical and logistical data systems. In addition, the performance of the manufacturing operation must be monitored and measured. This involves data collection and reporting activities.

These are the general areas of activity that one finds in any manufacturing operation. There are many detailed tasks below this level as well as activities that may be unique to a particular type of manufacturing. It is the sum total of these activities that must be tied together to achieve an integrated manufacturing system.

COMPUTER TOOLS FOR MANUFACTURING

Throughout this book, we have been introduced to a wide variety of computer tools which can be used to automate manufacturing tasks. An integrated manufacturing system is possible if these tools are applied wisely and planned so that they can ultimately be tied together. The major types of computer tools available for manufacturing to use are:

1. Computer technologies. Aside from powerful central processors, computer hardware includes I/O devices, microprocessors, and minicomputers. Computer systems also include a wide range of software, from large operating systems to programming languages and application programs. A key to system integration is data communication. This requires the use of communication controllers and networks. Other essential ingredients are database management and systems such as relational databases. An emerging computer technology that can be used to perform many tasks in manufacturing is artificial intelligence (AI). In the most general terms, this includes sensors, machine vision, voice systems, adaptive control, natural language, logic programming, and expert systems.

2. Computer-automated engineering (CAE) tools. These use computer graphics technologies to perform engineering tasks. They include computer-aided design (CAD) techniques such as geometric modeling and finite element analysis (FEA). Also included are simulation tools for kinematic analysis and line modeling.

3. Manufacturing automation technologies. Computer-based technologies can be used to automate both the data and physical operations in production. NC is, of course, a fundamental computer technology for the automation of manufacturing processes. Materials handling techniques can be used to automate the physical movement of parts and products through the line. Data collection can be automated with a variety of computer-based methods. In addition, robotics has emerged as a key technology for providing flexible automation in manufacturing.

4. Management systems. These are applications of computer systems to automate the management tasks involved in manufacturing, such as GT, MRP, and MRP II.

If we apply these tools to the tasks involved from the design through the manufacturing of a product, we can see how they can fit into a potential "system" (Fig. 19-24). Since they are all computer-based, it should be possible to link them. Since many of them need to use the same data, it should be desirable to link them. These computer tools automate the tasks that people and machines perform. In a traditional factory, these tasks would be integrated by an organizational structure. In an automated factory, the automated tasks must be integrated in a similar, computer-based structure. In either case, the structure is the "system" that runs the factory.

REQUIREMENTS FOR INTEGRATING MANUFACTURING SYSTEMS

Even if each of the major tasks in manufacturing were automated, it might not be practical, or perhaps even possible, to integrate them into an overall factory management system. There are several basic prerequisites to making the integration of manufacturing systems feasible:

1. Product model. The key to linking the design and

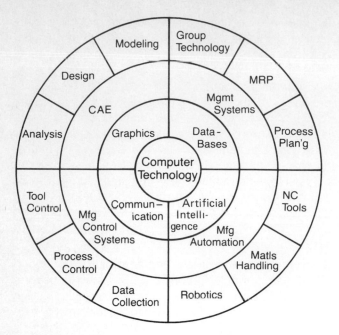

Fig. 19-24 Computer tools for manufacturing.

manufacturing activities is a computer model of the product. If those functions are to communicate automatically, design data must be available for both to use in a shared system. This means that a complete geometric model (and in electronics, a logical model as well) must be established.

2. Data extraction. Even when a product model exists, the communication link between the design and manufacturing activities is often a human—typically the process planner, who must interpret the design and extract the information required to generate process plans. To automatically link manufacturing to the product model, the computer system must interpret the information and extract the data needed. This can be accomplished by feature recognition techniques using logic programming. A computer must be taught rules of logic that would be used by an expert to sort through the facts or information contained in the model.

3. Database. For information to be communicated and shared, it must be easily accessible to those who need it. If most of the activities are automated, then the systems that need the information must have access to information that exists in other systems. This requires that such information reside in a database that can be accessed by those systems. It can be a large common database where all the information required is kept. In most cases, this is not practical or desirable. If the database is distributed, there must be a database management system that permits access to the information among the systems. This requires sophisticated techniques in data communication and information retrieval, such as relational databases.

4. System architecture. To tie together many automated systems, an overall architecture for the integrated system must be developed. For most manufacturing operations

this will be a hierarchical structure made up of many subsystems. In a complex structure, it will most likely have to be a distributed architecture which permits local systems to operate independent of the central system. The keys to integrating these systems are:

Standard interfaces

Common data communication links

Common database management system

5. Discipline. Computers can only process data. They perform the functions that they are taught—nothing more. For an automated manufacturing system to work, a great deal of discipline and control are required throughout the management process. To establish an integrated database, all the necessary data must be stored and updated in a standard format. The use of standard parts and processes, such as GT techniques, can minimize variations and maximize efficiency. Computer systems that are to communicate with each other must be designed with standard interfaces and compatible software protocols. An integrated system, of necessity, is a disciplined system. Such control and discipline may, by itself, have some benefits.

6. Manufacturing environment. Integrated systems may not be necessary or practical for all manufacturing operations. Those operations that can usually benefit the most from an integrated approach to CAM are characterized by:

Batch manufacturing processes

Short product life cycles

Competitive cost pressures

High quality and reliablity requirements

These, in general, mean that the manufacturing operation must be efficient and able to respond to change. Approaches to automation that use fixed and isolated systems cannot compete in that environment. Integrated manufacturing systems may prove to be a necessity for those that have to operate in an environment with such characteristics.

FUNCTIONS OF AN INTEGRATED SYSTEM

The functions of an integrated manufacturing system can be thought of in terms of the levels of control that it performs, the types of data that it manages, and the information that it provides:

1. Levels of system control. In a hierarchical structure, each level of the system is responsible for performing a different type of control (Fig. 19-25). This is similar to the structure of an organization in which each level of management deals with a different scope and level of business decisions. Such a division of responsibilities can be thought of as a four-level system structure:

☐ The lowest level controls the operation of machines. This involves tool and process control functions as well as data collection.

	Manufacturing activity	Control functions
Level 4	Plant management	Production planning Resource planning Database management
Level 3	Supervisory	Production scheduling Performance tracking
Level 2	Group of machines ("cell")	Machine loading Materials handling Monitoring
Level 1	Machine	Tool and process control

Fig. 19-25 Levels of system control in manufacturing.

☐ The next level controls a group of machines. This may be a manufacturing cell or line. The control functions include materials handling, machine loading, and monitoring the status of the line's operation.

☐ The next level performs supervisory functions over the production operations. It generates production schedules and tracks manufacturing performance. It is also usually the source of operational data that is needed by the lower-level systems (e.g., routings, NC programs).

☐ At the highest level, the control functions are focused more on planning than on the operational aspects of manufacturing. This includes forecasts, resource plans, and overall measurements. Central data processing functions may also be performed at this level, such as database management and data communications control. It also serves as the source for major common databases, such as engineering design information and production plans.

2. Types of data. An integrated manufacturing system must provide a wide variety of data to the production operation and support functions. This data generally falls into two categories:

☐ Manufacturing uses a number of different types of source or reference data, including information about the product design (e.g., geometry, specifications, bills of materials), production plans (demand and supply), and resource plans (e.g., labor, capacity).

☐ Most of the other data used by manufacturing is operational. This includes NC programs, routings, production schedules, and process control parameters.

3. Information. The manufacturing system must provide reports on the performance of the production operation. These may be real-time feedback to production operators, daily operational reports to supervisors, or high-level performance reports to management. The types of information typically reported include:

☐ Production status (e.g., shipments, back orders, WIP)

☐ Manufacturing performance (e.g., cycle time, production rates, labor)

☐ Equipment performance (e.g., availability, utilization, throughput)

☐ Quality (e.g., losses, rework, rejects, failures)

☐ Costs (e.g., spending, unit costs, inventory)

THE INTEGRATED MANUFACTURING SYSTEM

What will it look like when all of the pieces are put together? An integrated manufacturing system can be very complex. This makes it difficult to picture or describe in simple terms. One approach is to look at the relationships between the major subsystems or functions, such as:

Product design

Process planning

Production planning

Production control

Process control

Quality control

Distribution

Within each of these functions there are a number of activities which may use automated systems. For example, the process control system may include a number of subsystems, such as:

Tool control

Process monitoring

Materials handling

Data collection

To see how these functions relate to one another, one could look at the flow of product and materials through the plant. This might follow a sequence such as:

Receiving (e.g., raw materials and parts)

Stocking (warehouse)

Distribution (to production lines)

Manufacturing operations (process steps)

Inspections and testing

In-process inventory (buffers and queues)

Packaging

Finished goods inventory

Shipping

From an overall system viewpoint, one could also look at the flow of data. Since the manufacturing system is driven by data, this is the best way to see how the pieces are tied together. The data systems integrate the manufacturing operations—not the other way around. The physical flow of the manufacturing process can be automated only if the data handling system is automated. An integrated manufacturing system is a data processing system that drives the physical process activities. Figure 19-26 shows the relationship between data flow and the tasks in manufacturing.

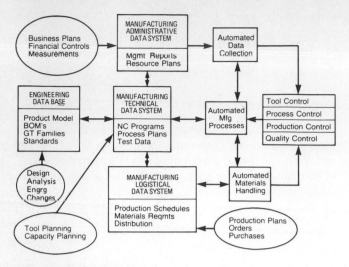

Fig. 19-26 An integrated manufacturing system.

19-7 SUMMARY

Automating individual manufacturing processes may improve the productivity of those particular operations. However, improvement from such an approach will not be as great as when the manufacturing systems that control the processes, tools, materials handling, and shop floor data are integrated. The key to such integration is not physical automation. It is automating the handling, storage, and retrieval of data. A common database can make it possible to share information between operations when it is needed. This may not be easy to implement. A large manufacturing operation may require a sophisticated database management system as well as a structured approach to formatting and storing data. To provide timely and accurate data requires an automated data collection system. The approach selected for data collection will depend somewhat on the manufacturing environment as well as on the type of data to be collected.

Materials handling is a major part of any manufacturing operation. In an automated manufacturing process, it is essential that there be an efficient materials handling system to minimize the work in process (WIP) and cycle time. Many different types of automated materials handling equipment are available. Such equipment should be integrated into a materials handling system which is designed to fit the nature and needs of the manufacturing process.

A flexible manufacturing system (FMS) is a small automated production operation which has been integrated. FMS's tie together automated tools, materials handling systems, and computer control systems. They are best suited for batch-type processes where flexibility is important. Integrating the manufacturing systems of an entire plant goes well beyond the scope of an FMS. It requires substantial time and effort, but this can be worthwhile for the additional efficiencies that can result. An integrated system uses computer tools to automate and tie together the major activities in manufacturing. The architecture of an integrated manufacturing system is described by the flow of the data that drives and manages the process and its relationship to the tasks performed in manufacturing.

REVIEW QUESTIONS

The answer to each question can be found in the section indicated at the end of the question.

1. Define an integrated manufacturing system. [19-1]

2. Identify some of the advantages of a common database. [19-2]

3. Describe how a database management system works. [19-2]

4. What are the primary reasons for collecting data on the manufacturing floor? [19-3]

5. Identify the major types of data collection systems and their principal differences. [19-3]

6. Why is a materials handling system important to an automated manufacturing operation? [19-4]

7. Identify the key elements of an integrated materials handling system. [19-4]

8. Describe some of the major types of automated materials handling equipment and how they are used. [19-4]

9. Why are robots used in materials transfer applications? [19-4]

10. Define an FMS and describe how it can be applied in manufacturing. [19-5]

11. Describe the major elements and features of an FMS. [19-5]

12. Identify some of the advantages of a typical FMS. [19-5]

13. What are some of the advantages of integrating the manufacturing systems of an entire factory? [19-6]

14. Identify the major areas of activity in manufacturing that must be automated in an integrated system. [19-6]

15. What are some of the basic requirements for integrating manufacturing systems? [19-6]

PART SIX

COMPUTER-AUTOMATED MANUFACTURING

CHAPTER 20
AUTOMATED MANUFACTURING

CHAPTER 21
IMPLEMENTING CAM

In the previous chapters, we have discussed the technologies that can be used to automate manufacturing operations. We have also discussed examples of these technologies actually being used in production applications today. To take full advantage of the capabilities of computer-automated manufacturing (CAM), however, requires that these technologies and applications be integrated into a total manufacturing system to run the factory. This is the underlying concept of the factory of the future. The question is whether this is truly achievable.

The factory of the future is not only achievable: It already exists! Many companies in a variety of industries have put the pieces together to automate entire manufacturing operations. To illustrate how and where CAM can be implemented, Part 6 will present examples of automated factories. In each case, all the basic elements of computer automation are used and integrated into a CAM system. Computer technologies, computer-automated engineering (CAE), robotics, and manufacturing systems are each essential ingredients that must work together to achieve computer-automated manufacturing.

Part 6 is divided into two chapters. Chapter 20 presents actual examples of CAM in each of the major types of production operations. Chapter 21 addresses the implementation process to provide a general understanding of what is involved and recommended approaches to assure its success. This last part of the book is intended to be an illustration and guide to implementing all the technologies and concepts that make up CAM, which were covered in earlier chapters. It is therefore written more as a wrap-up and summary of the subject than as a tutorial on new material.

AUTOMATED MANUFACTURING

20-1 INTRODUCTION

Factories do not automate their manufacturing operations just for the sake of automation. They do it to compete in their industries—and in some cases, perhaps even to survive. Automation should be used for specific reasons which result in benefits and improvements. These typically include such factors as:

☐ Reducing direct and indirect labor costs

☐ Improving product quality

☐ Reducing inventory and cycle time

☐ Avoiding exposure of people to hazardous or unpleasant work

In some manufacturing operations, such as those involving complex, high-technology products and processes, automation may be necessary just to make the manufacture of the product practical. Computer automation is also often used to provide flexibility in manufacturing. A flexible manufacturing system (FMS) can adapt to changes and is less likely to become obsolete in the near future.

When we speak of computer-automated manufacturing (CAM), we must think about the total manufacturing system. CAM is the automation of information as well as the physical manufacturing process. This involves automating tools and materials handling along with data collection and control activities. Provision of timely and accurate information can be even more important than the automation of the physical operations in the plant. The value of CAM cannot be judged by merely looking at the automated equipment on the factory floor. It must be evaluated on the basis of the efficiency of the total operation. This involves the planning, control, and management of all the manufacturing resources—people, equipment, and materials.

CAM is the integration of all the major functions in the factory. It ties the production operations together with the planning, scheduling, and inventory control activities into a total manufacturing system. This system can be thought of as having three major subsystems:

1. **Manufacturing.** Running the production operations.

2. **Engineering.** The interaction of the design and manufacturing engineering activities.

3. **Management.** The information required to control the total plant.

There are several approaches to implementing CAM. How it is done may depend on whether an existing factory is involved. An old facility or manufacturing line has to be converted to automation. A new one being built can be designed for CAM. In many cases, the most practical approach to achieving CAM is a gradual one. This means evolving the operation through several stages of automation (Fig. 20-1). One can start by automating individual tools and then groups of tools. However, to then tie all these groups together into an integrated system, a factorywide manufacturing control data system is needed.

This chapter will discuss approaches to CAM in three major types of production operations. Although these are not the only types of manufacturing operations that lend themselves to automation, they represent a major portion of the manufacturing industry:

1. Machining

2. Assembly

3. Process

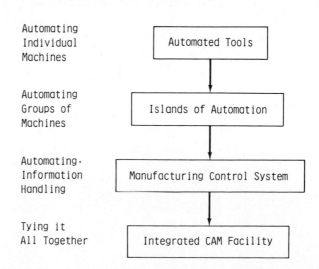

Fig. 20-1 Evolutionary approach to computer-automated manufacturing.

In each case, actual automated factories will be discussed as examples of how automation can be applied to those types of operations. Although each of these examples can be considered a "showcase" of automated manufacturing, they are not meant to exclude the many others that exist in industry or the new ones that are being implemented all the time. They are meant to illustrate what is feasible and practical and to demonstrate that CAM is not just a concept—it is real!

20-2 MACHINING

NATURE OF MACHINING OPERATIONS

Machining is a basic fabrication process. Many complex and precision metal parts are made by machining operations. The machining industry manufactures products for some of the most demanding performance applications, such as automobile engines, electric motors, airframes, and jet engines. A typical machining operation is characterized by:

Batches of production jobs

Many different part designs

Precision specifications

Variety of materials and properties

Large and heavy parts as well as small parts

Variety of large and complex machine tools

The machining industry was the first to use computers in manufacturing applications. It started with the numerical control (NC) of machine tools and evolved to the FMS's that are used today to automate machining operations. The machining process has all the basic elements that lend themselves to CAM:

☐ Complex, precision, custom product designs
☐ Low-quantity production of a large number of different parts
☐ Computer-controlled manufacturing tools
☐ Materials handling requirements
☐ Hazardous, undesirable, and monotonous tasks

This type of operation can be automated and integrated into a manufacturing system by building on the experience base that already exists in the machining industry. This means tying a computer-based design system together with a computer-controlled machining operation and manufacturing control system. This is a natural extension of the FMS's discussed in Chap. 19.

AN AUTOMATED MACHINING FACTORY IN THE HEAVY EQUIPMENT INDUSTRY

One example of a large machining operation that has been automated sucessfully is the General Electric loco-

motive plant in Erie, Pennsylvania. The factory was over 70 years old when it was refurbished and converted into an automated manufacturing operation. The objective was to establish a facility which could survive as a leading producer of locomotives for the world. This required increasing capacity, reducing manufacturing cycle time, and improving product quality while lowering costs. It took a significant investment (over $300 million) to make this a reality.

The principal machining operation in the plant fabricates motor frames from formed and cast metal parts. Some of the characteristics of this operation are:

☐ More than 100 different part types
☐ Average part size is 30 inches × 30 inches × 45 inches and 2500 pounds
☐ More than 140 machined surfaces
☐ Tolerances of 0.001 to 0.003 inches
☐ Surface finishes of 125 to 250 microinches

The entire operation is conducted on an automated machining line which is made up of the following major elements (Fig. 20-2):

☐ Nine computer numerical control (CNC) machine tools
☐ Fixture set-up station
☐ 21 load/unload stations
☐ 212-foot-long transporter (chain-driven shuttle cart on

Fig. 20-2 Automated machining operation at General Electric's Erie locomotive plant. *(General Electric Co.)*

a track) to move motor frames between load/unload stations (Fig. 20-3)

☐ Robots to load machines and change tools

☐ Automatic system for the detection and replacement of worn tools

☐ Laser inspection tools to check dimensions

☐ Central computer room which controls the operation (Fig. 20-4)

The line works as follows:

☐ The central control system displays the sequence of parts to be loaded at the fixture set-up station. This sequence is determined to maximize the throughput and utilization of the total line.

Fig. 20-3 Transporter for moving motor frames between machining stations. *(General Electric Co.)*

Fig. 20-4 Computer control room at General Electric's Erie locomotive plant. *(General Electric Co.)*

☐ The system directs the transporter to deliver the frame to the next available machine tool that can perform the required operations.

☐ The control system downloads the NC program to the machine tool selected for the next operation.

☐ The frame is returned to the fixture set-up station for repositioning so that other surfaces can be exposed for the next machining operation.

The central control system also isolates machines that are down for maintenance and schedules production around them. In addition, it collects and reports data on the performance of the system (e.g., machine and tool usage, failures). The entire system requires only two operators per shift. It has reduced the manufacturing cycle time for motor frames from 16 days to 16 hours.

This factory also uses several other computer tools to improve the productivity of the overall operation. For example, an interactive graphics system is used to "nest" different-shaped parts to be cut from steel plates. It can maximize materials usage and automatically generate NC programs. An expert system is used for diagnostics and equipment maintenance (Fig. 20-5). To use this trouble-shooting system, the operator responds to questions about the malfunction. This information is used by the system to identify the cause through a reasoning process based on expert knowledge. The system can also retrieve detailed drawings and display repair procedures on a video monitor. The result of all these applications of CAM

Fig. 20-5 Expert system for equipment diagnostics at General Electric's Erie locomotive plant. *(General Electric Co.)*

is a paperless operation that provides flexible automation of low-volume, multiple-part designs.

AN AUTOMATED MACHINING FACTORY IN THE DEFENSE INDUSTRY

Another machining operation that has been fully automated is at the Electro-Optical and Data Systems Group facility of Hughes Aircraft Co. in El Segundo, California. This operation fabricates precision aluminum housings for laser range finders and aircraft and missile optical systems. The production facility is an FMS comprised of the following features (Fig. 20-6):

☐ Nine 4-axes CNC machining centers (Fig. 20-7)

☐ Automatic pallet shuttles at each machine to interface with the materials handling system

☐ A coordinate measurement machine (CMM)

☐ Towline carts to move workpieces between machines (Fig. 20-8)

☐ A central computer control room which supervises the entire operation (Fig. 20-9)

A typical part produced on this line has approximately 400 machined surfaces, many with tolerances of only 0.001 inches. To hold such tolerances on large aluminum workpieces requires compensation for temperature changes. To accomplish this, each machine is equipped with a spindle probe in the tool changer which automatically adjusts the machine when it drifts out of tolerance. Each part is serialized and tracked through the process. The system controls all the aspects of scheduling, movement, and machining operations. It even records every specific machine and tool used to fabricate each feature on the parts.

Fig. 20-7 CNC machining center at Hughes Aircraft FMS. *(Hughes Aircraft Co.)*

The FMS facility provides substantial savings in equipment, space, cycle time, and labor over conventional NC machining alternatives. Only three people are required to run the system. To avoid problems in implementing such a complex system, its performance was first simulated. This permitted the layout and materials flow to be optimized prior to installation.

Fig. 20-6 Overview of flexible manufacturing system at Hughes Aircraft. *(Hughes Aircraft Co.)*

Fig. 20-8 Towline cart delivers workpieces between machining centers at·Hughes Aircraft FMS. *(Hughes Aircraft Co.)*

Fig. 20-9 Central computer control room at Hughes Aircraft FMS. *(Hughes Aircraft Co.)*

<u>20-3</u> ASSEMBLY

NATURE OF ASSEMBLY OPERATIONS

Assembly may be the largest single type of operation in the manufacturing industry. Most end products must go through some stages of assembly before they are complete. Assembly operations have the following general characteristics:

☐ Many different types of parts

☐ Materials handling requirements

☐ Several stages to the assembly process (e.g., subassembly, final assembly)

☐ Variety of features and models

☐ Frequent changes in design and bills of materials

☐ Test and inspection operations

Assembly operations usually involve batch-type production and have traditionally been labor-intensive. These factors, together with the general characteristics above, make them natural candidates for CAM. In addition to the physical assembly operations, there is usually a lot of data handling involved which should also be automated. This includes such activities as:

☐ Planning and ordering parts

☐ Controlling parts, work in process (WIP), and finished goods inventory

☐ Scheduling assembly operations

☐ Generating manufacturing routings and assembly and test instructions

☐ Collecting inspection and test data

Many different types of assembly operations are found in industry. However, most of them can be grouped into one of three major categories:

1. Mechanical. Most of the parts involved and the functions performed by the product are mechanical, as in automobiles or household appliances.

2. Electromechanical. Both electrical or electronic and mechanical parts and functions are involved, as in office equipment (e.g., typewriters, printers, disk drives, tape drives).

3. Electronics. All the parts and functions involved are electronic, as in printed circuit board assemblies.

To illustrate how CAM can be applied to assembly operations, we will look at examples from each of these categories.

MECHANICAL ASSEMBLY

1. Automated mechanical assembly in the automotive manufacturing industry. In recent years, the automobile industry has been undergoing a dramatic change in its manufacturing operations. To stay competitive, most companies have had to refurbish their old plants or build new ones. Automation is the key to achieving low cost and high quality in the production of automobiles in large quantities. One automated factory is the Chrysler assembly plant in Windsor, Canada. This facility was 55 years old when it was completely refurbished and transformed into one of the most automated plants in the industry. It took three years of planning and $660 million of investment to accomplish this. Now over 900 minivans are produced every working day in the facility, which incorporates the following features:

☐ Computer-aided design (CAD) systems are used to analyze and test designs through simulation prior to production

☐ An automated line with 58 robots performs more than 97 percent of the spot welds required on each vehicle (Fig. 20-10)

☐ A just-in-time program maintains a daily supply of high-quality parts for the line

☐ All the automated processes are linked by a 10-mile-long overhead and floor-level conveyor system

Fig. 20-10 Computer-controlled automated welding line at Chrysler's plant in Windsor, Canada. *(courtesy of Chrysler Canada Ltd.)*

☐ Large robots are used for materials handling tasks involving heavy payloads (Fig. 20-11)

☐ Robots are used to paint the interiors of the bodies by positioning the bodies diagonally on the conveyor (Fig. 20-12).

☐ A factory information system monitors and reports equipment performance and maintenance

☐ 125 robots are used for a variety of tasks, including the application of sealers and adhesives (Fig. 20-13)

☐ The dimensions and alignment of doors are measured automatically by a laser optical system

It takes features like these to manufacture automobiles competitively. Although the investment is very large, the payoff is substantial. It is nothing less than survival!

2. Automated mechanical assembly in the household appliance industry. Another automated mechanical assembly operation is the General Electric plant in Louisville, Kentucky. It is a large facility devoted to the production of household appliances. Substantial investments have been made in recent years to refurbish the production lines in order to improve their productivity and competitiveness as well as the quality of the products they manufacture. One such line was rebuilt and automated, with an investment of $39 million, to assemble dishwashers. It is made up of two subassembly lines that feed a final assembly and test operation. Some of the key features of the line are:

Standardized product design. The dishwasher product line was redesigned for automation. A standard plastic tub replaced the metal tub structure. Parts were designed to snap together instead of using screws and bolts. The number of unique features was minimized and the total number of parts involved was reduced by approximately 85 percent.

Automated subassembly lines. One line assembles the tub structure while another assembles the inner door (Fig. 20-14). Both use standard injection-molded parts that are fabricated in the plant and delivered by conveyor

Fig. 20-11 Large robot lifting floor pan of van. *(courtesy of Chrysler Canada Ltd.)*

to the lines. In an automated 21-step process, the tub-structure line assembles the plastic tub with steel structural parts. In an automated 13-step process the line for the inner door assembles accessories, such as dispensers and gaskets, onto the plastic door liner. These auto-

Fig. 20-12 Robots painting interior of vans with diagonally positioned conveyor. *(courtesy of Chrysler Canada Ltd.)*

Fig. 20-13 Robot applying sealer to car window. *(courtesy of Chrysler Canada Ltd.)*

mated operations require only one-tenth the number of operators involved in the previous manual operations.

Point-of-use manufacturing. Many of the parts used in the subassembly operations are fabricated in the plant as needed and fed by conveyor directly to the point on the line that uses them. This reduces the inventory involved, but it also requires consistently high-quality parts. This is an excellent example of just-in-time manufacturing.

Nonsynchronous final assembly line. A 3-mile-long overhead conveyor system moves the completed subassemblies to final assembly workstations. The product is held at each station until the final assembly is completed rather than moving continuously. Since the workers control the speed of the operation, it is easier to avoid defects and repairs.

Robots. Robots are used to perform heavy materials handling tasks, such as unloading tubs from conveyors.

Computer control system. The entire operation is controlled from a central room above the factory floor (Fig. 20-15). The central computer receives process data from the programmable controllers (PCs) throughout the line. Color graphics terminals can display the status of every manufacturing operation. If a problem on the line requires special attention, the unit is rerouted to a repair area along with instructions from the system. Lasers scan bar codes on the conveyor hangers to track units through the assembly process.

Vision system. A vision system is used to align the door hinges and the tub assembly.

Testing. Every unit undergoes an automatic final test for half an hour to demonstrate that it can perform all of its normal operational functions.

The results of this automation program have been significant:

☐ Manufacturing cycle time was reduced from 5 or 6 days to 18 hours

Fig. 20-14 Automated subassembly line at General Electric's Louisville plant. *(General Electric Co.)*

Fig. 20-15 Computer control room at General Electric's Louisville plant. *(General Electric Co.)*

- ☐ Service calls were cut by more than half
- ☐ WIP was reduced by more than 60 percent
- ☐ Capacity was increased by 20 percent
- ☐ Operator productivity was increased by more than 25 percent

The approach taken to automating this line and the basic concepts which were incorporated can be applied to many types of mechanical assembly operations with similar success.

ELECTROMECHANICAL ASSEMBLY

1. Electromechanical assembly in typewriter manufacturing. One automated electromechanical assembly operation is the IBM plant in Lexington, Kentucky. IBM recently invested approximately $350 million to automate the production of typewriters and printers in an existing 1-million-square-foot factory. The objective of the project was to provide a high-volume flexible manufacturing capability that could produce products with higher quality at lower costs.

The manufacturing process involves five major subassembly lines as well as automated final assembly, test, and packaging operations (Fig. 20-16). The output of the subassembly lines is fed to the final assembly stations by a conveyor system. An integrated manufacturing control system synchronizes the operation of the subassembly and final assembly areas to assure that there is a contin-

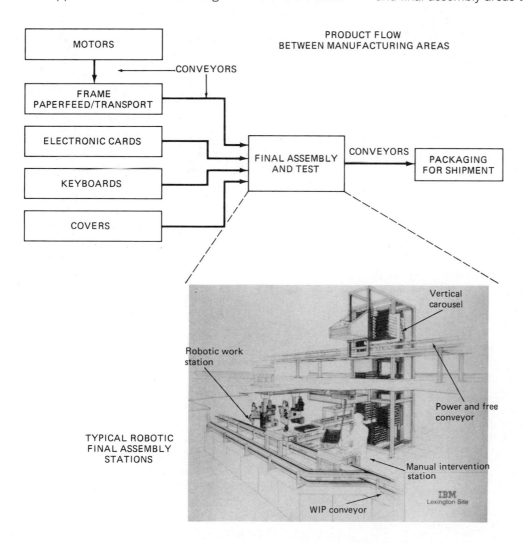

Fig. 20-16 Automated electromechanical assembly process at IBM's Lexington typewriter plant. *(International Business Machines Corp.)*

281

ous, closely scheduled flow of product through the factory. Each automated line is managed like a separate plant, with a production manager responsible for the schedule and quality on a daily basis. Some of the key features of the operation are:

Design for manufacturing. The products were designed using standard design guidelines so that they would be compatible with flexible automation. The automated facility was designed to handle any similar product that can fit into a standard work envelope.

Robots perform a variety of assembly tasks. The tasks selected involve a few simple, short moves in order to avoid complexity (Fig. 20-17). Multipurpose grippers are used in order to minimize gripper changes. A typical robot workstation includes a production-line conveyor and parts handling system as well as the robot system.

Materials handling. The overall flow of parts, subassemblies, and products is completely automated. A combination of automated materials handling systems is used, including overhead and workstation-level conveyors, automated storage and retrieval systems (ASRS's), vertical carousels, and automated guided vehicles (AGV's). Subassemblies and parts are delivered in containers by an overhead conveyor system to the vertical carousels that feed the robotic assembly workstations (Fig. 20-18).

Workstation ownership. Each operator is responsible for the quality and throughput of his or her workstation. Even though the workstation is automated, the operator has increased responsiblities. These include monitoring the operation of the equipment and the supply of parts, making adjustments to correct the operation of the equipment, and performing minor maintenance activities.

Flexibility. Multiple models can be processed simultaneously or model changes made quickly. The control system reads the bar-code identification on each carrier to determine to which final assembly station to deliver the product.

Manufacturing control system. A distributed hierarchical

Fig. 20-18 Vertical carousels feed robot assembly stations at IBM's Lexington typewriter plant. *(International Business Machines Corp.)*

computer system is used to control all the operations. It is organized into a four-level structure, from the plant level down to the individual workstations. The schedule for each manufacturing area is loaded for several days ahead, which permits the local control system to operate without the central host computer. There is also communication between manufacturing areas to anticipate potential problems, such as parts shortages or bottlenecks. At the workstation level, minicomputers interface with the controllers on production machines, robots, and conveyors in order to control the assembly operation.

Just-in-time materials control. The WIP and final assembly inventory are kept to a minimum. Only one shift's worth of subassemblies and two shifts' worth of final assemblies are maintained in inventory.

Each of the subassembly lines is a unique automated factory. One molds plastic covers. Another assembles small electric motors. Others assemble printed circuit boards and keyboards. Overhead transporters move the subassemblies between vertical transfer stations. Automated test stations perform functional tests on every final assembly. Together, these automated operations form a fully integrated process from parts fabrication through the shipping of the end product.

2. Electromechanical assembly in computer manufacturing. Another automated electromechanical assembly operation is Apple Computer's factory in Fremont, California. The facility was built for $20 million to assemble the Macintosh computer in high volumes. As a result of automation, production capacity was increased substantially while the number of workers required was actually reduced from what would have been required by manual assembly. The approach taken to automating this plant was not to introduce a lot of robots and automated tools. Automation focused mainly on the information and materials handling processes. Some of the features are:

☐ Component parts are stored and retrieved by an ASRS (Fig. 20-19).

Fig. 20-17 Robotic assembly at IBM's Lexington typewriter plant. *(International Business Machines Corp.)*

Fig. 20-19 Automatic storage and retrieval system at Apple Computer's Fremont plant. *(Apple Computer, Inc.)*

Fig. 20-21 Automated "burn-in" towers fed by automated elevator and conveyor system at Apple Computer's Freemont plant. *(Apple Computer, Inc.)*

☐ Inventory levels are kept to a minimum by a manufacturing resources planning (MRP) system and a "just-in-time" program.

☐ A bar-code system is used to track the components and computers through the assembly process.

☐ Printed circuit boards are assembled on a semiautomated line on which automated machines, robots, and people insert the various types of components (Fig. 20-20).

☐ Minicomputers monitor and control all the processes, providing automatic data collection and a paperless environment.

☐ Every computer is subjected to 24 hours of testing after final assembly. A special conveyor and elevator system carries the computers to the "burn-in" towers for the tests and then delivers them to the automated packing operation (Fig. 20-21).

The results of this automation have been significant:

☐ The quality of the product has been improved to a failure rate of less than 0.5 percent

☐ The cost of labor has been reduced to less than 1 percent of the manufacturing cost

☐ The inventory turnover has been increased to about 25 times a year

Fig. 20-20 Semiautomatic printed circuit assembly line at Apple Computer's Fremont plant. *(Apple Computer, Inc.)*

ELECTRONICS ASSEMBLY

There are many examples throughout industry of automated electronics assembly lines. Some are high-volume operations using mostly fixed-automation techniques. Others automate only a few of the steps in the total assembly process. Following are three examples.

1. Automated electronics in the defense industry. One automated electronics assembly factory that incorporates many CAM features is the Westinghouse plant in College Station, Texas. It builds special printed circuit board assemblies for the aerospace and defense industries. The nature of this business is small-quantity production runs of high-density electronics assemblies. This has traditionally been a manual operation that was subject to human error, causing a great deal of rework. Westinghouse built a new, automated factory, investing approximately $50 million, with a primary objective of reducing costs by eliminating such errors. Some of the key features are:

Materials supply. All of the various electronic component parts required for production are acquired by a central remote operation. This materials acquisition center does all the purchasing, receiving, inspecting, and testing of components for the production facility. The center is linked directly to the plant's manufacturing control system by a commmunication network. This permits component requirements to be scheduled automatically as they are needed. The system also manages the coordination of engineering changes between the materials acquisition center and the automated production operations.

Materials handling. The printed circuit board assemblies use a wide variety of electronic components. Each production job requires that a kit of parts be prepared for the assembly operation. An automated kitting system was developed for this task (Fig. 20-22). A miniload ASRS carousel stores the inventory of hundreds of different types of components in trays. When a production job is scheduled, the system withdraws the appropriate trays

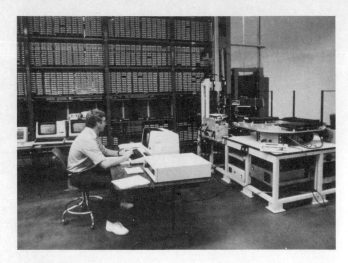

Fig. 20-22 Automatic kitting system at Westinghouse's College Station electronics assembly plant. *(Westinghouse Electric Corp. and the U.S. Air Force Get Price Program)*

Fig. 20-24 Robotic assembly station at Westinghouse's College Station electronics assembly plant. *(Westinghouse Electric Corp. and the U.S. Air Force Get Price Program)*

and presents them to a robot. The robot selects the component parts required, cuts their leads to the proper length, and places them in a pallet as a kit for production (Fig. 20-23). Each component is positioned in the sequence in which it will be assembled. The kits are then transported by a conveyor to another carousel which feeds the assembly operations.

Robotic assembly. Many of the individual components are small, delicate, and expensive. They must be precisely positioned on the printed circuit board before soldering. Robots are used at automatic assembly stations to orient, position, and insert the components (Fig. 20-24). Television cameras are used to determine the orientation and straightness of the leads on the components. Strain gauges are used to sense the pressure applied to the components during assembly in order to prevent damage.

Manufacturing data collection. Bar codes are used to track the components from initial acquisition through the assembly process. Optical scanning wands record the movement of the production jobs through each operation, including operator identification and any repair actions that may have been performed. This data makes the history of each assembly traceable, and the materials supply system is automatically updated with requirements.

2. Automated electronics assembly in the computer industry. Another automated electronics assembly operation is at IBM's plant in Poughkeepsie, New York. One of its manufacturing operations is the assembly of the printed circuit boards for its large computers. These boards are very large (24 × 27 inches) and heavy (up to 90 pounds). IBM installed a flexible assembly system to eliminate the manual handling of these boards as well as improve their quality and the manufacturing cycle time. The automated manufacturing line was installed on three floors of an existing facility. Features of the system include the following:

☐ Bar-code labels are applied to each board to track it through the process

☐ A car-on-track conveyor network automatically delivers the boards to workstations (Fig. 20-25)

☐ Boards are held in ASRS buffers until they are released to the conveyor system

☐ Each board is tested and analyzed for defects by a robotic system that probes all the interconnecting points (Fig. 20-26)

☐ Wires for overflow and engineering changes are assembled by automatic wire-bonding machines (Fig. 20-27)

☐ Robots automatically assemble small parts (such as resistors and capacitors) to each board (Fig. 20-28)

Before the line was installed, its operation was simulated to debug the design of the materials handling system.

Fig. 20-23 Robot kitting station at Westinghouse's College Station electronics assembly plant. *(Westinghouse Electric Corp. and the U.S. Air Force Get Price Program)*

Fig. 20-25 Car-on-track conveyor system delivers printed circuit boards to process operations. *(International Business Machines Corp.)*

Fig. 20-26 Robotic test systems at IBM's Poughkeepsie plant. *(International Business Machines Corp.)*

Fig. 20-27 Automatic wire-bonding machines on IBM's printed circuit board assembly line. *(International Business Machines Corp.)*

A central "spine" design which feeds all the assembly operations was chosen for maximum flexibility. It permits boards to be moved randomly between the process steps and can provide for future changes or expansion.

3. Electronics assembly in the electronic components industry. Another automated electronics assembly operation is the Allen-Bradley plant in Milwaukee, Wisconsin.

Fig. 20-28 Robotic assembly of small parts at IBM's Poughkeepsie plant. *(International Business Machines Corp.)*

The company invested $15 million to install a fully automated line in a 70-year-old facility to manufacture motor contactors and control relays. The objective was to establish an efficient production capability for a new line of high-quality products that could be marketed competitively worldwide. The alternatives were to locate the operation in a country with low-cost labor or to lose the market entirely. A computer-integrated approach to automated assembly made it possible to maintain production in the United States and still compete in the world marketplace. Some of the features of the line are:

☐ The product and the manufacturing line were developed simultaneously for an automated process by a multidisciplinary team. This involved not only designing the product for manufacturability, but also designing special automated equipment and a computer control system (Fig. 20-29).

☐ The operation is fully automated, with raw materials entering at one end of the line and finished products

Fig. 20-29 Special automated equipment designed for the Allen-Bradley contactor assembly facility; in this case, a contact insertion machine. *(photo courtesy of Allen-Bradley Co., Inc., a Rockwell International Co.)*

exiting at the other end. Only a few specialists are required to maintain the equipment and control the operation (Fig. 20-30).

☐ The line is designed to be flexible so that it can handle up to 999 variations in the product and lot sizes as small as one unit.

☐ The line is capable of producing 600 units per hour with a total cycle from start to finish in less than one hour (ship to customers within 24 hours). This minimizes the WIP inventory.

☐ Orders from field sales locations are entered and scheduled on the floor daily by the production control system. The production control system is tied into the manufacturing control system which drives all the machines on the factory floor.

☐ Orders are tracked with a bar-code system that applies a label to each unit that can be read throughout the process. The bar-code labels also provide information to the automated equipment on what specific parts to assemble on each unit (Fig. 20-31).

☐ High quality and low scrap are maintained by a process control system that monitors 3500 data collection points and 350 assembly test points.

☐ The entire operation is monitored by a computer control room with color graphics displays, diagnostic routines, and statistical process control programs (Fig. 20-32).

Experience with this line showed Allen-Bradley that it could manufacture a broader product line at less than half the cost of manual assembly methods.

20-4 PROCESS

NATURE OF PROCESS OPERATIONS

There are many different types of process manufacturing operations. Chemicals, pharmaceuticals, metals, and

Fig. 20-31 Bar-code information on each contactor allows automatic machines to assemble the right size covers and housings on the Allen Bradley contactor assembly line. *(photo courtesy of Allen-Bradley Co., Inc., a Rockwell International Co.)*

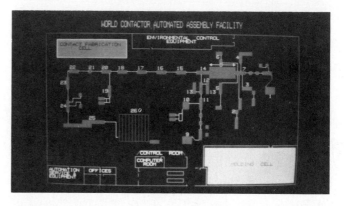

Fig. 20-32 The computer system which controls the Allen-Bradley automated contactor assembly facility uses color graphics terminals to monitor the operation of the line. *(photo courtesy of Allen-Bradley Co., Inc., a Rockwell International Co.)*

electronic components, for example, all involve process-type operations. Although these products are quite different, the nature of the manufacturing process involved has similar characteristics:

☐ It starts with a supply of raw materials

☐ Materials flow continuously through the process steps

☐ Many complex process steps are involved

☐ Complex and expensive production equipment is involved

☐ Material and process parameters must be controlled precisely

☐ Products are sensitive to small defects and contaminants

☐ High quality and reliability are required

As these types of processes become more complex and the product requirements become more stringent, manual operations become less practical. The capabilities of CAM are often essential to making the production of such products feasible on a large scale. The automation of such processes is often not as visible as it may be

Fig. 20-30 Overview of the Allen-Bradley automated contactor assembly line with systems maintenance specialists overseeing the operation. *(photo courtesy of Allen-Bradley Co., Inc., a Rockwell International Co.)*

in other types of manufacturing operations. The individual process steps often do not involve any movement or physical activity. They frequently involve sophisticated chemical and electrical processes that must be conducted under computer control. Usually a great deal of data is required to run and control the process. Process operations, by their nature, require a CAM approach to production.

AUTOMATED SEMICONDUCTOR FABRICATION

One example of a process operation that can be automated is semiconductor fabrication. The earliest fully automated semiconductor line was established by IBM at its East Fishkill, New York, facility. It was called the QTAT line (for "quick turn-around time"). As the name implies, its primary objective was to reduce the manufacturing cycle time for semiconductors in order to provide small quantities of many different designs of customized integrated circuit (IC) logic devices. QTAT made possible the quick prototyping of new designs as well as rapid introduction of production volumes and engineering changes early in the product program. The process started with a "master-slice" wafer which incorporated a standard array of transistors, diodes, and resistors. It then "personalized" the wafer by interconnecting these components into customized logic circuit designs. This involved 140 individual process steps. Automating this process made it possible to fabricate the ICs in less than half the normal cycle time.

Following are some of the key features of the QTAT line:

Automated wafer handling and storage. The traditional approach to semiconductor manufacturing placed the entire process area in a clean room, but QTAT enclosed only the wafers in a clean environment (Fig. 20-33). The wafers were moved automatically between tools on jets of air along enclosed tracks (Fig. 20-34). These "air tracks" eliminated manual handling and allowed faster and cleaner processing. The contamination levels could be controlled to a level below that of typical clean rooms

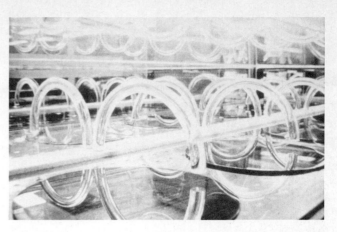

Fig. 20-34 Air-track wafer handling system at IBM's East Fishkill semiconductor plant. *(International Business Machines Corp.)*

inside the tracks while operators worked in a relatively normal environment. When the wafers were not moving between tools or being processed, they were stored at buffer stations in sealed cartridges which interfaced directly with the air track.

Computer-controlled wafer routing. QTAT was a single-wafer processing operation, rather than the batch approach that has traditionally been used in semiconductor manufacturing. Each wafer was routed between tools by computer control. Since many individual wafers were being processed simultaneously on the line, it was necessary to keep track of them automatically. This was accomplished by reading a laser-inscribed identification number on each wafer as it entered the air-track system. Single-wafer processing reduced the transport time and queues between operations. It also allowed for each individual wafer to be assigned a priority on the basis of its position in the process and its schedule for completion.

Manufacturing control system. The entire QTAT operation was under the control of a distributed hierarchical computer system (Fig. 20-35). It had a four-level architecture, from the central host system down to the individual tool controllers. When a wafer entered the line, all the

Fig. 20-33 QTAT line at IBM's East Fishkill semiconductor plant. *(International Business Machines Corp.)*

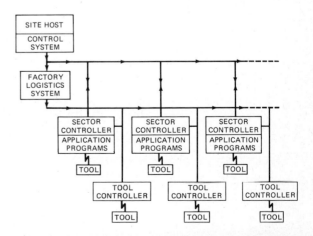

Fig. 20-35 Distributed hierarchical control system for IBM's QTAT line. *(International Business Machines Corp.)*

information necessary to process it was automatically distributed to each level of the system. Each level of the system could then operate without depending on the availability of the computer above it in the hierarchy. This made it possible for the system to operate at close to 100 percent effective availability. The control system integrated the automation of the tools, logistics, and data handling. It was the source of all the technical, logistical, and administrative information involved in the operation of the line.

Automated tools. The QTAT line was made up of approximately 100 automated tools that were arranged into eight groups or sectors (Fig. 20-36). These tools were all under the control of the central system and interconnected by air tracks. Each sector had a central distribution system for the supply of chemicals, gases, process materials, electricity, and air. The line was designed to operate 24 hours a day, 7 days a week. Operators and technicians were required only to monitor and maintain the equipment. Some of the tools used robotic arms to handle the wafers (Fig. 20-37). The "wet stations" filled and drained themselves of chemicals. Personalization patterns were "written" directly on the wafer by electron-beam tools which received instructions from the product design system (Fig. 20-38). This eliminated the need for the optical masks and manual exposure operations that were normally involved. These "e-beam" systems had to handle an enormous amount of data. The computer control systems converted detailed design information on each of the high-density ICs into NC programs which controlled the operation of the tool.

Process control. An automated process of such complex-

Fig. 20-37 Robot arms handling wafers at wet station in IBM's QTAT line. *(International Business Machines Corp.)*

Fig. 20-38 Direct-write electron-beam exposure system in IBM's QTAT line. *(International Busines Machines Corp.)*

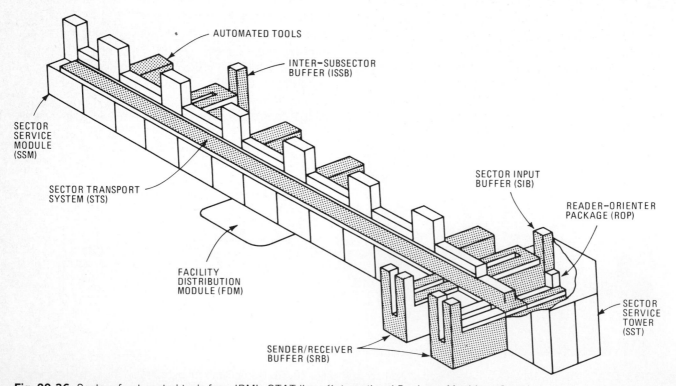

Fig. 20-36 Sector of automated tools from IBM's QTAT line. *(International Business Machines Corp.)*

ity required an automated control system. Automatic sensing and measurement tools were distributed throughout the process to monitor critical parameters (e.g., temperature, pressure, voltage, thickness, deposition rates). This data was collected by the system and tied to the wafer and tool identification numbers for analysis. The system was even capable of feedforward process control. It could send automatic instructions on process set points based on measurements from the previous step.

QTAT incorporated many of the computer tools we have addressed in this book. It was a truly integrated and automated manufacturing system. All the data and control functions in the design and manufacture of the product were tied together. The product was manufactured by a paperless and peopleless process.

QTAT is not the only example of automated manufacturing in the electronics industry. Other process operations outside of the semiconductor world have also been automated. Many high-technology products of the electronics industry today involve complex process operations during their manufacture. These include magnetic and optical disks, magnetic tape, ceramic substrates, and printed circuit boards. As technology advances, more and more products involve process operations in their manufacture. This will drive the need to use CAM approaches to produce them.

20-5 SUMMARY

Manufacturing operations are automated to improve overall productivity—not to replace people with machines. Both the physical manufacturing process and the information handling process are automated in a computer-automated manufacturing (CAM) environment. CAM is an integrated system which ties the production, engineering, and management functions together.

Three major types of manufacturing operations that have implemented CAM are machining, assembly, and process. The machining industry has a broad base of experience with numerical control (NC) to build on for automating its factories. CAM is a natural extension of flexible manufacturing systems (FMS's). Assembly operations have traditionally been both labor- and data-intensive. Flexible automation is the key to improving the productivity of these types of applications. Mechanical, electromechanical, and electronics assembly operations all lend themselves to a CAM approach. Process manufacturing operations have unique characteristics that not only benefit from computer control but, in some cases, require it. Manual operations become less practical as processes become more complex.

CAM is not just a concept—it is real. Factories of the future exist today. CAM is being applied throughout industry in a variety of different types of manufacturing operations.

REVIEW QUESTIONS

The answer to each question can be found in the section indicated at the end of the question.

1. What are some of the benefits of automating manufacturing operations. [20-1]

2. Identify some basic characteristics of a manufacturing operation that lend themselves to computer-automated manufacturing. [20-2]

3. Describe the general characteristics of a machining operation. [20-2]

4. Identify some of the typical elements of an automated machining factory. [20-2]

5. What are some of the basic characteristics of assembly operations? [20-3]

6. Describe some of the data handling tasks that are typically found in assembly operations. [20-3]

7. Identify some of the typical elements of an automated assembly factory. [20-3]

8. Describe the basic characteristics of a process-type manufacturing operation. [20-4]

9. Identify some of the elements of an automated process plant. [20-4]

IMPLEMENTING CAM

21-1 INTRODUCTION

If technologies exist today that can make computer-automated manufacturing (CAM) feasible, then one may ask, "Why are not all factories automated?" We have seen what CAM is capable of doing as well as actual examples of it working successfully. So we know that CAM is real. But it is also obvious that the entire manufacturing industry has not yet implemented it.

This chapter will conclude this book with a discussion on the implementation of CAM. It is intended to be a brief review of factors which should be considered and approaches which are recommended in pursuing a CAM project. Although it will identify important concepts and steps, it is not a comprehensive guide or manual.

The chapter starts with a summary of the basic reasons why CAM should be implemented. It then identifies some of the barriers which may be encountered when pursuing a CAM project and discusses the major changes that CAM will bring to a manufacturing organization. The balance of the chapter will deal with how to implement a CAM project. It will cover the key elements of a successful project, how to justify it, planning the project, and the implementation process itself.

The reader should not expect to be able to implement a CAM project as a result of reading this chapter. The chapter is intended to give an idea of what is involved so that the reader will have an appreciation of the factors which can influence the success of a CAM project. Such projects involve a substantial amount of time and effort on the part of many people. This book will not make anyone an expert in CAM. It takes a great deal of experience and skill to implement such a complex system. This chapter will deal with the basic knowledge that can prepare the reader for participating in that process.

21-2 WHY IMPLEMENT CAM?

The basic reasons for implementing CAM should be obvious to the reader. For some businesses, it may even be necessary for survival. There is an expression that is sometimes used in the manufacturing industry to refer to the alternatives: "Automate, emigrate, or evaporate!" In recent years some businesses chose to seek cheap labor around the world as a means to keep their costs low.

Others could not compete at all and went out of business altogether. Automation is an alternative which can allow a business to compete by improving its productivity.

Some of the benefits that can be realized by implementing CAM are:

☐ Reduced manufacturing costs

☐ Shorter manufacturing cycle times

☐ Lower inventory levels

☐ Improved product quality

☐ Reduced space requirements

☐ Improved productivity of individuals from the use of computer tools

☐ Shared use of design and manufacturing data

☐ Improved validity and consistency of data

☐ Flexibility in design and manufacturing operations

☐ Faster introduction of new products and changes into manufacturing

☐ More efficient data collection and communication

☐ Improved management control

This is not a complete or detailed list, but it includes some of the key benefits. Not all of them are obvious or even quantifiable, but each one can be important to a manufacturing operation. With such a list of benefits, it may not be clear why all manufacturing operations are not automated. CAM is not easy to implement. It not only requires a significant effort; it also requires overcoming a number of obstacles.

21-3 BARRIERS TO CAM

CAM has many potential benefits. However, responsible managers cannot assume it will be successful and the right solution to a problem without addressing the issues and exposures involved. Technical, financial, and organizational considerations must be dealt with in any manufacturing operation considering the implementation of CAM. Following are some of the barriers that are usually faced.

INERTIA

CAM involves significant change. Any manufacturing operation will have momentum to continue in its present mode; this must be overcome. This inertia comes in several different forms:

1. **Organization.** The structure of the manufacturing organization, as well as the roles and responsiblities of individuals, will have to change to accommodate CAM. Workers and managers tend to resist change in order to avoid the uncertainties involved.

2. **Financial system.** The traditional methods used to make financial decisions and measure manufacturing performance may cause automation to appear more expensive, even if the overall costs are truly lower. For example, such common indicators as direct to indirect ratios, cost per unit labor hour, and fixed versus variable cost all make a capital-intensive operation look undesirable.

3. **Perceptions.** People's feelings about CAM are based on what they have heard or experienced. These feelings, which bias people's opinions and affect their attitudes toward CAM, may not be based on facts or up to date. Even a positive opinion about CAM may be misguided. For example, people often associate automation with physical equipment only, such as robots. They do not consider the implications of automating data and integrating the total manufacturing system.

RESOURCES

CAM affects all the resources of a manufacturing operation. The facility, the equipment, and the people are all involved. Implementing CAM can also be very expensive. Some of the resource barriers that must be overcome are:

Large investment (capital and expense)

Lack of skills and experience in automation

Incompatible hardware and software

Inadequate data systems

Impact of automation on existing skills and work force

FEAR

Fear is a barrier that is very difficult to deal with and cannot be quantified, but it is real. The people involved in a manufacturing operation may have many reasons to fear CAM and therefore oppose its implementation:

Potential loss of jobs

New technologies

Changing job and skill requirements

Perceived threats to status and power

Potential failure

RISK

Any change brings with it some risk of problems or even failure. CAM is no exception. A large CAM project can involve a lot of risk that some managers may not be willing to take:

☐ Long-term payback

☐ Impact of start-up on current commitments and performance

☐ Potential for shutdowns and for problems that may be difficult to fix

This is a long list of barriers, and yet it is not complete. Convincing people to take the risk and implement a major CAM project is not easy. The first step in implementing a CAM project is to overcome these barriers and "sell" the project to all those involved in the decision and implementation process.

21-4 CHANGES CAUSED BY CAM

Implementing CAM will cause many changes in the plant. It will affect what people do and how the plant is run. Before implementing a CAM project, these changes should be anticipated. To prepare for them may require new skills, procedures, measurements, and management processes. These changes will generally affect the following areas:

Employees

☐ Operators will have new responsibilities (i.e., workstation "ownership")

☐ New skills will be required (e.g., use of computers and automated tools)

☐ Fewer jobs will involve "hands-on" work

☐ New, advanced techniques will be used to solve problems—for example, computer-automated engineering (CAE) tools

☐ There will be less opportunity for hands-on training (e.g., computer simulation and instruction will be used)

Managers

☐ Fewer employees will be in easily measurable jobs (i.e., "direct work")

☐ More employees will be in less structured jobs (e.g., debugging, problem solving, maintenance)

☐ More time will be spent in changing people's jobs and retraining them to acquire new skills

☐ New management tools will be available (e.g., decision support systems, simulation models)

☐ Routine jobs will be done by computer systems (e.g., expert systems, robots)

☐ More timely and accurate data will be available

☐ More time will be spent planning than reacting

Management process

☐ Operating procedures will change

☐ Measurements and methodologies will change (e.g., financial guidelines)

☐ There will be fewer layers of management

☐ Operational decisions will be made at lower levels

□ Team approaches will be taken to CAM implementation and operation

□ There will be more standardization

□ Communication will be faster and more efficient (e.g., on-line computer reports, electronic mail, video conferences)

□ Organizations will be tied more closely together (i.e., with interdependencies and team approaches)

21-5 KEYS TO SUCCESSFUL IMPLEMENTATION OF CAM

It is not easy to overcome all those barriers and cope with all those changes. An approach to implementing CAM must be developed which can help assure its success. No one formula will work in all cases. In fact, CAM's history is not long enough to give us a broad base of experience throughout industry. However, there have been enough attempts to implement automation projects to make possible some general observations about the key elements of a successful approach.

PLANNING REQUIREMENTS

A number of factors must be built into the plans and objectives for a CAM project if it is to be successful:

□ Knowledgeable and experienced people to develop and implement the plan

□ Preparation for changing skill requirements and job responsibilities

□ Adequate time and effort to prepare for and debug the system

□ A control system to manage the overall project

□ Measurements to track the performance of the operation after it is implemented

□ Stages for gradual or phased implementation

□ Involvement and support of all organizations affected by the plans and decisions

□ Compatibility with the goals and plans of the business

□ Use of computer tools for planning, evaluating, simulating, training, and decision making

□ Preparation for organizational implications of integrated systems and automation

SYSTEM REQUIREMENTS

The design and specification of the CAM system should be derived from some basic principles. It should have:

□ A common database management system

□ A standard protocol for data communications

□ A common language for tool control programs

□ A product model for engineering data

□ Standard design guidelines for manufacturing

□ A standard classification and coding system for parts

□ High quality in materials, parts, equipment, and process control

□ Provision for flexibility to respond to changes in products, processes, and volumes

MANAGEMENT ATTITUDES

The attitudes and perceptions of the management responsible for decision making and planning of a CAM project can have a significant effect on its success. In particular, management should:

□ Think of CAM as an opportunity to integrate operations, not just as the installation of a bunch of automated tools

□ Think of the CAM project as a strategic effort to improve the productivity of the entire manufacturing operation, not just as a solution to some immediate operational problems

□ Expect that a long-term return on the large investment will be required

□ Confront attitudes of fear or lack of priority with the threat of competition

□ Anticipate the problems of employees, such as retraining and displacement

These recommendations are by no means a guarantee of success. They are also very basic and should be fairly obvious. However, many automation projects have failed, or have never even been implemented, because some of these basic ingredients were missing.

21-6 JUSTIFYING CAM

A critical step in the early stages of a CAM project is to justify it. Automation projects can be very expensive and should not be implemented unless there are good business reasons to do so. No company can afford to automate just for the sake of automation. There must be some benefits or the company will lose money. Even though the benefits may seem obvious, each application should be thoroughly analyzed to assure that it can be justified before substantial investments are made. The justification process often reveals more attractive alternatives, errors in assumptions, or factors that were not previously considered.

This section is not a manual for preparing a project justification or for financial methods for evaluating business cases. It presents some basic factors that should be considered when trying to justify a CAM project.

RISKS

One reason for going through the justification process is to make sure that the project makes good business sense after considering all the significant factors that may be involved. Deciding whether to implement an automation project has some risks associated with it:

☐ The project may not result in an adequate return on investment (ROI)

☐ The project may not be the best application or alternative available

☐ A decision not to implement the project may result in a poorer return

COSTS

It is important to try to identify all the potential costs involved in the project. If some costs are overlooked, the project may turn out to be a poorer investment than was anticipated. Typical costs that must be evaluated include:

1. **Implementation costs**
   Planning efforts
   System design
   Hardware and software development or procurement
   Database development
   Facilities rearrangement or construction
   System installation and debugging
   Interruption or impact on current operations
   Training, education, and communications

2. **Operational costs**
   Direct and indirect labor
   Maintenance of equipment and systems
   Management and controls
   System support (i.e., changes and enhancements)
   Operational expenses (e.g., utilities)

BENEFITS

It is not unusual for a project justification to do a more thorough job in identifying the costs than the benefits. This often results in projects not being implemented that should have been. Aside from the obvious savings that may result from reduced labor costs, a number of other factors should also be considered. These may not be as easy to quantify, but their benefits could be even more important:

☐ Reduced product design costs

☐ Reduced manufacturing cycle time and inventory

☐ Improved product quality

☐ Establishing commonality in hardware, software, or procedures

☐ Providing flexibility in manufacturing and product capabilities

☐ Meeting technical requirements in the application (e.g., process control, contamination)

☐ Competitive posture (e.g., market share potential)

An attempt should be made to quantify as many of these factors as possible. A thorough financial analysis should be conducted in order to estimate the potential ROI of the project. Computer programs can be used to help make this financial analysis easier, particularly when considering several alternatives. However, the ROI should not be the only consideration in the justification if other benefits cannot be easily quantified. The decision to implement a CAM project is a strategic one which will require some judgment beyond the standard financial formulas. There are even some truly "hidden" benefits which should be realized from any successful CAM project. Not only is it impossible to quantify them; it is also difficult to assess their importance. However, they can be real benefits that should not be overlooked, such as:

Accuracy and timeliness of data

Enrichment of jobs

Improved communications

Common goals and measurements

Team involvement in the project and the operation

21-7 PLANNING FOR CAM

CAM projects can be very large and complex. They may involve many people over a long period of time as well as significant investments in capital and expense. Even if a project has been justified and a decision made to implement it, a great deal of work is required to first prepare a thorough plan. Without proper planning, the project may experience significant problems during implementation and may even fail. Following are some of the key ingredients that should be built into the planning process.

WHERE TO START

A good plan must start with a solid foundation. This means that the planner must understand the manufacturing operation today and what it should be like in the future. To define this starting point requires the following basic elements:

☐ A statement of the objectives of the project (in terms of costs, productivity, quality, etc.)

☐ A description of the current manufacturing environment (e.g., nature of the business, production operation, competitive posture)

☐ A description of the future manufacturing environment

☐ A definition of the products and processes that will be involved

☐ An identification of any constraints on the project (e.g., investment, scope, magnitude)

☐ An inventory of the current manufacturing resources (i.e., people, skills, tools, processes, facilities)

Although these basic factors are important to every manufacturing operation, they are not always obvious and thoroughly understood. It is therefore worth spending some time at the start to establish this base before proceeding in a direction that may turn out to be misguided.

The type of business will influence what type of manufacturing best fits it. For example, an environment of complexity and change can affect the productivity of a manufacturing operation (Fig. 21-1). As the complexity and degree of change increase, the manufacturing operation must become more flexible. A manufacturing environment of low complexity and little change will, by its nature, look and operate differently than one at the other extreme. Where the business is and where it is headed in this spectrum can influence the type of automation to implement.

INGREDIENTS OF A PLAN

A comprehensive plan must cover all the critical elements of the project:

☐ Architecture of the system

☐ Hardware (process tools, automation equipment, computers)

☐ Software (system and application)

☐ Materials handling systems

☐ Management systems (e.g., manufacturing resources planning, or MRP)

☐ Manufacturing control systems (i.e., tool, process, and floor control)

☐ Manufacturing-line architecture

☐ Data communication system

☐ Physical facilities

☐ Design systems

☐ Computer tools

☐ Personnel and skills

☐ Organization

☐ Human factors (e.g., safety and ergonomics)

These elements must then be organized into a planning and implementation process. This usually involves several major stages of activity:

1. Defining the project
2. Developing preliminary designs of the system
3. Preparing detailed designs and specifications
4. Installing and debugging the system
5. Scaleup and operation

PLANNING TOOLS

Since the planning process for a large project can get very complex, tools have been developed to make the job easier. Many of these tools are computer-based for ease of use and flexibility. One example of such an approach is the "ICAM Definition" (IDEF) language developed by the U.S. Air Force Integrated Computer-Aided Manufacturing (ICAM) project. It is a standard method for the design and specification of manufacturing systems. The structure of IDEF permits it to be used to describe the entire architecture of a complex system. It can be used to understand the manufacturing operation and determine whether the proposed CAM design can actually integrate all the elements of the system.

The approach is based on a hierarchy of logical building blocks (Fig. 21-2). Each one describes some function in the system and its relationships to others in terms of inputs, outputs, and controls. These building blocks are tied together into a diagram of the system. Such a diagram can be expanded to several levels of detail where higher-level functions are broken down into lower-level ones. In Fig. 21-3, an IDEF model is used to describe a portion of a project plan. It identifies the major functions involved, such as defining the system architecture and equipment requirements. Controlling such factors such as manufacturing objectives and equipment specifications affect these functions. Inputs are required, such as the definition of the process and capacity requirements. The result of all these relationships is an output—in this case, a capital plan. The IDEF approach forces the project planners to identify all the important functions of the system and their relationships to each other. This describes the system and reveals links and dependencies

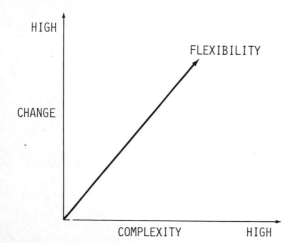

Fig. 21-1 Nature of the manufacturing environment.

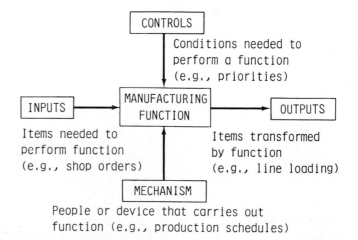

Fig. 21-2 IDEF building block.

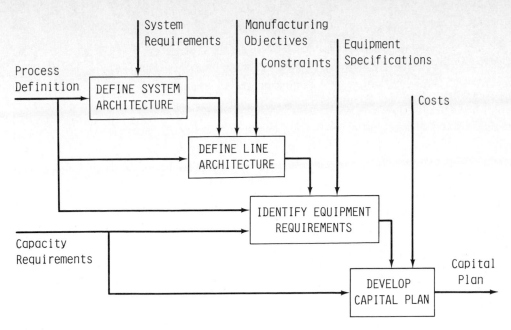

Fig. 21-3 Example of a simplified IDEF diagram of a portion of a project plan.

that need to be addressed in the design. This tool can even be used to design the planning process itself.

21-8 THE IMPLEMENTATION PROCESS

SCOPE

The implementation of CAM can affect a broad scope of the operations of a production facility. Implementation may involve not only the manufacturing organization but also the development and support organizations (e.g., finance, facilities, personnel). CAM integrates their activities with computer systems to automate both the factory floor and the management process which runs the facility. This integration occurs in three different forms:

1. **Information systems.** These are the data systems that support all the activities required to run the production operation. They include computer resources, communication links, and software.

2. **Physical automation.** This includes all the automated machines and materials handling systems in the production operation.

3. **Management process.** This includes the procedures and interfunctional relationships that determine the roles, responsibilities, and activities of organizations.

The implementation process must therefore address each of these areas. Unless all three are tied into the CAM system, it will not work. CAM is more than just robots and computers. It is the integration and automation of an entire business unit. If the CAM project addresses only the physical activities on the factory floor or belongs only to the information systems organization, it may fail. As we

have seen, most of the people and most of the costs involved in running a production facility are not directly associated with the physical manufacturing process. For CAM to improve the productivity of the entire operation, it must encompass all of the major activities involved.

KEY ELEMENTS

The implementation process involves four major stages of activity. For each of these, there are several key elements that are important to its success:

1. **Planning.** All CAM projects require thorough planning. The larger and more complex the project, the bigger this job becomes. Some guidelines for the planning process are:

☐ Schedule the implementation into stages rather than attempting to implement the ultimate system all at once

☐ Divide the plan into pieces which are manageable by individual teams for each area or system involved

☐ Provide adequate time and money for the installation and debugging phase

☐ Build flexibility into the plan so that changes and problems can be accommodated

☐ Involve all the organizations that will be affected by the implementation

2. **Preparation.** Once a plan is established, a lot of work is necessary to prepare for the installation of an automated system. Some of the key elements involved in this preparation phase are:

☐ Changing and standardizing operating procedures

☐ Developing databases

☐ Designing products for automated manufacturing

☐ Training employees and managers

☐ Establishing implementation teams

3. Installation. The actual installation of the systems is obviously a critical stage of the project. A lot of effort is involved and a lot can go wrong. Since this stage is the last one before full operation, manufacturing is dependent on its succeeding on schedule. Some of the key factors to assuring that success are:

☐ Applying an adequate number of experienced personnel and critical skills to the installation and debug activities

☐ Providing management attention and priority to the project

☐ Closely monitoring and reviewing the progress

☐ Having backup and alternative approaches to critical parts of the system

☐ Early warning of delays and problems

☐ Agreed-upon procedures for acceptance and qualification of equipment and systems

4. Operation. The final stage is, of course, the actual operation of the CAM system. Even though the project may appear to be completed at this stage, this stage can also be critical to success. Some of the key elements that should be built into the early stages of operation are:

☐ Close monitoring and measurement of performance

☐ Support of early learning and problems

☐ Gradual increases in production commitments

☐ A disciplined approach to system changes and improvements

☐ Provision for backup production capabilities

☐ Expanded and follow-on training of personnel

PREVENTING FAILURE

Many things can cause a complex project to have problems or even fail. Some of them are purely technical, such as the operation of a new automated machine. However, a number of basic elements in the implementation process itself can lead to failure. Some of the classic traps to avoid when implementing a CAM project include:

☐ Using inexperienced personnel and suppliers

☐ Not having the full support of upper management

☐ Committing to optimistic schedules

☐ Not planning the project thoroughly

☐ Inadequately training and educating personnel

☐ Not providing for flexibility and change

☐ Using traditional measurements of performance

☐ Becoming totally dependent on a single solution

☐ Not anticipating employee attitudes and problems

☐ Basing the project solely on financial considerations

These fundamental traps can be fatal, but they can easily be avoided. These are certainly not all of the possible traps, however. The success of a CAM project depends as much on how the implementation process is managed as it does on the hardware and software selected for the system.

21-9 SUMMARY

Although computer-automated manufacturing (CAM) is real, it has obviously not yet been implemented throughout industry in all manufacturing operations. CAM projects can be extremely complex, time-consuming, and expensive. The benefits of automation should be significant enough to justify the expense and effort. Despite the potential benefits, barriers to implementing CAM must be overcome before a project can be started. These barriers include a variety of technical, financial, and organizational considerations. CAM involves changes and risks which should be anticipated so that the project is adequately prepared to cope with them.

A successful CAM project starts with a thorough plan. The larger and more complex the project, the greater the planning effort required. The planning process should have the participation of all the organizations that will ultimately be affected by the implementation of CAM. The justification process is important to assure that the project makes good business sense and that better alternatives do not exist. Both quantitative and qualitative factors should be considered in the decision to implement CAM.

CAM integrates the activities of all the major functions involved in the manufacturing operation. Computer systems can automate data handling, physical production operations, and the management process. For a CAM project to improve the productivity of the entire factory, its scope must encompass all of these activities. There are many factors which could cause problems or even failure in a CAM project. The management of the implementation process can be as important as the technical solutions in preventing these problems.

GLOSSARY

accumulator A special temporary storage register in the CPU, used during arithmetic or logical operations. It handles the transfer of data by holding the data to be operated on by the arithmetic logic unit until it is ready and by receiving the results when it is done.

ADA A programming language developed by the U.S. Department of Defense and named after Ada Augusta Byron Lovelace, who is believed to be the first programmer in history. It is highly structured, modular, and relatively simple so that it is easy to use and maintain.

adaptive control A type of computer control system that compensates for sources of variability to optimize performance.

administrative data The types of information used to manage the manufacturing operation as a business on a daily basis, such as materials and labor costs.

AGV An acronym for automatic guided vehicle. This is a battery-powered vehicle that can move and transfer materials by following prescribed paths around the manufacturing floor without being physically tied to the production operation or being driven by an operator.

AI An acronym for artificial intelligence. AI is a field of computer science that deals with computers performing human-like functions, such as reasoning and interpretation. AI normally takes the form of a set of software that permits a computer to deal with very high level languages, adapt to sensory inputs, interpret data, and "learn" from experience.

algorithm A set of mathematical equations that has been developed based on some theoretical solution to a problem.

ALU An acronym for arithmetic logic unit. This is the section of the CPU of a computer which performs the logical and computational functions.

application software Computer programs developed to perform specific tasks for which the computer is being used, such as tool control or management reports.

APT An acronym for automatically programmed tool system. This is a symbolic language for NC applications. It uses relatively simple statements as instructions for common machine tool operations.

architecture The basic structure of the data flow of a computer system.

ASCII An acronym for the American Standard Code for Information Interchange. This is a seven bit binary code which was developed as a standard for data communications.

ASRS An acronym for automatic storage and retrieval system. It is an automated materials handling system for bulk or in-line storage of parts, materials, or products. It includes several different types of equipment, such as stacker cranes, narrow-aisle, and mini-load equipment.

assembler A program that translates a program written in an assembly language into machine language by replacing the symbolic code with its binary equivalent.

assembly language A machine-oriented, symbolic programming language which is somewhat easier to write programs with than machine language, but which is almost as efficient in directing the operation of a computer.

automation The mechanization and control of the physical movement, fabrication, or data handling operations in manufacturing.

auxiliary storage A computer storage device which provides high memory capacity for storing large amounts of data that is not currently being processed by the CPU. This is also known as secondary storage. Typical devices of this type include magnetic tape units and magnetic disk files.

backward chaining A technique used to solve problems in the logic process of expert systems. It selects a rule first and tries to match its consequence to the initial data or objective.

bar code A standard format of bars and spaces that can be printed on labels to identify parts or products which can be read automatically by a machine for data collection.

baseband A mode of data transmission involving only one stream of data on a line that connects a network of computer equipment.

BASIC An acronym for Beginner's All-Purpose Symbolic Instruction Code. This is a programming language designed for ease of use and interactive operation with very simple statements. It is used primarily for personal computers.

batch manufacturing The manufacture of parts or products in groups at each step of the process. It usually involves small quantities and a variety of types.

batch mode A mode of operation for a computer in which each program is executed sequentially, or one "batch" at a time.

BCD An acronym for binary coded decimal. This is the basic seven digit binary coding system used to represent numbers and characters in a form compatible with the on/off switching signals of a digital computer.

binocular vision A technique used for 3D vision systems which uses two cameras to obtain images from two different positions. This permits the system to compute the distances to all the edges and features on the object's surface. It is also known as stereo vision.

bipolar A solid state circuit technology which uses high speed, current-driven transistor devices.

bit slice A special type of microprocessor used for high performance applications. It is made up of small, high speed arithmetic logic units (e.g., 4- or 8-bit) that are connected in parallel to handle larger word sizes.

B-REP An acronym for Boundary REPresentation. It is a type of solid modeling technique which defines surfaces that bound solid objects.

broadband A mode of data transmission that allows multiple streams of data to be transmitted simultaneously using a system similar to cable TV.

bus A communication link that carries data between sections of a computer system or processor.

C language A programming language designed to be compatible with structured programming to make the program development job easier. It is a high-level language that is relatively easy to read and maintain while providing some functions that permit faster operations with less programming code than other languages.

CAD An acronym for computer-automated design or computer-aided design. This is an approach to using computers and computer graphics to automate engineering design tasks.

CAE An acronym for computer-automated engineering or computer-aided engineering. It means the technologies involved in using computers for engineering tasks such as design and analysis.

CAM An acronym for computer-automated manufacturing or computer-aided manufacturing. These are terms used to describe the general category of advanced approaches to manufacturing which use the power of the computer to automate the handling of data as well as the physical operations in the process. A formal definition developed by an industry group (Computer-Aided Manufacturing International or CAM-I) is: "The effective utilization of computer technology in the management, control, and operations of the manufacturing facility through either direct or indirect computer interface with the physical and human resources of the company."

CAPP An acronym for computer-assisted part programming or computer-automated process planning. This is an approach to using a computer to perform the computational work involved in programming tools to perform manufacturing tasks.

channels Computer control devices which direct the transfer of I/O data to minimize the impact of this operation on the CPU so it can concentrate primarily on computation.

CIM An acronym for computer-integrated manufacturing. It involves the use of the computer to tie together or "integrate" all the movement of data and product in the factory under the control of one complete manufacturing system.

closed-loop MRP An enhancement to materials requirements planning which adds capacity planning functions to adjust schedules or manufacturing resources in response to over or under scheduling conditions.

CMM An acronym for coordinate measuring machine. Numerically controlled tools that probe and record positions or coordinates on a workpiece.

CMOS An acronym for complementary metal oxide semiconductor. This is a low-power, high-density integrated circuit technology based on FET devices.

CNC An acronym for computer numerical control, which is a form of numerical control that uses a dedicated computer as the tool controller.

COBOL An acronym for COmmon Business Oriented Language. This is a programming language developed for large scale data processing tasks in business type applications. It uses statements that are relatively easy to read, but not easy to program.

communication protocol A standard data structure which permits different computer systems to exchange data.

compiler A program that converts an entire program from a high-level language into machine language. Once it completes the entire translation, the program can then be executed.

compliance device A spring-mounted fixture on the end of a robot arm that allows parts with chamfered edges to be grasped quickly and easily even when they are misaligned by several millimeters.

computer A machine that consists of an ALU, a memory, a control unit, and I/O devices that, through the use of a stored program, can process data.

computer control The use of a computer to control the operation of a machine, process, or tool. The computer translates the instructions of the programmer, monitors data from the machine or process, performs the necessary computations, and generates coded instructions to the controller.

concurrent processing The use of more than one processor operating in parallel to speed the process of computation in a computer system. This is also known as parallel processing, which can be accomplished by several different approaches to computer system architecture.

continuous manufacturing process A process which involves a continuos flow of materials through a series of process steps that eventually form a finished product.

continuous path NC An NC system which can simultaneously control more than one axis of motion and continuously control the path of a machine tool to generate the desired geometry of a workpiece.

contouring A continuous path NC operation, such as milling or turning, involving more than one axis of motion.

control-driven machine This is a form of parallel processing computer architecture. It uses multiple processors and memories to handle multiple streams of data through a complex, centrally controlled switching network. It is also known as "pipelining."

controlled path A mode of operation for a robot which uses a detailed control program and servo-control system to provide coordinated control of all the axes in terms of their position, velocity, and acceleration.

control system A system of hardware and software which controls the operation of a machine, such as a robot or NC machine tool. For motion control it may use either non-servo techniques, which control end points only, or a servo control of the path and speed.

conveyor A materials transport mechanism that provides a moving link between delivery points in a process. There are several types of conveyors, including continuous belt, roller, track, overhead trolley, and air tables.

coordinate system A geometric scheme to define positions in space, such as the length, width, and depth of a cube. Robots may be designed to use one of several different coordinate systems (i.e., rectangular, cylindrical, spherical, or revolute).

CPU An acronym for central processing unit. This is the heart of a computer which consists of an ALU, memory, and control unit.

CRT An acronym for cathode ray tube. This is the traditional technology used for computer displays using a vacuum tube and electron gun similar to a television monitor.

CSG An acronym for constructive solid geometry. This is a type of solid modeling technique. It uses a building block approach to create a 3D model out of simple shapes or primitives such as blocks or spheres.

CSMA/CD An acronym for carrier sense multiple access with collision detection. It is a type of communication control mechanism used in local area networks. It is a scheme based on avoiding conflict or contention between devices using the network by forcing each device wanting to use the network to wait until the line is clear.

DASD An acronym for direct access storage device. This is a memory in which the data can be stored and retrieved randomly but requires a combination of direct access and sequential searching to reach the specific storage location. A DASD is usually an auxiliary or secondary storage which is typically some form of magnetic disk storage device.

data administration The task of organizing data into a structured format so that it can be exchanged in a database system.

database A central file of information that can be used for many different purposes by many different people.

database management system The programs used to control central data files or databases.

data collection The task of providing information about the activities and results of production operations. A variety of different types of systems can be used to perform this task, either manually or automatically.

data communication The exchange of data between computer equipment connected by communication links, such as telephone lines or special cables.

data-driven machine A type of parallel processing computer architecture which uses a complex control scheme that permits instructions to be executed whenever sufficient data is available for them. It is also known as a "data flow" architecture.

data flow machine A type of parallel processing computer architecture which uses a complex control scheme that permits instructions to be executed whenever sufficient data is available for them. It is also known as a "data-driven" architecture.

debug routines Programs that permit the user to try out new programs or changes interactively with the computer.

decision support system (DSS) A form of artificial intelligence which uses knowledge representation to model the process and provide a rule-based expert system to predict or react to the performance of a production operation.

degrees of freedom (DOF) The number of axes or independent types of movement a robot manipulator can make. It takes three coordinates to locate the center of gravity of an object and three more to determine its orientation.

demand-driven machine A type of parallel processing computer architecture which uses different processors to handle specific types of

operations to optimize their efficiency. It is also known as a "reduction" architecture.

dextrous robot hand A general purpose robot gripper with features like the human hand to provide dexterity, agility, and sensing capabilities.

digitizer An analog input device used to trace drawings. It may use one of a number of different techniques, including electromagnetic, touch sensitive, or sonic.

DMA An acronym for direct memory access. A technique intended to improve the efficiency of the data transfer process by permitting data to be read or written directly to the memory from I/O devices without passing through the CPU. This is accomplished with a special bus line and DMA controller.

direct view storage tube This is a display device that uses a storage tube and a stroke writing approach. It can create high-resolution, flicker-free images at relatively low cost.

discrete manufacturing process A process which involves a series of individual and separate steps to form a finished part or product.

display An I/O device able to display alphanumeric characters or graphics. It is typically a video monitor much like a television.

distributed data processing (DDP) A computer system architecture which provides processing and storage capability at a low level in the system hierarchy using small computers.

distributed system A computer system configuration or architecture which uses small computers to reduce the dependency and demand on the large central host computer.

DNC An acronym for direct numerical control, which is a form of numerical control in which the tool control program is loaded by a direct link to a computer.

dynamic analysis The simulation of the behavior of objects when they are in motion. It is also known as kinematic analysis. Geometric models and FEA are used to create animated images of the movements of mechanisms and complex structures.

dynamic visual feedback This refers to computer graphics techniques that allow the user to see the effects of actions to create or modify images, such as rubberbanding or dragging.

EBCDIC An acronym for Extended Binary-Coded Decimal Interchange Code. This is a standard binary coding system which uses an 8 bit code plus a parity bit to define more characters (256) than the normal BCD code.

edge finding A technique used in vision systems to detect objects by recognizing the edges of their image.

EDIF An acronym for electronic design interchange format. It is a graphics standard which specifies a format for communicating design data for integrated circuits. It uses symbolic representations similar to the LISP language.

end effector The tooling at the end of a robot arm, such as a gripper or tool which permits it to perform tasks, such as picking up objects.

EPROM An acronym for erasable programmable read only memory. This is a type of ROM which can be reprogrammed by erasing the stored data with either ultraviolet light of electrical signals and then writing in new instructions.

Ethernet A control mechanism for data communications in a LAN based on a standard scheme called CSMA/CD. It was developed by Xerox.

event driven A mode of operation for computer systems when they must respond directly to signals generated by control devices in the process or tools being controlled.

expert system A form of AI in which a computer system provides decision-making capabilities for specific applications that require expert knowledge.

fault tolerance The ability of a computer system to continue to operate without losing data when a failure occurs. This can be accomplished by redundancy in the hardware or error-recovery software.

feature matching An image analysis technique used in vision systems that involves gray scale interpretation to match objects to design rules and shape characteristics.

feedback control A type of process control technique in which an error is detected at the output of a process step and the correction is made at the input.

feedforward control A type of process control technique in which the control signal is sent ahead in the process sequence to compensate for the deviation that was detected.

FET An acronym for field effect transistor. A type of solid state circuit technology that uses voltage-driven semiconductor devices that are capable of high-density and low-power operation.

finite element analysis (FEA) A mathematical technique that is used to calculate stresses in mechanical structures. It involves separating an object into many small, uniform pieces or elements to describe the behavior of each one when forces are exerted on the entire structure using stress and deflection equations.

firmware Programs that are stored permanently in hardware in the form of ROM.

fixed automation The mechanization and control of a fixed or repetitive sequence of physical operations.

flexible automation The mechanization and control of a sequence of operations which can be reprogrammed or changed.

FMS An acronym for flexible manufacturing system. A computer control system that ties together multiple manufacturing operations into an integrated production process which fabricates a finished product. A formal definition of an industry group (The Machine Tool Task Force) is: "A series of automatic machine tools or items of fabrication equipment linked together with an automatic materials handling system, a common hierarchical digital pre-programmed computer-control, and providing for randomly fabricating parts or assemblies that fall within predetermined families."

FORTRAN An acronym for FOrmula TRANslation. This is a high-level programming language which uses common mathematical notation for programming engineering and scientific problems.

forward chaining A technique used to solve problems in the logic process of expert systems. It starts with the initial data or objective and searches for a rule whose consequence matches.

fuzzy logic A problem solving technique used in AI applications. It uses a line of reasoning which is developed from uncertain or partial evidence, similar to human reasoning. For example, the term "usually" defines a relationship that is not precise, but may be helpful in solving a problem.

geometric model A mathematical representation of the geometry of objects which is used by a computer to display and manipulate images as data.

geometric transformation The process in a computer graphics system which converts data from a geometric model of an object into an image which can be displayed on a graphics screen.

geometry engine A special computer processor which performs complex geometric computations using floating point arithmetic and parallel processing techniques.

graphics The hardware and software technology involved in generating graphical representations of objects and data on a computer display device.

graphics processor A special set of hardware and software which is used in computer graphics systems to create and display the geometric model of an object. It converts the digital signals from the CPU into graphics commands for function generators that produce standard shapes, such as circles or lines.

gray scale imaging A technique used in vision systems to analyze and interpret the details of 3D shapes using shades of gray.

gripper A mechanism on the end of a robot arm which permits it to grasp and pick up objects. It may be a simple clamp or a complex device with sensors and multiple "fingers" like a human hand.

GT An acronym for group technology. An approach to classifying standard parts based on their physical characteristics so that they may be grouped together with other similar parts that could be manufactured using the same process.

hardware The "visable" computer. It is comprised of the physical technologies, such as the electronics, that make up the various types of data processing equipment.

heuristic A type of reasoning process based on experience or the use of empirical data (e.g., "rules of thumb").

hierarchical system A computer system architecture which is made up of several levels of processors and controllers between the points at which data is collected or generated and the central host computer which stores the database and controls the system.

high-level language A computer programming language that uses simple codes similar to normal speech to communicate complex instructions.

host computer A large computer in a system of computers which stores the common database.

human factors Considerations in the design of a workstation or piece of equipment that deals with the effects on humans. This includes operator safety, convenience, and efficiency. It may involve making the operator more comfortable, making the task easier, or preventing hazards.

IDEF An acronym for Icam DEFinition language. It is a standard method used for the design and specification of manufacturing systems developed by the U.S. Airforce ICAM project. The structure of IDEF permits it to be used as a planning tool to describe the entire architecture of a complex system.

IGES An acronym for Initial Graphics Exchange Specification. It is a computer graphics standard which specifies a file structure and language format for communicating geometric data.

image analysis or processing The process by which a computer describes a camera image in terms that can be related and compared to known objects. This is a form of AI used in vision systems.

inference engine The logic processor in an AI system. It solves problems by deducing conclusions using rules and facts available in the knowledge base. It is a computer that is programmed to process symbols that represent objects and their relationships.

integrated circuit A semiconductor device that incorporates many individual electronic components into a single chip.

integrated manufacturing system A computer system which ties together or integrates all the technical, logistical, and administrative data used to control a manufacturing operation.

intelligent robot A type of computer-controlled robot that is not restricted in its actions to only the preprogrammed instructions. Intelligent robots are automated manufacturing systems which integrate ma-

nipulation, sensing, computation, and control functions to perform complex tasks.

interactive computer graphics The computer technology which permits graphics displays to be created and modified interactively with the user.

interactive input Techniques used in computer graphics which permit the user to create and change images frequently. These include the use of high-level commands and interactive devices, such as light pens and function keys.

interpreter A program that both translates and executes programs which are written in a high-level language. It converts each instruction of the program, one at a time, as it is being executed.

interrupt A method for an I/O device to communicate with a CPU in which the operation of the CPU is interrupted to transfer data. The CPU must keep track of where it stopped so that it can return to the next program instruction after it has completed the data transfer.

I/O An acronym for input/output. The process of providing data to the computer (input) or to people or other machines (output).

I/O devices Computer equipment that is used to provide data to the computer (i.e., input devices such as a keyboard) or to people or other machines (i.e., output devices such as a display or printer).

I/O interface Circuits that control the transfer of data between I/O devices and other computer equipment. The I/O interface is needed to convert signals due to differences between the I/O devices and the CPU in terms of voltages, parallel versus serial data transfer, or data transfer rates.

I/O routines Programs that interpret data transfer instructions (e.g., READ and WRITE) and control the movement of data between the I/O devices and memory.

JIT An acronym for just in time. A production scheduling method used to optimize the flow of the manufacturing process. It requires that the parts and materials get to the proper workstation when they are needed—not too early or not too late.

job shop Manufacturing operations that deal with small quantities of parts that may often be customized for specific applications.

kan ban A technique developed in Japan to optimize production scheduling. It uses a reorder card for every workpiece or batch of parts as a simple way to identify when more parts are needed.

keyboard A computer input device which uses keys to represent specific functions such as alphanumeric characters, directions or special commands.

kinematic analysis The simulation of the behavior of objects when they are in motion. It is also known as dynamic analysis. Geometric models and FEA are used to create animated images of the movements of mechanisms and complex structures.

knowledge base The database used in an expert system in which all the rules and facts are stored to solve a particular problem.

knowledge engineering A field of AI involved in expert systems. It includes the development of problem-solving techniques and the acquisition and maintenance of knowledge bases.

knowledge representation The forms in which knowledge is stored so that it can be used in the problem-solving process of expert systems. They include constraints, assertions, rules, and certainty factors.

ladder diagram A symbolic programming technique which is used for programmable controllers. It simulates the electrical circuits of switches or relays that were used for many years to control electrical equipment.

LAN An acronym for local area network. A data communication system which connects computer equipment in a restricted geographical area. It uses techniques which provide high-speed data communication over relatively inexpensive transmission lines that would not be possible for a wide area or long distance network.

languages Coding schemes which are used to communicate a set of instructions or program to the logic circuits of a computer.

lead-through programming A technique for programming robots which uses a control panel, called a "teach pendant," with buttons or switches that control the motion of the robot through a cable connected to the control system. The operator or programmer can lead the manipulator through a task, one step at a time, recording each incremental move along the way.

light pen A computer input device used in interactive graphics to point at objects on a display screen.

linear programming An iterative method of problem solving using a computer. The problem is described in mathematical form and different solutions are tried until one is found that yields the best results.

LISP An acronym for LISt Processing. This is a logic programming language used for AI applications. The programs and data are structured as lists so that the statements are very simple and easy to program.

lithography A technology used to create microscopic images during the fabrication of semiconductor integrated circuit devices. Optical, electron beam, and x-ray exposure techniques may be used.

logic programming A type of programming technique used in AI applications. The programs reach a conclusion by matching the objects and rules with the goals or objectives of the problem.

logical design The stage of the electronics design process for integrated circuit devices which defines the functional requirements in terms of logic diagrams and circuit schematics.

logistical data Information that controls the flow of materials and products through the manufacturing process, such as quantities, schedules, and routings.

logistics system The process of controlling the flow of materials and products on the factory floor. A logistics system deals with such information as quantities, types of parts, schedule dates, priorities, and order status.

machine cell A set of complementary, computer-controlled machine tools that together perform a series of machining operations on a workpiece.

machine language The lowest level programming language. It uses a binary code which directly relates to the on/off switching operations of the computer logic circuits.

machining center A single machine tool which is designed to perform a variety of machining operations such as drilling, milling, boring, reaming, and tapping.

macroassembler A technique used when programming in assembly language to save time for the programmer. It assigns a name to a sequence of instructions that needs to be repeated in a program so that it can be assembled as an entire block of code.

magnetic stripe reader A type of data collection system which uses a strip of magnetic tape to store information that can be read automatically into a computer by scanning with a magnetic wand.

main memory The storage section of a CPU which stores the program instructions and data being processed by the computer at that time.

management system The process involved in managing the technical, logistical, and administrative data in a production operation. It includes functions such as process planning, production control, and management reporting.

manipulator The base and arm assembly of a robot which performs the movements and tasks.

manufacturing control system A computer system used to handle the technical data for process or tool control applications in manufacturing.

manufacturing technology The processes, equipment, and tools which are used to manufacture products.

MAP An acronym for manufacturing automation protocol. A token ring–based data communication protocol compatible with the International Standards Organization model of a standard system architecture. It was developed by General Motors for manufacturing applications to permit many different types of equipment to be connected together in a common manufacturing control system.

materials handling system The automation of the steps in a manufacturing process which involve the movement or handling of the product or materials (e.g., machine loading and unloading, machine-to-machine transfer, in-line movement of WIP, and storage).

memory A storage function in a computer system which retains data for use during the operation of a computer program.

microcomputer A small computer whose CPU is a microprocessor.

microinstructions The programs which control the internal operations of a microprocessor. They are stored in a microprogram memory such as ROM or PROM.

micromanipulator A special type of robot end-of-arm device used to perform precise alignments or accurate probing tasks.

MPU An acronym for microprocessor unit. A single, integrated circuit semiconductor device that contains all the arithmetic logic function of a CPU.

microprogramming A programming technique used to increase the performance of microprocessors. The user can specify a special instruction set and word size which is different from those provided with a standard microprocessor.

minicomputer A machine with all the basic processing and storage functions of a computer in a small package with limited capabilities.

miniload system A type of automatic materials storage system which is relatively small and used to store and retrieve materials or small parts within a production line. It usually takes the form of a rotating carousel with a picker mechanism.

modeling In computer applications, it means using mathematical equations to represent objects or events, such as geometric models or simulation models of manufacturing lines.

modem An acronym for MOdulator/DEModulator. A device which translates the digital signals of a computer to audio signals that can be transmitted through the telephone.

mouse A type of computer input device used as a positioning tool for graphics applications.

MRP An acronym for materials requirements planning. A computer-based, comprehensive approach to planning, scheduling, and controlling production activities. The production schedules are based on actual manufacturing status as well as planned requirements.

MRP II An acronym for manufacturing resources planning. A total manufacturing management system which ties in the resource and financial implications of production decisions. It involves the control of factors which cause schedule changes to occur, such as product design changes, production problems, or changes in customer requirements.

MTBF An acronym for mean time between failure. A measure of the availability of computers or production equipment which identifies how long a period of time can be expected between failures occurring.

MTTR An acronym for mean time to repair. A measure of availability of computers or production equipment which identifies how long a period of time can be expected to repair a failure when it occurs.

multipoint network A type of computer network in which a number of terminals or satellite computers share one line to the host computer.

multiprocessing A type of computer system architecture in which more than one processor is used simultaneously to increase the speed of computation for a particular task.

multiprogramming A mode of operation of a computer system in which the operating system must control the execution of multiple programs at the same time. This technique was developed to increase the efficiency and utilization of the CPU.

natural language The native language of human speech. This is the ultimate form of making computer programming and the human interface easy to use.

NC An acronym for numerical control. A form of programmable automation for machine tool control applications. An NC system is comprised of a control program, a control unit, and a machine or tool.

network A group of computer systems which are tied together with data communication links which permit them to share or exchange data.

noncontact sensors A type of sensing device that generates signals using transducers that are not in physical contact with the objects involved. Examples include optical and magnetic sensors.

object code The basic binary code which is used to communicate with the on/off switching signals of the computer logic. It is also known as "machine language," since it is the only code that the computer understands.

object hierarchy An approach to creating geometric models from building blocks to improve the efficiency of use of computer resources as well as save design time. The model is made up of a hierarchy of primitive graphics commands and transformations.

operating system A set of special programs which manage all of the hardware resources of a computer system during its operation and provide the primary interface to the user.

OCR An acronym for optical character recognition. A computer input device which can read written or printed characters from paper or forms and convert them into data which can be stored in a computer.

PIA An acronym for parallel interface adapter. Bidirectional data transfer circuits which are used for direct communication between the I/O devices and the CPU.

parallel processing A computer system architecture in which more than one processor is used to perform computations simultaneously to increase overall system performance.

part manufacturing A fabrication process that usually involves discrete operations on batches of objects to create parts.

part programmer An individual who prepares a program of instructions which is used to operate an NC machine.

part programming The process of planning and specifying every step and movement of an NC machine into a complete process sequence in the form of an NC program.

PASCAL A programming language named after the famous French mathematician Blaise Pascal. It is a very structured and easy to read language which is used on small computers.

PC An acronym for programmable controller. Small, dedicated computers with logical, but not computional capabilities, used to control machine tools.

PDDI An acronym for product definition data interface. This is a standard for the exchange of design data developed by the U.S. Airforce ICAM project.

physical design The stage of the electronics design process which involves the physical layout of the functional design. This includes the design and placement of circuit elements and their interconnections.

pick and place Tasks involving relatively simple loading and unloading operations which can be mechanized by either fixed or programmable automation.

pipelining A type of parallel computer processing in which multiple processors and memories handle multiple streams of data through a complex, centrally controlled switching network.

plotter A type of computer output equipment used for graphics applications. It may use pens or electrostatic techniques to trace the graphics image as an outline drawing.

point-to-point network A type of computer system network in which each terminal or satellite computer is connected directly to the host computer on its own communication line.

point-to-point programming A type of NC programming in which the tool is moved to a predetermined position with no control over the tool's speed or path.

printed circuit An electronics packaging technology which incorporates printed wiring patterns into a planar substrate which is used to interconnect electronic devices. The printed circuit may be made of organic or ceramic materials and have a number of internal layers.

process control The monitoring, comparing, and controlling operations of a manufacturing process to maintain its operation.

process manufacturing A manufacturing operation which involves a continuous flow of materials through a series of process steps that eventually form a finished product.

process planning The interpretation of information about the design of a part to be fabricated (such as from an engineering drawing) into a description of manufacturing process steps (such as a routing).

product manufacturing The final process which assembles parts and subassemblies into a functional end product, such as a machine.

program A set of coded instructions in an ordered sequence which tell a computer to carry out certain arithmetic or logical operations.

program development system Software tools which are used by a programmer to develop computer programs. Such a system typically includes an assembler or compiler, a text editor, and a debugging program.

programmable automation Using computer control to reprogram or change the sequence or control of a mechanized operation.

programmable control The use of a computer to control and change the operation of a machine through programmed instructions.

programming Developing a set of coded instructions which tell a computer to carry out certain arithmetic or logical operations.

programming languages Symbolic coding schemes used to communicate with the logic circuits of a computer.

PROLOG An acronym for PROgramming in LOGic. This is a programming language used for AI applications which is a derivative of LISP. It is based on the symbolic logic of inference processing and consists only of declarative statements that define relationships between objects and data.

PROM An acronym for programmable read only memory. A type of ROM which can be programmed permanently with electrical signals to store unique instructions for a particular microcomputer application.

protocol A standard data structure which permits different computers to exchange data through a communications network.

PROWAY An acronym for PROcess data highWAY. This is a token bus data communication standard which was developed by the U.S. committee of the International Electro-Technical Committee for industrial use.

pull system A method of controlling inventory which does not allow WIP to move until it is needed at the next process step. This approach can prevent excessive buildups of inventory which create bottlenecks in the process flow.

push system A method of managing a production operation which allows work to move to the next process regardless of the conditions at that step. This may cause WIP to build up in front of steps that have slow throughput or experience problems.

quality control Activity and procedures designed to assure that the manufacturing process does not produce defective products. It is usually performed by a quality control organization whose job is to acquire and analyze data which will determine whether the process is under control and the product being manufactured will meet its specifications.

queue A waiting line or backlog. In manufacturing, it refers to the magnitude of jobs waiting to be processed at a workstation.

RAM An acronym for random access memory. A memory in which data can be stored and retrieved directly from addressable locations. RAMs are typically internal memories using high-speed semiconductor devices.

range finding A technique used in 3D vision systems which uses light reflected from an object's surface as a depth cue to determine its distance and shape. It is similar in concept to radar.

raster scan CRT A type of graphics display technology which generates images by scanning a CRT screen with an electron beam, as in a TV, and illuminating specific picture element positions.

real time system A type of computer system environment in which the operating system is designed to respond to inputs from multiple sources very rapidly so that the user perceives that there is no time lost.

reduction machine A type of parallel-processing computer architecture in which different processors are dedicated to handle specific types of operations to optimize their efficiency. This is also known as a "demand-driven machine."

register Small memory device that can receive, store, and transfer data. Registers are used in the CPU for temporarily storing data during computer operations.

relational database A type of computer database which uses the relationships that exist between sets of data to retrieve information. The data is stored as 2D tables of rows and columns that the user can select from with a simple query language or menu-driven interface.

ring network A type of data communication architecture in which each device is connected to a common communication line and gains access by some control scheme in the data structure (e.g., by passing a "token").

RISC An acronym for reduced instruction set computer. A type of computer architecture in which the number of instructions used in a program is reduced to increase the speed in which the computer can execute those instructions.

robot An automated tool that can be programmed to perform a wide variety of manufacturing tasks using human-like capabilities. A formal definition developed by an industry group [Robot Institute of America (RIA)] is: "A reprogrammable, multifunctional manipulator designed to move material, parts, tools or specialized devices, through a variety of tasks."

robotics A technology comprised of a mechanical manipulator, sensors, software, and computer control which provides programmable automation capabilites.

ROI An acronym for return on investment. A method to evaluate the economics of alternative solutions to business decisions. It is a calculation which determines the rate of return and payback period for an investment.

ROM An acronym for read only memory. A memory in which the data is stored permanently and can only be read out. ROMs are used primarily as internal memories to store small control programs that can be accessed frequently at high speeds.

secondary storage A memory which is external to the CPU which is used to store data that is waiting to be processed or record data for future use. This is also called "auxiliary storage."

semantic network A technique used to represent knowledge in expert systems. It represents relationships between objects and provides a mean for efficient logic processing.

sensor A device which is designed to detect, measure, or record physical phenomena, such as pressure or temperature. Sensors are used to provide machines, such as robots, with capabilities similar to the human senses, such as touch or vision.

servo control A motion control system which is capable of controlling the velocity, acceleration, and path of motion as well as the end points. It uses control programs, electronic controllers, and sensors.

shop floor control system The control of the flow of the product and materials on a factory floor involving the quantities, types of parts, schedule dates, priorities, and the status of jobs and orders.

silhouette matching An image analysis technique used in 2D vision systems which uses the light reflected off an object to create a silhouette or outline that is compared to a reference image or template stored in memory.

silicon compiler A software tool that automatically generates detailed logical and physical designs of an integrated circuit from high-level functional specifications.

simulation The use of a model of an operation to evaluate its behavior under varied conditions.

software Computer programs that have been developed to accomplish specific tasks.

software development system A set of software tools which help the programmer develop computer programs. Such a system typically includes an assembler or compiler, a text editor, and a debugging program.

solid model A true mathematical representation of a solid object. It includes all the information necessary to define both the surface and interior of an object as well as its mass properties.

stacker crane A type of automated storage system which uses a crane-like mechanism to store and retrieve materials in high racks or "stacks" in a warehouse. It uses the "air space" to provide higher density storage that would normally be achieved with conventional warehousing techniques, such as forklift trucks.

star network A type of data communication system which connects all the devices through a central computer. Each device must have its own cable and communicate all messages through the central computer.

stereo vision A technique used in 3D vision systems which uses two cameras to obtain images from two different positions which are used to compute the distances to all the edges and features on the object's surface. This is also known as binocular vision.

straight cut NC The moving of a cutting tool parallel to an axis at a controlled rate of speed, such as in an NC milling operation.

structured programming A technique used to deal with large programming tasks. It uses a very organized program structure which is designed to have a form which matches the meaning of the program. This approach makes it easier for the programmer to complete the whole job without getting lost in the complexity of the program.

synchronous/asynchronous Modes of data communication which permit computer equipment to send and receive data from each other. The synchronous mode transmits data in a continuous stream of characters which are divided into blocks of time that are synchronized between the equipment. The asynchronous mode uses stop and start bits before and after each character.

symbolic programming A type of programming used in AI applications which manipulates symbols that represent objects and their relationships.

system A data processing or computer system comprising all the hardware, software, and data tied together to perform a particular set of tasks. A basic computer system is made up of a CPU, main memory, input, and output peripheral equipment, with operating system software and applications programs.

system software The programs that direct the internal operations of the computer while it is executing instructions from the user. They include functions such as translating languages, managing computer resources, and developing programs.

tactile sensor A type of contact sensor which detects objects by touch. It may use a microswitch or pressure-sensitive elements.

teach pendant A control panel which can be held by an operator or programmer of a robot which is used to program the motion of the robot. It has buttons or switches which are connected through a cable to the control system. The manipulator is led through a task, one step at a time, recording each incremental move along the way.

technical data system A computer system which controls the technical information about the design and manufacture of a product. It drives the tools and physical manufacturing activities on the factory floor.

teleprocessing The transmission of data between remote locations over telephone lines.

test data generation The stage of the electronics design process where programs use design data to generate test patterns which can be used by production test equipment to detect faults if they occur.

text editor A program that helps a programmer make changes to software.

time-sharing A type of computer architecture which permits the CPU to handle data from several sources at the same time. It uses delays between memory access and data transmission cycles to process other available data.

token ring A type of data communication control mechanism used in LAN. It is a ring network architecture which provides access for devices to use the network by passing a "token" or control bit around the ring.

tool changer Mechanisms which automatically change tools on a machine under program control. It may take the form of a carousel with a variety of cutting tools for an NC machining center or a special device at the end of a robot arm which provides for quick changes of the end-effector or tool.

tool control The control of the operation of a piece of manufacturing equipment or tool by a predetermined program of instructions.

transducer A device that converts energy from one form to another, such as pressure into electrical signals. Transducers are the basic mechanism used to provide sensing functions for machines.

Unix A standard operating system used by small computers developed by AT&T. It is a highly structured hierarchical system which can easily be expanded and modified for a wide variety of functions.

utility programs The system software which performs a variety of func-

tions to help the user write and execute programs. It includes I/O routines, text editors, and debug routines.

vector refresh CRT A type of computer graphics display device which uses a refresh tube and stroke writing to draw interactive images on a CRT screen. The image is regenerated at a high frequency to sustain it so that it does not appear to flicker.

virtual storage A computer system architecture which significantly expands the effective size of the computer's main memory without actually increasing it physically. Only those instructions that are needed at the specific time of execution of each portion of a program are stored in the main memory.

vision system An application of AI techniques in which a computer system is used to analyze and interpret 2D camera images to infer 3D objects.

voice recognition system A technique for automatic data collection that allows operators to record data by talking directly to the computer. Voice recognition systems can translate a limited vocabulary of numbers and words into digital signals that can be transmitted to a computer.

walk-through programming A method used to program robots which involves counterbalancing the manipulator arm so that it can be moved manually through the intended motions while its path is being recorded by the control system. It is also known as the "guiding" or "playback" method.

wireframe model A type of geometric graphics modeling technique which uses a collection of lines and arcs that are connected to represent the boundary edges of an object.

work envelope The maximum reach or range of movement of a robot arm. It will vary in shape and size depending on the configuration and size of the manipulator.

workstation When used in relation to computers, it refers to an intelligent terminal or small computer which provides the user with local computational capability to reduce the dependency on remote processing on a mainframe computer. When used in relation to manufacturing, it refers to a single process operation which is run by a production operator.

world coordinates Graphics data which identifies the location of an object in terms of measurements that relate to the real world of the object.

world modeling A function in robot programming which specifies the position of the robot in terms relative to other objects. The position data may be obtained from sensors or geometric models.

REFERENCES

CAM/CIM—IN GENERAL (Chapters 1–5)

Illinois Institute of Technology: *APT Part Programming*, McGraw-Hill, New York, New York, 1967.

Kutcher, M. M.: *How CAD/CAM Has Been Applied to Customized Mass Production*, Society of Manufacturing Engineers, MS72-919, 1972.

Bradford, R., and J. E. Stuehler: "Using Process Data to Control and Optimize Manufacturing Processes," *IEEE Transactions on Manufacturing Technology*, June 1972.

Childs, J. J.: *Numerical Control Part Programming*, Industrial Press, New York, 1973.

Harrington, Joseph P., Jr.: *Computer-Integrated Manufacturing*, Industrial Press, New York, 1973.

Rembold, Ulrich, Mahesh K. Seth, and Jeremy S. Weinstein: *Computers in Manufacturing*, Marcel Dekker, New York, 1977.

Groover, Mikell P.: *Automation, Production Systems and Computer-Aided Manufacturing*, Prentice-Hall, Englewood Cliffs, NJ, 1980.

Besant, C. B.: *Computer-Aided Design and Manufacture*, John Wiley, New York, 1980.

Gunn, Thomas G.: *Computer Applications in Manufacturing*, Industrial Press, New York, 1981.

Krouse, John K.: *What Every Engineer Should Know About Computer-Aided Design and Manufacturing*, Marcel Dekker, New York, 1982.

Computers in Manufacturing: Execution and Control Systems, Auerbach Publishers Inc., Princeton, NJ, 1982.

"Data-Driven Automation," special issue of the *IEEE Spectrum*, May 1983.

Hegland, Donald E.: "The Automated Factory—A Progress Report," *Production Engineering*, June 1983.

Groover, Mikell P. and Emory W. Zimmers Jr.: *CAD/CAM: Computer-Aided Design and Manufacturing*," Prentice-Hall, Englewood Cliffs, NJ, 1984.

"The Promise of Automated Manufacturing," special issue of *Production Engineering*, September 1984.

Gunn, Thomas G.: "CAD/CAM/CIM: Now and in the Future," *Industrial and Control Systems*, April 1985.

"Computers in Manufacturing," special issue of the *IBM Journal of Research and Development*, vol. 29, no. 4, July 1985.

"Integrated Manufacturing," special section of *Production Engineering*, September 1986.

COMPUTER TECHNOLOGY—IN GENERAL (Chapters 6 and 7)

Gear, C. William: *Computer Organization and Programming*, McGraw-Hill, New York, 1980.

Shell, Gary B. and Thomas J. Cashman: *Introduction to Computers and Data Processing*, Anaheim Publishing Co., Anaheim, CA, 1980.

Bohl, Marilyn: *Information Processing*, Science Research Associates, Chicago, 1980.

Stone, Harold S. (ed.): *Introduction to Computer Architecture*, Science Research Associates, Chicago, 1980.

"Computer Technology," special issue of the *IBM Journal of Research and Development*, vol. 25, no. 5, September 1981.

Vles, John M.: *Computer Fundamentals for the Non-Specialist*, AMACOM, New York, 1981.

Bernhard, Robert: "Giants in Small Packages," *IEEE Spectrum*, February 1982.

Schaefer, David H. and James R. Fisher: "Beyond the Supercomputer," *IEEE Spectrum*, March 1982.

Graham, Alan K.: "Software Design: Breaking the Bottleneck," *IEEE Spectrum*, March 1982.

"Computer Software," special section of the *IEEE Spectrum*, August 1982.

Manuel, Tom: "Computers People Can Count On," *Electronics*, January 27, 1983.

Douglas, John H.: "New Computer Architectures," *High Technology*, June 1983.

Jurgen, Ronald K.: "Our Computers Aren't Speaking," *IEEE Spectrum*, September 1983.

Torrero, Edward A. (ed.): "Next Generation—Tomorrow's Computers," special issue of the *IEEE Spectrum*, November 1983.

Alexander, Tom: "Reinventing the Computer," *Fortune*, March 5, 1984.

Manuel, Tom: "Computer Security," *Electronics*, March 8, 1984.

Lerner, Eric J.: "Data Flow Architecture," *IEEE Spectrum*, April 1984.

Davis, Dwight B.: "Super Computers: A Strategic Imperative," *High Technology*, May 1984.

Manuel, Tom: "The Coming Surge in Data-Base Systems," *Electronics*, May 17, 1984.

Evanczuk, Stephen: "New Tools Boost Software Productivity," *Electronics Week*, September 10, 1984.

"Computer Software," special issue of *Scientific American*, September 1984.

Bear, Jean-Loup: "Computer Architecture," *Computer*, October 1984.

Helms, Harry (ed.): *The McGraw-Hill Computer Handbook: Application, Concepts, Hardware, Software*, McGraw-Hill, New York, 1985.

Torrero, Edward A. (ed.): *Next Generation Computers*, IEEE Press, New York, 1985.

"Technology '86," special issue of the *IEEE Spectrum*, January 1986.

Freedman, David H.: "Programming Without Tears," *High Technology*, April 1986.

MICRO- and MINICOMPUTERS (Chapter 8)

Sloan, M. E.: *Introduction to Minicomputers and Microcomputers*, Addison Wesley, Reading, MA, 1980.

Ramirez, Edward V. and Melvyn Weiss: *Microprocessing Fundamentals—Hardware and Software*, McGraw-Hill, New York, 1980.

Lines, M. Vardell and Boeing Computer Services Co.: *Mini-Computer Systems*, Winthrop Publishers, Inc., Cambridge, MA, 1980.

Bennett, William S. and Carl F. Evert: *What Every Engineer Should Know About Microcomputers*, Marcel Dekker, New York, 1980.

Capice, Raymond P. and John G. Posa (eds.): *Microprocessors and Microcomputers—One Chip Controller to High End Systems*, McGraw-Hill, New York, 1981.

Bernhard, Robert: "Less Hardware Means Less Software," *IEEE Spectrum*, December 1981.

Jikhambata, Adi: *Microprocessors/Microcomputers—Architecture, Software and Systems*, John Wiley, New York, 1982.

Lu, Cary: "Microcomputers: The Second Wave," *High Technology*, September/October 1982.

Hall, Douglas V.: *Microprocessors and Digital Systems*, McGraw-Hill, New York, 1983.

Guter, Fred: "Chip Architecture: A Revolution Brewing," *IEEE Spectrum*, July 1983.

Evanczuk, Stephen: "Bell Lab's Unix Operating System Spreads Powerful New Wings," *Electronics*, July 28, 1983.

Lu, Cary: "Dawn of the Portable Computer," *High Technology*, September 1983.

Davis, Dwight B.: "Super Micros Muscle Into Mini Markets," *High Technology*, December 1983.

Hindling, Harvey J.: "Software Portability," *Electronics*, December 1, 1983.

Jasany, Leslie C.: "Tying the Factory Together With PCs and Networks," *Production Engineering*, April 1984.

Evanczuk, Stephen: "Integrating the Engineer's Environment," *Electronics*, May 17, 1984.

Lu, Cary: "Making Computers Easy to Use," *High Technology*, July 1984.

Lu, Cary: "Integrated Software Bids for Center Stage," *High Technology*, November 1984.

Ohr, Stephan: "RISC Machines," *Electronic Design*, January 10, 1985.

Stanley, Robert C.: "Microprocessors in Brief," *IBM Journal of Research and Development*, vol. 29, no. 2, March 1985.

Zorpette, Glenn: "Computers That Are Never Down," *IEEE Spectrum*, April 1985.

ARTIFICIAL INTELLIGENCE (Chapter 9)

Barr, Avon and Edward K. Feigenbaum: *The Handbook of Artificial Intelligence*, William Kaufman, Los Altos, CA, 1982.

Kinnucan, Paul: "Artificial Intelligence: Making Computers Smarter," *High Technology*, November/December, 1982.

Rich, Elaine: *Artificial Intelligence*, McGraw-Hill, New York, 1983.

Kinnucan, Paul: "Machines That See," *High Technology*, April 1983.

Feigenbaum, Edward and Pamela McCorduck: "Land of the Rising Fifth Generation Computer," *High Technology*, June 1983.

Winston, Patrick: *Artificial Intelligence*, Addison Wesley, Reading, MA, 1984.

Kinnucan, Paul: "Computers That Think Like Experts," *High Technology*, January 1984.

Poggio, Tomaso: "Vision by Man and Machine," *Scientific American*, April 1984.

Zadeh, Lotfi A.: "Making Computers Think Like People," *IEEE Spectrum*, August 1984.

Lerner, Eric J.: "Why Can't a Computer Be More Like a Brain?," *High Technology*, August 1984.

Lenat, Douglas: "Computer Software for Intelligent Systems," *Scientific American*, September 1984.

Hayes-Roth, Frederick: "The Knowledge-Based Expert System: A Tutorial," *Computer*, September 1984.

Hayes-Roth, Frederick: "Knowledge-Based Expert Systems," *Computer*, October 1984.

Charniak, Eugene and Drew V. McDermott: *Introduction to Artificial Intelligence*, Addison Wesley, Reading, MA, 1985.

Dym, Clive L.: "Expert Systems: New Approaches to Computer-Aided Engineering," *Engineering With Computers*, no. 1, 1985.

Schindler, Max: "Expert Systems," *Electronic Design*, January 10, 1985.

Kinnucan, Paul: "Software Tools Speed Expert System Development," *High Technology*, March 1985.

Walker, Adrian: "Knowledge Systems: Principles and Practice," *IBM Journal of Research and Development*, January 1986.

CAE/CAD (Chapters 10–12)

Ryan, Daniel L.: *Computer-Aided Graphics and Design*, Marcel Dekker, New York, 1979.

Schaffer, George: "Computer Graphics Goes to Work," *American Machinist*, July 1980.

Posa, John G.: "The System/370 Processor Chip: A Triumph for Automated Design," *Electronics*, October 9, 1980.

Requicha, A. A. G. and H. B. Voelker: "An Introduction to Geometric Modeling and Its Applications in Mechanical Design and Production," in *Advances in Information Systems Science*, vol. 8, U. T. Tou (ed.), Plenum, New York, 1981.

Swerling, Stephen: "Computer-Aided Engineering," *IEEE Spectrum*, November 1981.

Foley, James D. and Andries van Dam: *Fundamentals of Interactive Computer Graphics*, Addison Wesley, Reading, MA, 1982.

House, William C.: *Interactive Computer Graphics Systems*, Petrocelli Books, Princeton, NJ, 1982.

Scott, Joan E.: *Introduction to Interactive Computer Graphics*, John Wiley, New York, 1982.

Kinnucan, Paul: "Computer-Aided Manufacturing Aims for Integration," *High Technology*, May/June 1982.

Boyes, John W. and Jack E. Gilchrist: "GM Solid: Interactive Modeling for Design and Analysis," *IEEE Computer Graphics and Applications*, March 1982.

Requicha, A. A. G. and H. B. Voelcher: "Solid Modeling: A Historical Summary and Contemporary Assessment," *IEEE Computer Graphics and Applications*, March 1982.

Kinnucan, Paul: "Solid Modelers Make the Scene," *High Technology*, July/August 1982.

Swerling, Stephen: "Computer-Aided Engineering," *IEEE Spectrum*, November 1982.

Goetsch, David L.: *Introduction to Computer-Aided Drafting*, Prentice-Hall, Englewood Cliffs, NJ, 1983.

Pao, Y. C.: *Elements of Computer-Aided Design and Manufacturing*, John Wiley, New York, 1984.

Krouse, John K.: "Automation Revolutionizes Mechanical Design," *High Technology*, March 1984.

Wallich, Paul: "On the Horizon: Fast Chips Quickly," *IEEE Spectrum*, March 1984.

Manuel, Tom: "Computer Graphics," *Electronics*, June 28, 1984.

Krouse, John K.: "Paperless Engineering: From Concept to Finished Part in the Computer," *Machine Design*, July 12, 1984.

van Dam, Andries: "Computer Software for Graphics," *Scientific American*, September 1984.

"Design Automation," special issue of the *IBM Journal of Research and Development*, vol. 28, no 5., September 1984.

Machover, Carl and Ware Myers: "Interactive Computer Graphics," *Computer*, October 1984.

Fichera, Richard: "Rendering Adds Realism to Graphics," *Electronics Week*, October 22, 1984.

Davis, Dwight B.: "Chip-Based Graphics," *High Technology*, February 1985.

Warner, James R.: "Standard Graphics Software for High Performance Applications," *IEEE Computer Graphics and Applications*, March 1985.

Goldberg, Andrew: "Approaches Toward Silicon Compilation," *IEEE Circuits and Devices*, May 1985.

Bairstow, Jeffrey N.: "Chip Design Made Easy," *High Technology*, June 1985.

Voisinet, Donald D.: *Introduction to Computer-Aided Drafting*, McGraw-Hill, New York, 1986.

Krouse, John K.: "Engineering Without Paper," *High Technology*, March 1986.

Lu, Cary: "Micros Get Graphic," *High Technology*, March 1986.

ROBOTICS TECHNOLOGY (Chapters 13 and 14)

Kinnucan, Paul: "How Smart Robots Are Becoming Smarter," *High Technology*, September/October 1981.

Susnjara, Ken: *A Manager's Guide to Industrial Robots*, Corinthian, Shaker Heights, OH, 1982.

Hunt, Daniel V.: *Industrial Robotics Handbook*, Industrial Press, New York, 1983.

Taylor, Russell H. and David D. Grossman: "An Integrated Robot System Architecture," *Proceedings of the IEEE*, July 1983.

Lozano-Perez, Tomas: "Robot Programming," *Proceedings of the IEEE*, July 1983.

Brooks, Rodney A.: "Planning Collision-Free Motions for Pick and Place Operations," *International Journal of Robotics Research*, vol. 2, no. 4, Winter 1983.

Latombe, Jean-Claude: "Survey of Advanced General Purpose Software for Robot Manipulators," *Computers in Industry*, vol. 4, 1983.

Tanner, William R. (ed.): *Industrial Robots: Volume 1—Fundamentals*, Robotics International of the SME, Dearborn, MI, 1983.

Togai, Masaki: "Japan's Next Generation of Robots," *Computer*, March 1984.

Schreiber, Rita R.: "How to Teach a Robot," *Robotics Today*, June 1984.

Jarvis, John F.: "Robotics," *Computer*, October 1984.

Zeldman, Maurice: *What Every Engineer Should Know About Robots*, Marcel Dekker, New York, 1984.

Kehoe, Ellen J.: "The Expanding Repertoire of Robot End Effectors," *Robotics Today*, December 1984.

Cardoza, Anne and Suzee J. Vlk: *Robotics*, TAB Books, Blue Ridge Summit, PA, 1985.

Nof, Shimon Y. (ed.): *Handbook of Industrial Robots*, John Wiley, New York, 1985.

Asimov, Isaac and Karen A. Frenkel: *Robots: Machines in Man's Image*, Harmony Books, New York, 1985.

Koren, Yoram: *Robotics for Engineers*, McGraw-Hill, New York, 1985.

Albus, James S.: "Research Issues in Robotics," *SAMPE Journal*, January/February 1985.

Brady, Michael: "Artificial Intelligence and Robotics," *Artificial Intelligence*, April 1985.

Nitzan, David: "Development of Intelligent Robots: Achievements and Issues," *IEEE Journal of Robotics and Automation*, March 1985.

Dario, Paulo and Danilo DeRossi: "Tactile Sensors and the Gripper Challenge," *IEEE Spectrum*, August 1985.

Edson, Daniel V.: "Giving Robot Hands a Human Touch," *High Technology*, September 1985.

Hoekstra, Robert L.: "Robotics and Automated Systems," SouthWestern Publishing Co., Cincinnati, OH, 1986.

ROBOTICS APPLICATIONS (Chapters 15 and 16)

Engelberger, Joseph F.: *Robotics in Practice*, AMACOM. New York, 1980.

Tanner, William R. (ed.): *Industrial Robots: Volume 2—Applications*, Robotics International of the SME, Dearborn, Michigan, 1981.

Jablonowski, Joseph: "Robots That Assemble," *American Machinist*, November 1981.

Warnecke, H. J.: *Industrial Robots Application Experience*, IFS Publications, Kempton, England, 1982.

D'Ignazio, Fred: *Working Robots*, Dutton, New York, 1982.

Stauffer, Robort N.: "IBM Advances Robotic Assembly in Building a Word Processor," *Robotics Today*, October 1982.

Hartley, John: *Robots at Work*, IFS Publications, Kempton, England, 1983.

Argote, Linda, et al.: "The Human Side of Robotics: How Workers React to a Robot," *Sloan Management Review*, Spring 1983.

Scheinman, Victor: "Current Successful Robot Applications," *Proceedings of the DOD Robot Application Workshop*, Sacramento, CA, October 4–7, 1983.

"The Specification and Application of Industrial Robots in Japan," Japan Industrial Robot Association, Tokyo, 1984.

Griffin, Kirby G.: "Safety for Industrial Robot Applications," *Professional Safety*, December 1983.

Potter, Ronald D.: "Safety for Robotics," *Professional Safety*, December 1983.

Harvard Business Review, January–February 1984.

Howie, Phil: "Graphic Simulation for Off-Line Robot Programming," *Robotics Today*, February 1984.

Jetley, Sudershan: "Arc Welding with Robots," *Production Engineering*, March 1984.

Bublick, Timothy J.: "Guidelines for Applying Finishing Robots," *Robotics Today*, April 1984.

Giacobbe, A. L.: "Diskette Labeling and Packaging System Features Sophisticated Robot Handling," *Robotics Today*, April 1984.

Stauffer, Robert N.: Robot System Simulation," *Robotics Today*, June 1984.

Kessel, David S.: *Robotic Safety*, Sandia National Laboratory, Albuquerque, NM, October 1984.

Stauffer, Robert N.: "Robotic Spray Painting Update," *Products Finishing*, October 1984.

Thompson, C. C.: "Robot Modeling—The Tools Needed for Optimal Design and Utilization," *Computer-Aided Design*, November 1984.

Peterson, Tom: "Anatomy of a Clean Room Robot," *Semiconductor International*, November 1984.

Addison, J. H.: *Robotic Safety Systems and Methods: Savannah River Site*, E. I. Dupont De Nemours & Co., Aiken, SC, December 1984.

Powers, John H.: "Robotics Applications in Electronics Manufacturing," *IEEE Circuits and Devices*, January 1985.

Fischetti, Mark A.: "Robots Do the Dirty Work," *IEEE Spectrum*, April 1985.

Graham, J. H. and Meagher, J. F.: "A Sensory-Based Robotic Safety System," *IEEE Proceedings*, vol. 132, no. 4, July 1985.

Kak, A. C. et al.: "A Knowledge-Based Robotic Assembly Cell," *IEEE Expert*, Spring 1986.

SYSTEM ARCHITECTURE (Chapter 17)

Rembold, Ulrich, Karl Armbruster, and Wolfgang Ulzmann,: *Interface Technology for Computer-Controlled Manufacturing Processes*, Marcel Dekker, New York, 1983.

Strole, Norman C.: "A Local Communications Network Based on Interconnected Token-Access Rings: A Tutorial," *IBM Journal of Research and Development*, September 1983.

"Process Control," special issue of *Computer*, February 1984.

Zadoff, Mike: "Networks: Building Blocks for Integrated Manufacturing," *Assembly Engineering*, June 1984.

Dixon, Tom: "Networks Bridge the Gap Between Islands of Automation," *Electronic Packaging and Production*, June 1984.

Spector, Alfred Z.: "Computer Software for Process Control," *Scientific American*, September 1984.

Graube, Maris and Michael C. Mulder : "Local Area Networks," *Computer*, October 1984.

Williams, Theodore J.: "The Development of Reliability in Industrial Control Systems," *IEEE Micro*, December 1984.

Komoda, Norishisan, Kazuo Kera, and Takeaki Kubo: "An Autonomous, Decentralized Control System for Factory Automation," *Computer*, December 1984.

National Research Council: "The Interface Challenge, Special Report of the Committee on the CAD/CAM Interface," *American Machinist*, January 1985.

Wilson, P.R., et al.: "Interface for Data Transfer Between Solid Modeling Systems," *IEEE Computer Graphics and Applications*, January 1985.

Eaton, Robert J.: "Networking: Its Importance to Modern Control," *Industrial and Control Systems*, February 1985.

Voelcker, John: "Helping Computers Communicate," *IEEE Spectrum*, March 1986.

MANAGEMENT SYSTEMS (Chapter 18)

Adams and Ebert: *Production and Operations Management*, Prentice-Hall, Engelwood Cliffs, NJ, 1978.

Houtzeel, Alexander: "The Many Faces of Group Technology," *American Machinist*, January 1979.

Bothroyd, Geoffrey, Corrado Poli, and Laurence E. Murch: *Automated Assembly*, Marcel Dekker, New York, 1982.

Chevalier, Peter W.: "Group Technology as a CAD/CAM Integrator in Batch Manufacturing," *International Journal of Operations and Production Management*, vol. 4, no. 3, 1984.

Hyer, Nancy L. and Urban Wemmerlov: "Group Technology and Productivity," *Harvard Business Review*, July/August 1984.

Waterbury, Robert: "MRP-II Orchestrates Harmonious Production," *Assembly Engineering*, September 1984.

Marsland, D. W.: "A Review of Computer-Aided Part Programming System Developments," *Computer-Aided Engineering Journal*, October 1984.

Mill, Frank and Stuart Spragtett: "Artificial Intelligence for Production Planning," *Computer-Aided Engineering Journal*, December 1984.

Shunk, Dan L.: "Group Technology Provides Organized Approach to Realizing Benefits of CIMs," *Industrial Engineering*, April 1985.

Suzaki, Kiyoshi: "Japanese Manufacturing Techniques" Their Importance to U.S. Manufacturers," *The Journal of Business Strategy*, Winter 1985.

INTEGRATED MANUFACTURING SYSTEMS (Chapter 19)

Brunini, A. J.: *On-Line Planning, Scheduling and Control in a Process Line Environment*, Electronics Manufacturing Technology and Systems Conference, Phoenix, AZ, February 1–3, 1983.

Kinnucan, Paul: "Flexible Systems Invade the Factory," *High Technology*, July 1983.

Markstein, Howard W.: "Materials Handling Systems Support the Automated Factory," *Electronics Packaging and Production*, August 1983.

"FMS," special report in *Manufacturing Engineering*, September 1983.

Hughes, Tom and Don Hegland: "Flexible Manufacturing," *Production Engineering*, September 1983.

White, David A.: "Single Unit Processing is Solved by Flexible Manufacturing," *Electronic Packaging and Production*, September 1983.

The Charles Stark Draper Laboratory, Inc.: "Flexible Manufacturing Systems Handbook," Noyes, Park Ridge, NJ, 1984.

Kochan, D.: "Integrated Information Processing for Manufacturing for CAD/CAM to CIM," *Computers in Industry*, no. 5, 1984.

Managaki, Masao, Kyoji Kawagoe, and Masaru Naniwada: "A Model and Its Implementation in a Practical CAD/CAM Database," *Computers in Industry*, no. 5, 1984.

Moseng, Bjorn and Bjarte Haaoy Nes : "Integration of CAD/CAM as Seen From a Production Planner's Point of View," *Computers in Industry*, no. 5, 1984.

Bourne, David A. and Mark S. Fox: "Autonomous Manufacturing: Automating the Job Shop," *Computer*, September 1984.

Melkanoff, Michel A.: "The CIMs Database: Goals, Problems, Case Studies and Proposed Approaches Outlines," *Industrial Engineering*, November 1984.

Robbins, J. H., R. Kapur, and G. L. Berry: "Shop Floor Information System Is Foundation and Communication Link for CIMs," *Industrial Engineering*, December 1984.

Downie, Donald: "Automatic Guided Vehicles Move into the Assembly Line," *Modern Materials Handling*, January 1985.

Ashburn, A. and J. Jablonski: "Japans Builders Embrace FMSs," *American Machinist*, February 1985.

Tompkins, James A.: "Without Materials Handling There Is No Automated Factory," *Modern Materials Handling*, May 1985.

CAM APPLICATIONS (Chapter 20)

Brunner, R. H., E. J. Holden, J. C. Luber, D. T. Mozer, and N. G. Wu: "Automated Semiconductor Line Speeds Custom Chip Production," *Electronics*, January 27, 1981.

Moore, R. D., G. Caccoma, H. Pfeiffer, E. Weber, and O. Woodward: "Electron Beam Writes Next-Generation IC Patterns," *Electronics*, November 3, 1981.

"Productivity's Role in the Reindustrialization of America," special issue of *Production Engineering*, January 1981.

Burgam, Patrick M.: "Flexible Fabrication Moves In at Hughes Aircraft," *Manufacturing Engineering*, September 1983.

Long-Fox, E. R.: "Computer-Aided Design, Manufacture and Test of Printed Circuit Boards at IBM," *Computer-Aided Engineering Journal*, February 1984.

Waterbury, Robert: "The Total Approach to Automated Assembly," *Assembly Engineering*, March 1984.

Freeman, Nancy Brooks: "At Windsor It's Not the Same Old Line," *American Machinist*, June 1984.

Phillips, L. W. and J. W. Pearson: *Computer-Integrated Manufacturing: AT&T Technologies' Works—Richmond, Virginia*, Electronics Manufacturing Technology and Systems Conference, Raleigh, NC, September 28, 1984.

Lineback, J. Robert: "Air Force Arms Itself to Slash PC Board Costs," *Electronics Week*, December 3, 1984.

Tanimoto, Hiromu: "Factory Automation: An Automatic Assembly Line for the Manufacture of Printers," *Computer*, December 1984.

Blumenthal, Marjory and Jim Dray: "The Automated Factory: Vision and Reality," *Technology Review*, January 1985.

Coleman, J. R.: "World Class CIM in America," *Tooling and Production*, January 1985.

Reichenbach, Ray: "IBM's Automated Factory—A Giant Step Forward," *Modern Materials Handling*, March 1985.

Sepehri, Mehran: "A Machine Builds Machines at Apple Computer's Highly Automated Macintosh Manufacturing Facility," *Industrial Engineering*, April 1985.

Davis, Dwight B.: "Renaissance on the Factory Floor," *High Technology*, May 1985.

"Integrated Manufacturing," special staff report in *Production Engineering*, February 1986.

Formicelli, Joseph: "Automated Materials Handling Boosts Flexible Assembly Safety," *Modern Materials Handling*, February 1986.

Pound, T. R. (ed.): "Design and Manufacturing Automation," special section of *Electronic Packaging and Production*, March 1986.

Powers, John H.: *Automating Electronics Manufacturing*, International Electronic Manufacturing Technology Symposium, San Francisco, CA, September 15–17, 1986.

IMPLEMENTING CAM (Chapter 21)

Schaffer, George H.: "Implementing CIM," *American Machinist*, August 1981.

Gerwin, Donald: "The Do's and Don'ts of Computerized Manufacturing," *Harvard Business Review*, March–April 1982.

Gold, Bela: "CAM Gets New Rules for Production," *Harvard Business Review*, November–December 1982.

Holden, Happy: *Implementing Computer-Aided Manufacturing in Electronics*, Electronics Manufacturing Technology and Systems Conference, Phoenix, AZ, February 1–3, 1983.

Blumberg, Melvin and Donald Gerwin, *Human Problems Associated With the Acquisition and Use of Computer Integrated Technology*, Electronics Manufacturing Technology and Systems Conference, Phoenix, AZ, February 1–3, 1983.

Mackvlak, Gerald T.: "High Level Planning and Control: An $IDEF_0$ Analysis for Airframe Manufacturers," *Journal of Manufacturing Systems*, vol. 3, no. 2, 1984.

Harrington, Joseph J., Jr.: *Understanding the Manufacturing Process*," Marcel Dekker, New York, 1984.

Miller, Steven M.: *Impacts of Robotic and Flexible Manufacturing Technologies on Manufacturing Costs and Employment*, Carnegie Mellon University, Pittsburgh, PA, March 1984.

Groover, Mikell P., John E. Hughes, Jr., and Nicolas G. Odrey: "Productivity Benefits of Automation Should Offset Workforce Dislocation Problems," *Industrial Engineering*, April 1984.

Curtin, Frank T.: "Planning and Justifying Factory Automation Systems," *Industrial Engineering*, May 1984.

Shaiken, Harley: "Automation In Industry: Bleaching the Blue Collar," *IEEE Spectrum*, June 1984.

Jelinek, Mariann and Joen D. Goldhar : "The Strategic Implications of the Factory of the Future," *Sloan Management Review*, Summer 1984.

Shewchuk, John: "Justifying Flexible Automation," *American Machinist*, October 1984.

Primrose, P. L., G. D. Creamer, and R. Leonard,: "Identifying and Quantifying the Company-Wide Benefits of CAD Within the Structure of a Comprehensive Investment Program," *Computer-Aided Design*, January/February 1985.

Carrie, A. S., E. Adhami, A. Stephens, and I. C. Murdoch: "Introducing a Flexible Manufacturing System," *International Journal on Production Research,* March 1985.

National Research Council: "Computer-Integrated Manufacturing: Barriers and Opportunities," *National Productivity Review*, Spring 1985.

Hill, Kenneth D. and Steven Kerr: "The Impact of Computer Integrated Manufacturing on the First Line Supervisor," *Canadian Data Systems*, April 1985.

Brody, Herb: "Overcoming Barriers to Automation," *High Technology*, May 1985.

National Research Council, report of the Committee on the Effective Implementation of Advanced Manufacturing Technology: "People and Automation," *American Machinist*, June 1986.

"Factory Automation," special report in *High Technology*, October 1986.

ADA (language), 74
Adaptive control, 23, 190, 226-227
Administrative data, 10-11, 219, 245-247
AGV (automated or automatic guided vehicle), 32, 263-264
Algorithm, 252
ALU (arithmetic logic unit), 49-50, 77-78
American Standard Code for Information Interchange (ASCII), 49-50, 56
Application Software, 63-64, 67
Architecture (see Computer system)
Arithmetic logic unit (ALU), 49-50, 77-78
Artificial intelligence:
    applications of, 92-93, 93-94, 108, 146, 191-192
    definition of, 92
    expert system (see Expert system)
    hardware for, 101-102, 107
    major elements of, 93
    process, 93
    in robotics, 190-192
    software, 100-101
    symbolic programming, 94
    system, 102
    trends in, 106-108
ASCII (American Standard Code for Information Interchange), 49-50, 56
ASRS (automated storage and retrieval system), 32-33, 261-262, 282-284
Assembler, 64, 85
Assembly language, 64, 68, 85
Assembly manufacturing, 8, 33-34, 198-199, 204-205, 278-286
Asynchronous transmission, 51, 56
Automated drafting, 114
Automated or automatic guided vehicle (AGV), 32, 263-264
Automated materials transport system, 263-264
Automated storage and retrieval system (ASRS), 32-33, 261-262, 282-284
Automation:
    applications of, 33-37, 195-206, 261-264, 268-269, 274-289
    barriers to, 290-291
    benefits of, 33, 209, 268, 274, 290, 293
    changes caused by, 291-292
    economics of, 207-208, 292-293
    implementation of, 207-215, 290-296
    justification of, 292-293
    planning for, 293-295
    technologies, 269

Backward chaining, 95
Bar code reader, 259
Baseband, 224
BASIC, 69, 71
Binary-coded decimal (BCD), 49, 64
Binocular vision, 187-188
Bit-slice microprocessor, 78
Boundary representation (B-rep), 127, 149
Broadband, 224
Bus, 78-79, 81-82, 223

C (language), 74, 84
CAD (computer-aided or computer-automated design):
    application of, 114-115, 128, 141-143
    basic elements of, 29-30
    benefits of, 30, 132, 133
    electronic design, 140-144
    expert system for, 102-103, 146
    implementation of, 134
    mechanical design, 137-140
    process, 132-134
    systems, 134-137
    techniques, 137
    trends, 144-146
CAE (computer-aided or computer-automated engineering):
    application of, 115, 128, 158-160, 161 269
    design analysis, 148-151
    dynamic or kinematic analysis, 151
    integrated systems, 157-160
    manufacturing analysis, 152-154
    modeling and simulation, 154-157
    process planning, 152-153
    system architecture, 157-158
    trends in, 160-161
CAI (computer-aided instruction), 103
CAPP (computer-aided or computer-automated process planning), 152-153, 237-238
Carrier sense multiple access with collision detection (CSMA/CD), 223
Central processing unit (CPU), 48-51
Closed loop MRP, 241
CNC (computer numerical control), 23
COBOL, 68-69, 71
Communication (see Data communication)
Compiler, 64, 67
Compliance device, 178-181
Computer:
    applications in manufacturing, 28-37, 43-44, 93-94, 108, 220, 232, 246-247, 268-269
    architecture of, 60-62, 79, 89
    basic elements of, 2, 48-49
    basic functions of, 48
    channels, 51, 56
    definition of, 48
    equipment, 51-55
    memory, 52-53
    microcomputer (see Microcomputer)
    minicomputer (see Minicomputer)
    operation of, 49-51
    process control computer, 35-36, 224-230
    reduced instruction set computer (RISC), 88-89
    role of, in manufacturing, 4-5, 44-45, 220-221, 268-269
    technology, 38-39, 59-60, 87-88, 232, 269
    tools, 269-270
    trends, 16, 38-41, 57-62, 87-90, 107, 230-232
    types of, 51-52
Computer-aided instructions (CAI), 10
Computer-aided or computer-automated design (see CAD)
Computer-aided or computer-automated engineering (see CAE)

Computer-aided or computer-automated process planning (CAPP), 152-153, 237-238
Computer graphics (see Graphics system)
Computer network (see Network)
Computer numerical control (CNC), 23
Computer program (see Program)
Computer system:
    architecture of, 3, 25-26, 40-41, 60-62, 89, 218-233, 230-231, 270-271
    data flow or data-driven, 61-62
    definition of, 3
    distributed, 3, 26-27, 221-222, 287-288
    hierarchical, 25-26, 220-222, 257
    operation of, 50-51
    pipelined or control-driven, 60-62
    reduction or demand-driven, 61-62
Concurrent processing, 60-62
Constructive solid geometry, (CSG), 126-127, 149
Continuous manufacturing process, 225
Continuous path, 21, 174
Contouring, 21
Control system:
    adaptive control, 23, 190, 226-227
    applications of, 232
    architecture of, 227-228, 270
    computer numerical control (CNC), 23
    direct digital control, 227
    direct numerical control (DNC), 22-23
    event-driven, 228
    feedback control, 226
    feedforward control, 226
    functions of, 225-226
    manufacturing control system, 3, 24-25, 51, 220-221, 254-272, 282, 287-288
    numerical control (see NC)
    process control, 35-36, 104, 225-226, 288-289
    production control, 240-244, 251
    reliability of, 228-230
    robot, 165, 168-169, 174-175, 188-190
    servo, 169
    shop floor control, 30-31, 102, 220, 242-243
    tool control, 28-29, 220-228, 287-288
    types of, 224-225
Controlled path, 174
Controller, 4, 22, 57, 82-84
Conveyor, 263, 284-285
Coordinate system, 166, 170
CPU (central processing unit), 48-51
CSG (constructive solid geometry), 126-127, 149
CSMA/CD (carrier sense multiple access with collision detection), 223

DASD (direct access storage device), 53
Data:
    administration, 257
    administrative, 10-11, 219, 245-247
    design, 143
    graphics, 121, 143
    logistical, 10-11, 219, 240-245
    security, 74, 255
    sources of, 12

Data (continued)
  technical, 10-11, 219, 234-240
  types of, 7-8, 10, 219, 271
Database:
  definition of, 67
  design, 143
  graphics, 121, 143
  integrated, 255-258, 270
  knowledge base, 95, 98
  management system, 67, 104, 256-257
  relational, 73-75, 94
Data collection, 200, 225, 258-259, 284
Data communication or transmission,
    55-57, 74, 80-81, 89, 225,
    230-232, 257-258
DDC (direct digital control), 227
Debug routines, 67
Decision support system, 252
Degrees of freedom, 166-168
Design:
  applications, 141-143
  database, 143
  of facilities, 248-249
  functional, 141, 142
  for manufacturing, 248-249
  physical, 141, 142
  process, 132-133
  rules, 143
  standards, 146
  system (see CAD)
  verification, 141, 143
Dextrous robot hand, 178-181
Discrete manufacturing process, 224-225
Direct access storage device (DASD), 53
Direct digital control (DDC), 227
Direct memory access (DMA), 80
Direct numerical control (DNC), 22-23
Displays, 54, 60, 116-117, 129-130
Distributed data processing or distributed
    system, 3, 26-27, 221-222, 287-288
DMA (direct memory access), 80
DNC (direct numerical control), 22-23
Dynamic analysis, 151

EBCDIC (Extended Binary-Coded Decimal
    Interchange Code), 49-50, 56
Economic justification, 207-209, 292-293
Edge finding, 185-186
EDIF (electronic design interchange
    format), 146
Electromechanical assembly, 281-283
Electronic design interchange format
    (EDIF), 146
Electronic design system, 140-144
Electronics manufacturing, 281-286
End-of-arm tooling or end effectors, 174,
    177-180
Engineering analysis, 128
EPROM (erasable programmable read-
    only memory), 80
Erasable programmable read-only
    memory (EPROM), 80
Ethernet, 223
Expert system:
  advantages of, 99
  applications of, 102-104, 146, 276
  architecture of, 98, 108
  backward/forward chaining, 95
  basic elements of, 94-95, 98
  definition of, 94
  fuzzy logic, 96
  heuristic reasoning, 94, 95
  if-then rules, 95

Expert system (continued)
  inference engine, 95, 98
  knowledge base, 95, 98
  knowledge engineering, 95
  knowledge representation, 95
  semantic networks, 95-97
Extended Binary-Coded Decimal
    Interchange Code (EBCDIC), 49-50,
    56

Fabrication manufacturing, 7
Facilities applications, 36, 248-249
Factory of the future, 43-44, 274-275
Fault tolerance, 74, 89, 229
Feature matching, 185
Feedback control, 226
Feedforward control, 226
Finance, 247
Finishing, 197, 280
Finite element analysis, 149-151
Firmware, 86
Flexible manufacturing system (FMS),
    28-29, 264-268, 275-278
FMS (flexible manufacturing system),
    28-29, 264-268, 275-278
FORTRAN, 68
Forward chaining, 95
Fuzzy logic, 96

Geometric model, 121, 123-128, 130,
    138-140, 148, 149-150, 269-270
Geometry engine, 130
Graphics system:
  advantages of, 114
  applications of, 115, 128-129, 130-131,
    158-160, 161, 214-215, 228-229, 239
  architecture of, 120, 130, 157-158
  basic operations of, 115
  chips, 130
  data, 121
  database, 121, 143
  definition of, 112
  displays, 116-117, 129-130
  geometry engine, 130
  hardware, 116-119
  input devices, 118
  interactive, 113, 114, 122, 137
  major elements of, 112-113
  output equipment, 119
  processors, 120
  programming, 122, 213-215
  software, 120-123, 135-136, 214
  trends in, 129-130
  types of, 114-115
Gray scale imaging, 185
Gripper, 177-178
Group technology, 153, 161, 235-237,
    257

Heuristic reasoning, 94, 95
Hierarchical system, 25-26, 220-222,
    270-271, 257
High-level languages, 64, 68, 100-101
Human factors, 156, 210, 294

ICAM definition language (IDEF), 294-295
IGES (Initial Graphics Exchange
    Specification), 122, 146, 258

Image analysis or processing, 185-188
Inference engine, 95, 98
Initial Graphics Exchange Specification
    (IGES), 122, 146, 258
Input/output (I/O):
  bus, 81-82
  communications, 80-81
  equipment, 53-55, 118-119
  interface, 80
  interrupts, 80-81
  routines, 67
Integrated manufacturing system, 4, 25,
    254-272, 274-289
Intelligent robotic systems (see Robotics)
Interactive computer graphics, 113, 114,
    122, 137
Interpreter, 67
Interrupt, 80-81, 228
Inventory management, 241-242, 251
ISO open-system interconnection model,
    230-231

Just-in-time manufacturing, 251

Kan-ban, 251
Kinematic analysis, 151
Knowledge base, 95, 98
Knowledge engineering, 97-98
Knowledge representation, 95

Ladder diagram, 83-84
Languages (see Programming)
Language translators, 67, 257
Light striping, 187
Linear programming, 252
Line architecture, 248-249
Line tracking, 184, 200
LISP, 69-70, 100
Local area network (LAN), 222-224
Logic programming, 94-95, 95-97, 100-101
Logistical data, 10-11, 219, 240-245

Machine language, 64, 68
Machine tools, 19, 202
Machining, 196-197, 265-267, 275-278
Machining cell, 28
Machining centers, 21, 28, 265
Macroassembler, 85
Magnetic stripe reader, 259
Maintenance applications, 36-37, 102,
    276
Management systems, 234-252, 269
Manipulator, 165-168, 178, 181, 192
Manufacturing automation protocol
    (MAP), 231
Manufacturing:
  complexity, 12
  cost, 254
  design for, 248-249
  interdependencies, 15-16
  line architecture, 248-249
  line modeling, 155-156
  nature of, 5-6, 14-15
  operations, 6-7, 249-252
  organization of, 6, 220
  performance measurement, 246
  productivity, 250

Manufacturing (continued)
  technology, 41-43
  types of, 7-8, 15, 196-199, 201-206,
      224-225, 275-288
  assembly, 8, 198-199, 204-205,
      278-286
  continuous, 225
  discrete, 224-225
  electromechanical, 281-283
  electronics, 281-286
  fabrication, 7, 196-198, 201-204,
      275-278
  mechanical, 278-281
  metalworking, 201-202, 275-278
  part, 15
  plastics, 202-204
  process, 7, 15, 205-206, 286-289
  product, 15, 281-283
  semiconductor, 287-289
Manufacturing control system, 3, 24-25,
    51, 220-221, 264-271, 282, 287-288
Manufacturing resources planning (MRP
    II), 241, 243-244
MAP (manufacturing automation
    protocol), 231
Materials handling:
  automated or automatic guided vehicle
      (AGV), 32, 263-264
  automated storage and retrieval systems
      (ASRS), 32-33, 251-262, 282-284
  automated systems, 31-32, 275-288
  basic elements of, 260
  benefits of, 32, 195
  considerations, 261, 264
  conveyors, 263, 280, 281-282, 284-285
  integrated systems, 32, 260, 267
  mini-load system, 261-262
  optimizing, 251, 261
  robotics in, 195-196, 276, 279-280,
      283-284, 288
  stacker crane, 261-262
  transfer systems, 264, 277
  transport systems, 263-264, 275-276,
      278
Materials requirements planning (MRP),
    241-243
Mean time between failure (MTBF), 229
Mean time to repair (MTTR), 229
Measurement and inspection, 34-35, 199,
    268
Mechanical design system, 137-140
Memory, 52-53, 65, 79-80
Metal working, 201-202, 275-278
Microcomputer:
  architecture of, 79, 89
  definition of, 2, 79
  I/O communications, 80-81
  memories, 79-80
  programming, 83-86
  software, 83-86, 88-89
  supermicro, 89
  technology, 87-88
  trends, 87-90
Microinstructions, 78
Micromanipulator, 178, 181
Microprocessor, 77-79, 87
Minicomputer:
  architecture of, 89
  bus configurations, 81-82
  definition of, 2, 81
  programming, 83-86
  software, 83-86, 88-89
  supermini, 89
  trends, 87-90
Mini-load system, 261-262

Modeling, 102, 121, 123-128, 130, 133,
    138-140, 154-157, 213-215, 269-270
MODEM (modulator/demodulator), 55,
    57, 80
Modulator, demodulator (MODEM), 55,
    57, 80
Molding, 202-203
MRP (materials requirements planning),
    241-243
MTBF (mean time between failure), 229
MTTR (mean time to repair), 229
Multiplexing, 57
Multiprocessing, 60-62, 89
Multiprogramming, 65

Natural language, 74-75, 98
NC (see Numerical control)
Networks:
  applications of, 55
  basic elements of, 55
  definition of, 3
  local area, 222-224
  of programmable controllers, 83
  types of, 57, 223-224
Noncontact sensors, 182
Numerical control (NC):
  advantages of, 22, 238
  applications of, 23-24
  computer numerical control (CNC), 23
  continuous path, 21, 174
  contouring, 21
  controlled path, 174
  definition of, 4
  direct digital control (DDC), 227
  direct numerical control (DNC), 22-23
  history of, 18-22
  machine tools, 19
  part programming, 20, 153-154,
      238-240
  point-to-point, 21, 174
  process, 20
  process planning, 20, 237-238
  program, 20-21, 153-154, 238-240
  straight cut, 21
  system, 19-20, 21-22, 227-228

Object code, 64
OCR (optical character recognition), 259
Operating system, 50, 64-66, 83-85
Optical character recognition (OCR), 259

Parallel interface adapter (PIA), 80
Parallel processing, 60-62, 107
Part programming, 20, 153-154, 238-240
PASCAL, 69-70
PC (see Programmable controller)
PDDI (process data definition interface),
    231
PIA (parallel interface adapter), 80
Pick and place, 174
Plastics manufacturing, 202-204
Point-to-point (see Numerical control or
    networks)
Process control, 35-36, 104, 225-227,
    288-289
Process data definition interface (PDDI),
    231
Process data highway (PROWAY), 231
Process manufacturing, 7, 35, 205-206,
    286-289

Process planning, 20, 152-153, 191,
    237-238
Production control, 102, 240-245, 251
Productivity, 250-252
Program or Programming (see also
    Software):
  in assembly language, 85
  definition of, 3, 64
  development techniques, 71-73, 75, 86,
      94, 100, 101, 102
  languages, 3, 64, 68-69, 74-75,
      100-101, 173-174
  lead-through, 172
  linear, 252
  logic programming, 94-97, 100-101
  microinstructions, 78
  microprogramming, 85-86
  multiprogramming, 65
  numerical control, 20-21, 153-154,
      238-240
  part, 20, 153-154, 238-240
  playback, 172
  process of, 69-72
  of programmable controllers, 83-84
  robot programming, 171-174, 190, 192,
      214
  symbolic programming, 94
  trends in, 75, 88-89
  types of, 64, 83-86
  utility, 65-67
  walk-through, 172
Programmable automation, 4
Programmable controller (PC):
  advantages of, 83
  basic elements of, 22, 82
  basic functions of, 22, 82-83
  definition of, 4, 82
  programming of, 83-84
  networks, 83
Programmable read-only memory
    (PROM), 80
PROLOG, 100
PROM (programmable read-only
    memory), 80
Protocol, 57, 74, 230-231, 257-258
PROWAY (process data highway), 231
Purchasing, 246-247

Quality control, 35, 246, 251
Queue, 246

RAM (random access memory), 52, 79
Random access memory (RAM), 52, 79
Range finding, 187
Read-only memory (ROM), 52, 80
Real-time system, 65, 228
Reduced instruction set computer (RISC),
    88-89
Register, 50, 78
Relational database, 73-75, 94
Reliability, 228-230
Return on investment (ROI), 208-209, 293
Ring network, 57, 223-224
RISC (reduced instruction set computer),
    88-89
Robot or Robotics:
  advantages of, 165, 177
  application of, 195-206, 207-215, 264,
      279-288
  for assembly, 33-34, 198-199, 204-205,
      278-286

Robot or Robotics (continued)
  basic elements of, 42, 164-165, 176-177
  capabilities of, 169-171
  compliance device, 178, 181
  control system, 165, 168-169, 174-175, 188-190
  definition of, 4, 164, 176
  dextrous robot hand, 178-181
  drive systems, 167-168
  end-of-arm tooling or end effectors, 174, 177-180
  features of, 171
  gripper, 177-178
  history of, 164
  human factors, 210
  implementation of, 207-215
  intelligent systems, 176-193
  justification of, 207-209
  line tracking, 184, 200
  manipulator, 165-168, 178, 180, 192
  materials transfer, 264, 279-280, 283-284, 287-288
  mechanical configurations, 166
  mobile robots, 192
  operation, 174-175
  programming, 171-174, 190, 192
  return on investment (ROI), 208-209
  safety, 211-213
  sensors, 173, 178, 180-184
  simulation, 213-215
  specifications, 170
  tool holder or changer, 177-180, 196-198
  trends, 192-193
  types of, 165-169
  vision system, 184-188, 191, 193, 284
ROI (return on investment), 208-209
ROM (read-only memory), 52, 80

Safety, 211-213
Semantic networks, 95-97
Semiconductor manufacturing, 287-288
Semiconductor technology, 38-39, 59, 87, 130, 142-143, 232
Sensors, 173, 178, 180-184
Servo control, 169
Shop floor control, 30-31, 102, 220, 242-243
Simulation, 102, 143, 145, 154-157, 213-215, 252
Silhouette matching, 185-186
Silicon compiler, 145-146
Software:
  application, 63-64, 67
  artificial intelligence, 100-101
  definition of, 64
  development systems, 71-73, 75, 86, 94, 100, 101, 102
  for robotics systems, 171-174, 188-190, 213-215
  standards, 122-123, 146
  system, 63-64, 73-74, 83-85
  trends in, 72-75, 88-89
  types of, 63-64
Solid model, 126-128, 138-140, 149, 214
Stacker crane, 261-262
Standards:
  communications, 57, 74, 230-231, 257-258
  graphics software, 122-123, 146
Star network, 57
Stereo vision, 187-188
Storage devices, 52-53
Straight-cut numerical control, 21
Super micro/mini, 89
Surface model, 125-126
Symbolic programming, 94
Synchronous transmission, 51, 56
System software, 63-64, 73-74

Tactile or touch sensors, 180-181
Teach pendant, 172
Technical data, 10-11, 219, 234-240
Teleprocessing (see Data communication)
Test data generation, 141, 144, 154
Testing and measurement, 34-35, 103, 141, 144, 154
Text editor, 67
Transport system, 263-264
3D vision systems, 185, 187-188
Time sharing, 60, 65
Token ring, 223-224
Tool control, 28-29, 220-228, 287-288
Tool holder or changer, 177-180, 196-198
2D vision systems, 185-187

Unix, 84-85
Utility program, 65-67

Virtual memory, 65
Vision system, 105, 184-188, 191, 193, 284
Voice recognition system, 259

Walk-through programming, 172
Welding, 197-199, 279
WIP (work in process), 251, 260
Wireframe model, 124-125, 149
Work envelope, 167-168, 170
Work in process (WIP), 251, 260
Workstation, 89-90, 113
Workstation ownership, 282
World modeling, 170, 172